Becoming a Sustainable Organization

A Project and Portfolio Management Approach

Best Practices and Advances in Program Management Series

Series Editor
Ginger Levin

RECENTLY PUBLISHED TITLES

Becoming a Sustainable Organization: A Project and Portfolio Management Approach
Kristina Kohl

Improving Business Performance: A Project Portfolio Management Approach
Ramani S

Leading and Managing Innovation: What Every Executive Team Must Know about Project, Program, and Portfolio Management, Second Edition
Russell D. Archibald and Shane Archibald

Program Management in Defense and High Tech Environments
Charles Christopher McCarthy

The Self-Made Program Leader: Taking Charge in Matrix Organizations
Steve Tkalcevich

Transforming Business with Program Management: Integrating Strategy, People, Process, Technology, Structure, and Measurement
Satish P. Subramanian

Stakeholder Engagement: The Game Changer for Program Management
Amy Baugh

Making Projects Work: Effective Stakeholder and Communication Management
Lynda Bourne

Agile for Project Managers
Denise Canty

Project Planning and Project Success: The 25% Solution
Pedro Serrador

Project Health Assessment
Paul S. Royer, PMP

Portfolio Management: A Strategic Approach
Ginger Levin and John Wyzalek

Program Governance
Muhammad Ehsan Khan

Becoming a Sustainable Organization

A Project and Portfolio Management Approach

Kristina Kohl

CRC Press is an imprint of the
Taylor & Francis Group, an **informa** business

AN AUERBACH BOOK

CRC Press
Taylor & Francis Group
6000 Broken Sound Parkway NW, Suite 300
Boca Raton, FL 33487-2742

© 2016 by Taylor & Francis Group, LLC
CRC Press is an imprint of Taylor & Francis Group, an Informa business

No claim to original U.S. Government works

Printed on acid-free paper
Version Date: 20160328

International Standard Book Number-13: 978-1-4987-0081-8 (Hardback)

This book contains information obtained from authentic and highly regarded sources. Reasonable efforts have been made to publish reliable data and information, but the author and publisher cannot assume responsibility for the validity of all materials or the consequences of their use. The authors and publishers have attempted to trace the copyright holders of all material reproduced in this publication and apologize to copyright holders if permission to publish in this form has not been obtained. If any copyright material has not been acknowledged please write and let us know so we may rectify in any future reprint.

Except as permitted under U.S. Copyright Law, no part of this book may be reprinted, reproduced, transmitted, or utilized in any form by any electronic, mechanical, or other means, now known or hereafter invented, including photocopying, microfilming, and recording, or in any information storage or retrieval system, without written permission from the publishers.

For permission to photocopy or use material electronically from this work, please access www.copyright.com (http://www.copyright.com/) or contact the Copyright Clearance Center, Inc. (CCC), 222 Rosewood Drive, Danvers, MA 01923, 978-750-8400. CCC is a not-for-profit organization that provides licenses and registration for a variety of users. For organizations that have been granted a photocopy license by the CCC, a separate system of payment has been arranged.

Trademark Notice: Product or corporate names may be trademarks or registered trademarks, and are used only for identification and explanation without intent to infringe.

Visit the Taylor & Francis Web site at
http://www.taylorandfrancis.com

and the CRC Press Web site at
http://www.crcpress.com

Dedication

I am dedicating this book to my husband, Morris. Thank you for your continuous support in all of my endeavors.

Contents

Dedication		v
Contents		vii
Preface		xiii
Acknowledgments		xv
About the Author		xvii
Chapter 1	**What Is Sustainability? Why Does It Matter?**	**1**
1.1	Defining Sustainability	2
1.2	History of Sustainability	6
1.3	Sustainability in Business	8
1.4	Social Capital	9
1.5	Natural Capital	9
1.6	Megatrends	10
1.7	Natural Resource Scarcity	11
1.8	Climate Change	12
1.9	Population Growth, Urbanization, and Demographic Change	15
1.10	Global Connectivity and Information Transparency	16
1.11	Stakeholder Interest in Sustainability	17
1.12	Portfolio, Program, and Project Management Impact	17
1.13	Conclusion	18
	Notes	18
Chapter 2	**Building the Business Case for Sustainability**	**21**
2.1	Sustainable Strategy Drives Value	22
2.2	CEO Perspective	22
2.3	Financial Performance/Competitive Advantage	26

2.4	Managing Risk	27
2.5	Climate Change Risk	29
2.6	Creating New Opportunities and Products	31
2.7	Access to Capital	32
2.8	Engaging Stakeholders	33
2.9	Competitive Ratings and Rankings	34
2.10	Operational Improvements and Energy Savings	39
2.11	Project and Portfolio Management Impact	40
2.12	Conclusion	41
	Notes	41

Chapter 3 Gaining Stakeholder and C-Suite Support and Sponsorship — 43

3.1	Creating Value Through Sustainable Strategy	44
3.2	Planning for Sustainability Success	44
3.3	The Sustainability Journey	45
3.4	Building the Case for Sustainability	47
3.5	Engaging the C-Suite	49
3.6	Finding a Champion in the C-Suite	52
3.7	Materiality	54
3.8	Barriers to Adoption	58
3.9	Project, Program, and Portfolio Management Impact	59
3.10	A Conversation About Engaging the C-Suite	60
3.11	Conclusion	62
	Notes	63

Chapter 4 Alignment of Business and Sustainability Strategy Creates Sustainable Strategy — 65

4.1	Alignment of Business and Sustainable Strategy	66
4.2	Creating and Implementing Sustainable Strategy	68
4.3	Organizational Alignment	73
4.4	Portfolio Management Selection to Promote Alignment	76
4.5	Alignment of Programs and Projects with Sustainability Vision	77
4.6	Materiality Supports Alignment	79
4.7	Developing a Framework to Promote Sustainability	83
4.8	Value Creation Drivers	87
4.9	Aligning Programs and Projects with Sustainable Strategy	88
4.10	Conclusion	88
	Notes	90

Chapter 5 Project Management Techniques Inform Sustainable Strategy Development — 91

5.1	Project Manager Barriers to Effectively Managing Sustainability Projects	91

5.2		Project Management Maturity and the Sustainability Journey	92
5.3		Sustainability Project Charter	94
5.4		Sustainability Drivers	97
5.5		Sustainability Assessment	99
5.6		Sustainability Continuum	103
5.7		Harness the Power of Project Management	105
5.8		Scope, Time, and Cost	106
	5.8.1	Scope	106
	5.8.2	Time and Cost	107
5.9		Building Sustainability into All Projects	110
5.10		Program Management	113
5.11		Conclusion	116
Notes			117

Chapter 6 Creating a Culture of Sustainability — 119

6.1	Organizational Culture	119
6.2	Culture of Sustainability	122
6.3	Defining Sustainability as a Core Value	124
6.4	Benefit Corporations	127
6.5	Creating a Culture of Sustainability	130
6.6	Misalignment of Values and Actions	131
6.7	Establishing Goals	135
6.8	Create a Team	136
6.9	Internal Stakeholder Adoption	137
6.10	Metrics to Drive Change	138
6.11	Best Practices	139
6.12	Portfolio, Program, and Project Manager Perspective	140
6.13	Conclusion	140
Notes		140

Chapter 7 Portfolio Management Supports Strategic Sustainability Alignment — 143

7.1	The Portfolio Analysis Process	144
7.2	Identifying the Portfolio Management Process	145
7.3	Sustainability Portfolio Assessment	147
7.4	Communication of Goals and Drivers to Stakeholders	148
7.5	Portfolio Management Alignment	150
7.6	Creating a Supportive Organizational Structure	150
7.7	Portfolio Component Selection	154
7.8	Types of Components	156
7.9	Incorporating Lessons Learned	158
7.10	Conclusion	163
Notes		163

Chapter 8   Identifying and Engaging External Stakeholders   165

- 8.1 Impact of Engagement — 166
- 8.2 Creating an Engagement Plan — 169
- 8.3 Identifying Stakeholders and Collecting Requirements — 170
- 8.4 Engaging Stakeholders to Identify Needs and Define Requirements — 172
- 8.5 The Stakeholder Engagement Process — 173
- 8.6 Identifying Key Stakeholders — 175
- 8.7 Portfolio, Program, and Project Management Impact — 178
- 8.8 Conclusion — 184
- Notes — 184

Chapter 9   Leveraging Organizational Relationships and Assets   187

- 9.1 Impact of an Organization's Sustainability Maturity on Stakeholder Engagement — 188
- 9.2 Managing Internal Stakeholders — 191
- 9.3 Engaging Function Business Leaders — 193
- 9.4 Internal Stakeholder Drivers for Sustainability — 196
- 9.5 Best Practices for Engagement — 202
- 9.6 Measuring Engagement — 206
- 9.7 Project, Program, and Portfolio Management Impact — 207
- 9.8 Conclusion — 207
- Notes — 207

Chapter 10   Leveraging Human Capital Professionals   209

- 10.1 Human Capital Relevance in Reporting Frameworks — 210
- 10.2 Human Resource Alignment — 211
- 10.3 Impacting Global Challenges — 215
- 10.4 Employee Engagement — 216
- 10.5 Creating Business Value — 218
- 10.6 Integrating Sustainability into Human Capital Management — 220
- 10.7 Relevance to Project Management Professionals — 222
- 10.8 Conclusion — 226
- Notes — 226

Chapter 11   Adopting a Culture of Change to Unlock the Benefits of Sustainable Strategy   229

- 11.1 Organizational Structure for Change — 230
- 11.2 Change Management Plan — 233
- 11.3 Barriers to Project Success — 237
- 11.4 Communication Plan — 239
- 11.5 Engaging Stakeholders in the Change Process — 245

11.6	Human Capital Advantage	247
11.7	Conclusion	248
Notes		249

Chapter 12 Sustainable Strategy as a Lever to Attract, Engage, and Retain the Workforce — 251

12.1	The Value of Engagement	252
12.2	Building a Sustainable Workforce	255
12.3	Creating a Sustainable Workplace	256
12.4	Employee and Team Engagement	257
12.5	Best Practices to Embed Sustainability and Maximize Engagement	262
12.6	Project and Portfolio Perspective	264
12.7	Case Study: How the City of Cambridge, Massachusetts, Engaged Employees on Commuting Habits and Effected Change	266
12.8	Conclusion	270
Notes		270

Chapter 13 Tools and Techniques to Embed Sustainability — 273

13.1	Resources	273
13.2	Benchmarking Organizational Sustainability	275
13.3	Opportunities for Employee Engagement	277
13.4	Continuous Improvement Plan	281
13.5	Impact of Maturing on the Sustainability Continuum	285
13.6	Promoting Plug-in Vehicle Adoption by Employees by Developing Workplace Charging Stations	289
	13.6.1 Project Manager Tools	289
	13.6.2 Impact of a Workplace Charging Station	289
	13.6.3 Determine Organizational Readiness	291
	13.6.4 Workplace Charging Station Success Stories	292
13.7	Creating a Sustainable Culture	293
13.8	Diversity as a Cornerstone of Sustainability	296
13.9	Programs to Support D&I	297
13.10	Conclusion	299
Notes		299

Chapter 14 Selecting Goals and Metrics that Matter — 301

14.1	Alignment of Strategy, Goals, and Metrics	301
14.2	Reporting Standards and Frameworks	303
14.3	Selecting Meaningful Metrics	306
14.4	Meaningful Metrics Drive Engagement	308
14.5	Data Quality and Accuracy	311
14.6	Role for Technology	311

14.7	Portfolio, Program, and Project Management Impact	315
14.8	UVM Case Study: STARS Project[21]	316
14.9	Conclusion	318
Notes		318

Chapter 15 Celebrating Success — 321

15.1	Impact of Incentives	322
15.2	Gamification	323
15.3	Employee Resource and Affinity Groups	324
15.4	Culture of Sustainability	325
15.5	Promoting Project Team Success	325
15.6	Sonoma County Winegrowers Celebrate Sustainability	326
15.7	The Sustainability Continuum Impact	327
15.8	Conclusion	331
Notes		332

Index — 335

Preface

Sustainable strategy is a business model that focuses on creating long-term financial value while preserving natural and social capital. Increasingly, management is adopting sustainable strategy as part of their business strategy in order to garner benefits such as opening new markets, creating new products and solutions, reducing costs, improving customer relations, reducing risk, and engaging employees. However, a gap exists between strategy and benefit realization, as the process of integrating sustainable strategy into people, processes, and policies is challenging. Leveraging project management and human capital professionals facilitates embedding sustainability into an organization's culture, unlocking the value creation potential of sustainable strategy on financial, social, and natural capital.

The target audience for this book includes organization leaders who wish to create a sustainable culture in order to reap the full benefits of a sustainable approach; sustainability champions who are driving an organization's sustainability agenda forward; project management professionals who have been tasked with managing portfolios, programs, and projects in order to integrate sustainability within their organizations; and human capital professionals who are seeking to add organizational value through developing a culture of sustainability. The case studies and interviews in this book include sustainability stories and projects from a wide variety of organizations in both function and size, such as family-owned businesses, higher-education institutions, NGOs, municipal and federal government agencies, and large global organizations. These cases are based on interviews with experienced sustainability and project management professionals who have not just "talked the talk" but also "walked the walk." The voices of these professionals provide inspiration and guidance to sustainability champions and to program and project managers seeking to move their sustainability portfolio components forward within their own organizations.

As a global society, we are facing megatrends such as population growth, natural resource depletion, climate change, divergent living standards, and biodiversity destruction. Leading global organizations have adopted sustainable strategy as an approach to create long-term value by addressing both the challenges and opportunities created by these megatrends. However, sustainable strategy is not just for global companies; increasingly, organizations of all sizes are being encouraged by customers, regulators, and communities to consider environmental, social, and governance standards as part of their business model.

Becoming a sustainable organization is a journey that requires significant change in the ways in which organizations conduct their operations, including identifying and focusing on material issues, engaging with diverse stakeholders, managing climate- and resource-related risks, creating new sustainability-driven opportunities, and reporting transparently on actions and performance. Understanding the stages of the sustainability journey and the indicators and drivers of sustainable strategy at each of the stages facilitates the process of sustainability visioning and organizational integration and adoption.

Driving sustainability into an organization's culture requires the C-suite to adopt sustainability as an organizational pillar. In order to maximize value creation, management must align business and sustainable strategies and create an organizational structure that promotes cross-functional collaboration. Leveraging organizational assets to drive sustainability includes incorporating all business functions and utilizing their expertise to drive transformative organizational change. Interconnectivity is key to a sustainable approach, and collaborations among business functions, industry, community, and government promote sustainable solutions. This approach requires complex, multifaceted stakeholder engagement and a broad array of tools and techniques to handle the large and increasingly influential number of stakeholders.

As drivers of change within an organization, project management professionals have an important role to play in planning and implementing sustainable change. While project management methodologies provide beneficial structure and techniques to manage the process, project management professionals require additional tools and techniques in order to maximize their effectiveness. We discuss tools such as the Sustainability Accounting Standards Board (SASB) model for selecting material issues, life-cycle assessment (LCA), and the circular economy. Sustainability reporting frameworks such as the CDP and Global Reporting Initiative (GRI) are discussed in terms of their organizational impacts. A variety of checklists, matrices, and assessment tools are made available to facilitate the process of embedding sustainability into an organization. For those interested in utilizing these tools and templates, downloadable copies are available at the companion website at http://www.becomingsustainable.org.

In order to promote broad-based understanding and acceptance of sustainability across the organization, we highlight the beneficial impact of leveraging the skill set of human capital management (HCM) professionals. As keepers of the organizational culture, experts in change management, communication, and compensation incentives, HCM professionals are well positioned to drive sustainability into the people, process, and policies of an organization.

In addition, a sustainable approach creates value for HCM professionals by impacting their core business mission of employee attraction, retention, and engagement through promoting a positive corporate image relative to natural and social capital. We include an in-depth look at tools and techniques available to human capital professionals to drive programs and projects that promote a sustainable workforce. It is people, not technology or equipment, that drive sustainability within an organization.

This book provides a roadmap for organizations during their sustainability journey by sharing case studies, best practices, and lessons learned, as well as tools and techniques to drive change. Effecting transformation into a sustainable organization begins with senior management support but requires embedding a culture of sustainability within the organization so that it becomes part of every employee's daily function, mindset, and actions. Best of luck with your organization's sustainability journey.

Acknowledgments

Thank you to all who have contributed to the creation of this book. Whether you contributed content or proofread pages, your many hours of time are greatly appreciated. I wish to extend my appreciation to the organizations that granted me permission to reprint their sustainability materials. Sharing of our lessons learned and best practices helps each of us grow in our sustainability journey. I would also like to thank the Wharton Initiative for Global Environment Leadership for promoting business sustainability by providing forums to generate dialogue among sustainability practitioners. In addition, I would like to thank the Office of Sustainability at the New Jersey Department of Environmental Protection for their work in creating a forum for New Jersey–based organizations to share ideas, best practices, and lessons learned as we all seek to become sustainable organizations. I am very grateful to my family for their support and patience throughout the lengthy process of researching and writing this book. Every day we can all make decisions that improve our world. Thank you to all who have shared their thoughts and experiences on how to make our world a more sustainable place. To my colleagues at HRComputes, thank your for your research, ideas, and proofreading. Without all of you, this book would not be possible. To my editors, thank you for your patience and guidance.

I would like especially to thank the following people for sharing their organization's sustainability stories with me so that I could share them with you. A heartfelt thank you to:

Irv Cernauskas, Co-founder, Irv & Shelly's Fresh Picks
Jennifer Haugh, CEO, Iconic Energy Consulting
Neil Hawkins, Corporate VP and Chief Sustainability Officer, The Dow Chemical Company
Trudy Heller, Ph.D., Executive Education for the Environment
Shelly Herman, Co-founder, Irv & Shelly's Fresh Picks
Paul Kohl, PE, Energy Program Manager, The Philadelphia Water Department
Jennifer Lawrence, Sustainability Planner, The City of Cambridge
Talia Manning, Founder and Chief Creative Officer, Chameleon Studios
Sarah Olexsak, Energy Efficiency & Renewable Energy, U.S. Department of Energy
Mieko Ozeki, Sustainability Officer, University of Vermont
Paul Reig, Associate II—The Aqueduct Project, World Resource Institute
Morris Yankell, Principal, HRComputes
Mark Weick, Director, Sustainability Programs and Enterprise Risk Management, The Dow Chemical Company

If I have overlooked anyone, please accept my personal thank you for all of your help. Best of luck with your sustainability journey!

About the Author

Kristina Kohl is the managing principal of HRComputes, a management consulting firm that provides strategic guidance, program, and project management to corporate clients in order to maximize their human capital investment through technology and change management. Kristina leads the Corporate Social Responsibility and Sustainable Strategy practice for the firm and provides strategic guidance to senior management on the value proposition of adopting a sustainable strategy. HRComputes is a registered New Jersey Sustainable Business, WBE, and SBE. They are recipients of the Gotham Green Award for sustainable strategy and a certificate of recognition from the New Jersey EPA for sustainability innovation.

Kristina speaks at industry events including the PMI Global Conference and the Society for Human Resource Management Global Conference. In addition, she has been a guest lecturer for the University of Pennsylvania, La Salle University, and the Project Management Institute. She has authored white papers and articles for online publications of Project Management.com, Sustainable Brands, and the Wharton Magazine.

Kristina brings over 25 years of executive business management experience in the corporate and not-for-profit areas. Her background includes serving as Vice President and Manager at JPMorgan Chase and President of the Moorestown Home and School Association. She has served on numerous boards including the La Salle University Human Capital Advisory Board, the Tender, an Alzheimer respite care center, the Moorestown Home & School Association, and the Moorestown Education Foundation. Kristina holds an MBA from the Wharton School, University of Pennsylvania. She is also a Project Management Institute member and a certified PMP.

Chapter 1
What Is Sustainability? Why Does It Matter?

Sustainability is defined as considering the short- and long-term environmental, social, and economic impacts of organizational decisions and actions. Sustainability is a megatrend impacting the way in which successful business currently is and will continue to be conducted in the future. Increasingly, management is tasked with steering their operation through a world of environmental, social, and economic constraints while seeking new opportunities to create value for stakeholders. Sustainable initiatives and policies are changing product design, material selection, resource allocation, supply chain management, logistics, waste management, and product disposal. A reminder of the importance of management addressing megatrends is the bankruptcy of Kodak, the one-time market leader that failed to adapt to the rise of digital film. Kodak's management was unable to adapt in a timely manner and the end result was bankruptcy, with significant losses for investors, employees, and the community. Sustainable strategy is the new paradigm for global business.

Global trends in natural resource scarcity, population growth, urbanization, climate change, and interconnectivity among organizations, communities, and nations are influencing how internal and external stakeholders view value creation. As a result, senior management is receiving increased pressure from stakeholders to consider both traditional financial indicators as well as environmental and social performance indicators as they guide their organizations. Investors view the inclusion of environmental, social, and governance metrics as a proxy for a well-managed firm. Because of global competition, consumer and investor demands, stakeholder requirements, and governmental regulation, the success of an organization is increasingly tied to its purpose-driven values and ethics as they relate to environmental and social capital. As this change in value creation unfolds, the strategies and processes of running an organization are changing. Leaders in their industries, such as Nestlé and Alcatel Lucent, are embracing sustainability as a competitive advantage. They are striving to create corporate cultures that embed sustainability into the core of their operations, including project, program, and portfolio management. One of Nestlé's goals is to address the challenge of world population growth and the corresponding increased demand for water by committing to a 40% reduction in direct water usage per tonne of product by 2015 from a 2005 baseline level.[1] They are engaging with internal and external stakeholders to address water stewardship opportunities within their own operation and throughout their supply chain. In addition, they are committed to establishing and implementing guidelines to address the fundamental human right of access to clean water and to ensure equitable access. As corporations develop major strategic initiatives around sustainability, it creates opportunities for portfolio, program, and project managers to use their skills for planning and integrating these strategies through program and project selection and implementation. Adoption of sustainable strategy is essentially managing a complex, highly visible program that is designed to create long-term change within the organization.

1.1 Defining Sustainability

Sustainability is sometimes referred to as corporate social responsibility (CSR), corporate responsibility (CR), product stewardship, or environmental compliance. When the topic of sustainability is raised, people ask: "What do you mean by sustainability?" Even among practitioners of sustainability, there is discussion about the meaning and factors that should be included. This book defines the concept of sustainability as considering the short- and long-term environmental, social, and economic impacts of organizational decisions and actions. This definition takes a holistic view and includes concepts such as greenhouse gas emissions, product stewardship, regulatory compliance, and supply chain management, but also community impacts, employee health and safety, and workforce diversity. Known as the *triple bottom line* (TBL), this three-pronged approach was created by John Elkington in 1994, as a way for organizations to consider the social, environmental, and economic impacts of their decisions and actions.[2] The concept is partially based on the call to action for business leaders in the publication of *Our Common Future* by the Brundtland Commission (now known as the UN World Council for Economic Development). Their recommendation is to use sustainable development, focusing on the

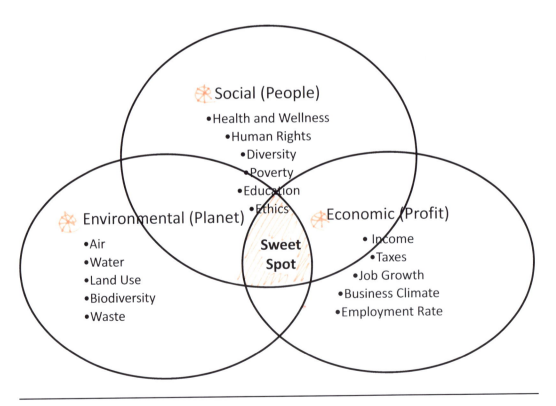

Figure 1.1 The Triple Bottom Line

three pillars of the economy, environment, and society, as a means of addressing global challenges such as resource depletion, poverty, and vastly different global living standards. (Another term for the concept is the "3 Ps," which stand for people, planet, and profit as the basis for measuring organizational success.) Rather than focus just on financial metrics such as net income or return on investment (ROI), the TBL approach seeks to understand the impact of organizational decisions on the world, including their impact on natural, social, and human capital. The TBL approach facilitates management focusing not only on shareholder value creation but also on environmental and social value creation or depletion. With the tremendous challenges facing our global society, a new paradigm is needed in order to more accurately assess and measure organizational value creation.

Figure 1.1 depicts environmental, social, and economic impacts as equally important aspects of determining an organization's performance. Each of the circles represents people, planet, and profit. The area of intersection of the three overlapping circles is the sustainability sweet spot, where outcomes are evaluated on all three types of criteria. While there is no universal standard for calculating TBL, the figure includes some examples of variables to consider in establishing metrics for measuring economic, environmental, and social impacts.[3] The size and type of organization, project scope, and geographic location are some of the factors that will determine the metrics selected. Economic measures might include profit, personal income, expenditures, investment, and employment. Environmental indicators look at air quality, water supply and quality, energy usage, natural resource depletion, waste, and land use. Social issues consider quality of employment, education, health and wellness, and quality of life in the local and regional community. Using this paradigm, management evaluates performance on metrics that include economic as well as social and environmental measurements known as externalities. Historically, financial statements and market valuation have not accounted for externalities. As a result, the full picture of performance and risk was incomplete.

The TBL approach moves beyond traditional financial metrics such as net income, ROI, cash flow, and earnings before interest, taxes, depreciation, and amortization (EBITDA) that are based solely on Financial Accounting Standard Board (FASB) compliant financial statements and considers other forms of capital such as natural, social, and human capital. As investors, customers, communities, and regulators evaluate future performance of an organization, they are looking for how externalities can impact this valuation. Financial statements focus on historical performance. They provide limited information on an organization's utilization of natural capital such as natural resources, energy, water, and pollution generation, and social capital such as human rights, health and wellness, diversity, and living wage. Metrics that might be considered in a TBL approach include environmental metrics such as fossil fuel consumption measured by absolute or relative greenhouse gas (GHG) emissions or social metrics on diversity, such as the percentage of women in senior management.[4] Sustainable strategy as measured by the TBL approach considers each of the economic, environmental, and social dimensions equally in order to better assess a long-term and holistic valuation.

As organizations have matured in their sustainability journey, they have collaborated with partners from government, academia, and others to develop and adopt a more encompassing definition of sustainability. The United Nations Global Compact (UNGC) is the largest corporate social responsibility program in the world, with 12,000 corporate, government, academic, and non-governmental organization (NGO) participants from over 145 countries. These corporations and institutions have agreed to align their strategies and operations with sustainable and responsible business practices based on 10 principles concerning human rights, labor, the environment, and anti-corruption.[5] As drivers of globalization, businesses that have adopted these principles help ensure that commerce, development, markets, technology, finance, sourcing, and manufacturing are conducted in a manner that benefits society, the environment, and the economy. Included in the list are market leaders such as Johnson & Johnson, 3M, Kimberley-Clark, and Citi.[6] A significant number of global businesses believe that adopting sustainable and responsible business practices provides a sound foundation for creating value for their stakeholders, and they are willing to state their commitment publicly.

> **The 10 UNGC Principles**[7]
>
> - **Human Rights**
>   Principle 1: Businesses should support and respect the protection of internationally proclaimed human rights; and
>   Principle 2: make sure that they are not complicit in human rights abuses.
> - **Labor**
>   Principle 3: Businesses should uphold the freedom of association and the effective recognition of the right to collective bargaining;
>   Principle 4: the elimination of all forms of forced and compulsory labor;
>   Principle 5: the effective abolition of child labor; and
>   Principle 6: the elimination of discrimination in respect of employment and occupation.
> - **Environment**
>   Principle 7: Businesses should support a precautionary approach to environmental challenges;
>   Principle 8: undertake initiatives to promote greater environmental responsibility; and
>   Principle 9: encourage the development and diffusion of environmentally friendly technologies.
> - **Anti-corruption**
>   Principle 10: Businesses should work against corruption in all its forms, including extortion and bribery.

These principles provide a framework through which to consider the broad-base impact of sustainable strategy on organizational values and culture. As drivers of globalization, business leaders who incorporate sustainability into their value systems through strategies, policies, and procedures are creating a positive impact on global economies and societies and a foundation for long-term sustainable development. The UNGC shares best practices, management tools, and resources to advance sustainable solutions to global challenges. The UNGC philosophy is that strong global markets and strong societies can coexist through strategic alignment with the 10 Principles and partnerships between public and private entities. The UNGC is an excellent resource for establishing a corporate sustainability program with tools, risk assessment frameworks, and compliance lists to facilitate integration into organizational programs, projects, and processes.

To understand sustainability from a slightly different perspective, one needs only look at the evaluation criteria for sustainability performance by the leading global ranking agency. The Dow Jones Sustainability Index (DJSI) is one of the most recognized and respected sustainability ranking indices in the world. Annually, RobecoSAM produces the DJSI by selecting top performers for inclusion and naming global businesses as leaders in their industry based on their responses to industry-specific sustainability questionnaires. The RobecoSAM Corporate Sustainability criteria are evenly distributed among economic, environmental, and social dimensions and are further segmented between industry-specific criteria (57%) and general criteria (43%).[8] As indicated in Figure 1.2, corporate governance, ethics, and human capital programs are evaluated and measured alongside more traditional environmental metrics such as GHG emission, product ingredients, and risk strategy to determine the sustainability leaders. We can clearly see that sustainability focuses not only on environmental stewardship but also on social, human, and ethical considerations.

Each of the categories—Economic, Environment, and Social—is further divided into subcategories such as corporate governance, climate strategy, and stakeholder engagement, requesting details on specific programs and actions. While participation in the DJSI Assessment is on an invitation-only

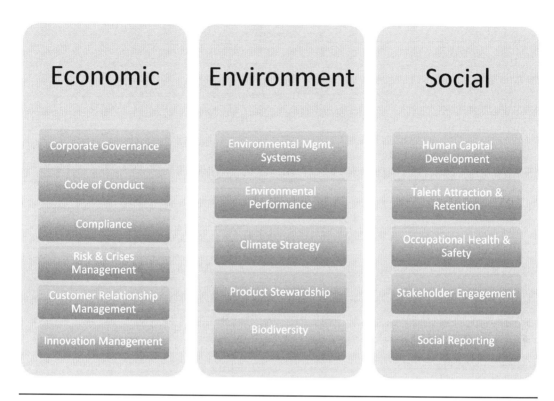

Figure 1.2 RobecoSAM Corporate Sustainability Assessment Criteria[9]

basis, reviewing the dimensions and categories provides sustainability leaders with a best-of-industry framework on which to begin to build their own organization's program.

To understand how a global corporation incorporates sustainability into its core operations, we will look more closely at Nestlé SA, which was the 2013 DJSI Food Products group leader. Nestlé's approach to sustainable development is outlined in the concept of "Creating Shared Value," which links the long-term prosperity of the business to the communities it serves. Here is a quotation from the letter introducing the 2012 Corporate Social Responsibility Report from Chairman Peter Brabeck-Letmathe and CEO Paul Bulke about Shared Value:

> We believe that we can create value for our shareholders and society by doing business in ways that specifically help address global and local issues in the areas of nutrition, water and rural development. This is what we mean when we speak about Creating Shared Value (CSV). We proactively identify opportunities to link our core business activities to action on related social issues.[10]

Their philosophy is that they can deliver better returns to shareholders by sustainably addressing social and environmental challenges, especially in the areas of nutrition, water, and rural development. Nestlé is a founding participant in the UN Global Compact LEAD, a select group of sustainability leaders from around the globe who represent the cutting edge of the UNGC, and is committed to incorporating the UNGC principles into its business model. In addition, the company engages in training and educating both internal and external stakeholders on key environmental, social, and governance issues and makes investments in technologies designed to lower their environmental impact. Management recognizes that leadership comes with responsibility, and they are committed to governance standards including compliance with international standards and national laws as well as their own internal ethics, values, and guidelines.[11] Their values are based on respect for people, cultures, environment, and the future of the world. As we can see from this example, sustainability at Nestlé is holistic and meets the broad-based economic, environmental, and social criteria outlined by RobecoSAM, including standards to address corporate governance, ethics, risk and compliance, and human capital development.

Sustainability goes beyond environmental stewardship and energy efficiency projects; it is a strategic approach that fundamentally changes the mission and vision of an organization. In order to effect this transformation, the sustainability champion, the person tasked with driving sustainability within an organization, must understand how her organization defines value creation and then develop a sustainable strategy that aligns with that value creation process. In the Nestlé example, 10 principles of business operation are used across global operations and provide the foundation for "Creating Shared Value."

Nestlé's 10 Principles[12]

1. Nutrition, Health and Wellness
2. Quality Assurance and Product Safety
3. Consumer Communication
4. Human Rights
5. Leadership and Personal Responsibility
6. Health and Safety
7. Customer and Supplier Relations
8. Agricultural and Rural Development
9. Environmental Sustainability
10. Water

These principles have been distributed to employees around the world and have been supported by employee education and training programs. Projects, programs, and people are evaluated on these principles in the areas of impacting consumers around nutrition, quality and communications, human rights and labor practices, leadership, personal responsibility, health and safety, supplier and customer relations, and the environment including water. In order to track performance to plan, key performance indicators (KPIs) are selected and tracked. Nestlé is committed to sustainable strategy and has transformed its operation to embed these principles into its core operation. Understanding the concepts of sustainability is important for project managers, program managers, and portfolio managers as they are well positioned within the organization to drive adoption of sustainability principles and standards into an organization.

1.2 History of Sustainability

The definition of sustainability traces back to the Brundtland Commission, which defined the concept of sustainable development as "to ensure that it meets the needs of the present without compromising the ability of future generations to meet their own needs."[13] In 1983, Javier Perez Cuellar, the Secretary General of the United Nations, asked Gro Harlem Brundtland, a medical doctor and former Prime Minister of Norway, to become the Chairwoman of the World Commission of Environment and Development (UNWCED), widely known as the Brundtland Commission after its first Chair. The Commission's publication of its report, *Our Common Future,* in April 1987 has provided the framework for the conversation on sustainable development and the foundation for future global summits such as Earth Summit/UNCED in 1992, the UNCED conference in Johannesburg, South Africa, in 2002, and the UNCSD conference in Rio, Brazil, in 2012. The concept of the triple bottom line was based on the Brundtland report's three pillars—environmental, social, and economic considerations.

As the Brundtland Commission was forming, nations around the globe were facing many environmental and social challenges as well as a diverse and divergent socioeconomic world. While strides had been made in terms of life expectancy, infant mortality, literacy rates, and per-capita food supply, the world still faced many challenges, such as poverty, population growth, urbanization, future economic growth given finite natural resources, and continued food security.[14] In addition, the commissioners were witnesses to numerous natural and man-made disasters that helped form their foundational beliefs. The following is a partial list of environmental and social catastrophes mentioned in the report that occurred from 1984 through 1987, between the formation of the commission and the issuance of its final report.

- In Africa, a drought-triggered, environment-development crisis killed 1 million and put 36 million people at risk.
- In Russia, the explosion of the Chernobyl nuclear reactor created nuclear fallout across Europe.
- During a warehouse fire in Switzerland, the Rhine River was polluted with chemicals, solvents, and mercury, killing millions of fish and threatening drinking water in Germany and the Netherlands.
- Globally, diarrhea from unsafe drinking water and malnutrition was responsible for the deaths of approximately 60 million people, the majority of whom were children.[15]

Reacting to their own experiences and these tragedies, it was apparent to the commissioners that significant change needed to take place in how we were developing the world and using its resources. In addition to the environmental and social challenges facing the globe, there were economic issues such as valuing natural capital used in production processes. While global society had generated economic gain, natural resource utilization fell into a paradigm of 20% of the countries utilizing 80% of the natural resources. This model is unsustainable as the Earth's population continues to grow. For all of the people of the globe to coexist, we were going to need to rethink our policies, processes, and priorities around resource distribution.

Despite the enormous challenges facing the commissioners, they believed that as a society, we would be able to utilize our human capability of innovative thought to create new technologies and solutions to address the global challenges of today and tomorrow. They said:

> Human progress has always depended on our technical ingenuity and a capacity for cooperative action. These qualities have often been used constructively to achieve development and environmental progress: in air and water pollution control, for example, and in increasing the efficiency of material and energy use. Many countries have increased food production and reduced population growth rates. Some technological advances, particularly in medicine, have been widely shared.[16]

The members of the Brundtland Commission believed that, through government, business, and NGO cooperation and partnerships, solutions could be developed to meet global challenges. In the years since the publication of *Our Common Future*, there have been a host of conferences, summits, and meetings to help the world better address the environmental, social, and economic challenges to our global society. In 2012, the UN Conference on Sustainable Development met in Rio de Janeiro, Brazil, to outline a proposal for "The Future We Want." While the attendees found that substantial progress had been made, the path to sustainable development continued to be fraught with numerous obstacles. In their report, eradicating poverty remained the greatest global challenge. In order to address these global challenges, the UN has called for increased collaboration among governments, NGOs, and the private sector. In addition, the UN recognized the importance of sustainability reporting and encouraged large public companies to integrate sustainability reporting into their reporting cycle.[17] In order to achieve the transformative global changes originally envisioned by the Brundtland Commission and subsequent meetings of the UNWCED, corporations need to partner with NGOs, academia, and governments to develop solutions that ensure long-term success of both the corporation and human society. While large corporations have been called upon to lead the initiative, small and mid-size organizations must take up the sustainability mantle in order to create the desired global sustainability impact.

1.3 Sustainability in Business

Businesses significantly impact natural and social capital through their global operations. Management's view on the importance of sustainability as a core strategy in their business has a significant impact on their success in preserving and enhancing natural, social, and human capital. Table 1.1 highlights corporate sustainability statements from a selection of companies named as industry leaders in the 2013 Dow Jones Sustainability Index (DJSI).

These firms are industry leaders, and they have adopted sustainable strategy as part of their core mission and values. They view sustainability as a way to create competitive advantage and to better manage their organizations. Their visions of sustainability are holistic and include environmental, social, and economic impacts. Alcatel, SAP, and Roche Holding specifically cite the use of technology and innovation as a tool to assist in the process of addressing challenges such as poverty, environmental protection, health and wellness, living standards, and socioeconomic disparity. While guiding current decisions, these statements are forward thinking and represent management's plan for future value creation. These visionary sustainable strategies seek to address significant global challenges using their own expertise and technologies while creating value for their more traditional stakeholders, such as shareholders, employees, and clients. In short, they are following a path for innovation and change that was originally envisioned by the Brundtland Commission.

Table 1.1 Corporate Vision of Sustainability: Selection of 2013 DJSI Group Leaders[18]

Company	Industry	The Company's Vision of Sustainability
Alcatel-Lucent	Technology hardware & equipment	"Our vision is to make communications more sustainable, more affordable and more accessible in line with our company's industrial focus on 'The Shift Plan.' The goal is to increase the capacity and reach of communications technology, making it more accessible and affordable for users worldwide."[19]
SAP AG	Software & services	"Our mission is to help every customer become a best-run business. We do this by delivering technology innovations that address the challenges of today and tomorrow without disrupting our customers' business operations."[20]
Roche Holding AG	Pharmaceuticals, biotechnology, & life sciences	"We use innovations in science to drive research and development of medicines and diagnostics that address some of medicine's most pressing challenges. At the same time, we must also deliver sustainable growth and value for our stakeholders, be they employees, investors, society or patients."[21]
Citigroup Inc.	Diversified financials	"Integrating environmental sustainability into our core business generates value for our clients, customers, communities and our firm."[22]

As indicated in Table 1.1, Alcatel-Lucent is seeking to bridge the digital divide by improving education and socioeconomic standards of those in less developed countries through new technologies. Through this process, Alcatel creates shared value for both itself, by identifying new markets and developing new solutions, and for the countries in which it operates, by raising living standards through education and access to technology. Each of these vision statements guides its organization forward through planning, creating, and implementing strategic change, further driving sustainability into their organizations.

1.4 Social Capital

Social capital includes issues such as the health and wellness of the population, education access and quality, safety and crime, community citizenship, human rights, standard of living and poverty rates, child labor, diversity and inclusion. In order to better understand social capital from a company perspective, we will analyze SAP's sustainability vision and how it relates to social capital. SAP's vision includes using its own innovation and expertise to develop new technology and solutions for its clients and prospective clients, including fast-growing emerging markets.[23] In order to build relationships within the developing world, SAP collaborated with PlaNet Finance, a NGO, to address poverty in Africa. This voluntary collaborative project dovetails with SAP's core business strategy to develop innovative solutions for customers. It also allows SAP to maximize its investment in human capital by driving employee engagement through meaningful volunteer opportunities.

Ghana, an impoverished West African nation, is a significant grower of shea nuts and producer of shea butter for use in cosmetics. In order to improve the standard of living in Ghana, SAP formed a joint venture with PlaNet Finance in 2009. The program was designed to provide a better business model to 1500 women who farm shea nuts and produce shea butter in Northern Ghana. Through the use of education, mobile technology, and microfinancing, the women were able to improve their processing techniques, purchase better equipment, maintain inventory, and gain access to the global market. Using cell phones, they have real-time access to market prices, allowing them to hold and then sell their shea butter to end users at the best price. In the past, they were subject to selling at below-market price to middlemen. Income levels of these women have risen 82%. The quality and stability of supply has also improved for the cosmetic companies that use the shea butter in their products. This venture has grown into a self-sustaining business serving over 10,000 women, and their StarShea network has become a worldwide market leader of organically produced shea butter.[24] More information about the StarShea Network program is available at http://www.sap-tv.com/video/#/7334/sustainable-business-in-ghana.

While the project has been a resounding success from the perspective of Ghanaian women, it has also benefitted SAP. This opportunity allowed SAP to develop innovative solutions using cloud and mobile technologies to meet the needs of other high-growth emerging markets, such as 500 million small farmers in developing countries who face similar challenges. From a human-capital perspective, this venture improved employee engagement and provided an opportunity to develop employee skill levels through 2000 volunteer hours and highly coveted fellowships.[25]

Management has embraced the concepts of sustainable development as a means to create value for their stakeholders through developing an innovative solution to meet global challenges of poverty and the digital divide. Using their products, skills, and expertise, they have aligned their business and sustainability strategy to address global challenges while creating and furthering their core business mission.

1.5 Natural Capital

Interestingly, it was the advent of space exploration that made us more aware of our own planet's natural resources. From the photos sent back of Earth, we gained a new perspective on its role in the planetary system, and we saw that Earth is dominated by interconnected natural systems such as oceans, clouds, mountains, deserts, and plains.[26] As a global society, we began to think about the impact of human activity on these natural ecosystems. We saw the interconnections of waterways and began to understand that in a watershed, each upstream farmer and factory impacts those communities downstream. We all depend on the Earth's resources for our survival, yet the distribution and usage of global natural resources is uneven between developed and developing nations.

Part of the Brundtland charter and those of subsequent UN conferences was to discuss a unified approach for sharing common space such as land, rivers, oceans, air, and space, and to address environmental stress as a source of conflict. A social problem such as poverty can lead to an environmental problem such as land degradation, as people misuse resources in an attempt to raise their living standard. Poverty and land degradation move in tandem. This concept is known as "the tragedy of the commons."

> **The Tragedy of the Commons**
>
> In 1968, Garrett Hardin wrote an article referring to "the tragedy of the commons," in which he highlighted the opposing interests of individuals and the long-term collective good of the group. The traditional example is a finite pasture area that is available to the community for shared grazing. If each household has 1 cow, the grassland is sufficient to support the community herd. Everyone's standard of living is relatively the same. However, a family that is, somehow, able to afford to have 2 cows on the pasture will have a higher standard of living through more milk production, meat, and additional calves to sell. If others in the community see this improved living standard and also add a second cow, the demand on the shared pasture will increase. As this process unfolds, the pasture eventually cannot support the additional grazing cows. The community loses the pasture, reducing everyone's standard of living.[27]
>
> A short YouTube clip depicting the inherent conflict can be found at https://www.youtube.com/watch?v=MLirNeu-A8I.

With open-access resources such as rivers, oceans, and air, it becomes very difficult to limit resource utilization. All parties are trying to maximize their own standard of living. If they do not seek to exploit the shared resource, someone else will. The traditional resolution has been either public or private ownership of open-access resources, but this remains a problem in globally shared resources such as oceans.

In order to feed expanding populations in the mid-twentieth century, governments around the globe created favorable policies, loans, and subsidies to support the expansion of large industrial fishing fleets, which supplanted smaller local fisherman. By 1989, we were taking 90 million tons of fish from the oceans. Since then, fish yields have been declining. Combining overfishing with pollution, climate change, and increasing acidity of the oceans, this once-abundant resource is projected to collapse by 2048.[28] Scientists believe that fisheries can be saved through better global fishery management, enforcement of laws governing catches, and improved aquaculture. Stocks of some species, such as bluefin tuna, are already collapsing. Greater international cooperation is needed among businesses, governments, and NGOs to meaningfully address this marine "tragedy of the commons." As organizations develop their long-term strategies, these are the types of challenges and risks that need to be addressed as part of the planning process.

1.6 Megatrends

As management teams create strategies to navigate an ever more complex global economy, they need to consider major trends that are developing in their competitive marketplace. Some of these "megatrends"

Table 1.2 Megatrends[29,30]

Megatrend	Highlights
Population growth and demographic changes	Growth of the global middle class Population growth in developing countries Increased governmental regulation Aging population in developed countries
Increased urbanization	Increased population density in urban areas Increased need for infrastructure for power, water, and services Change in delivery and usage of products and services
Climate change	Property damage and operations interruption from storms and flooding Loss of habitat and biodiversity Melting ice caps Food production limitations Rising sea levels
Natural resource scarcity	Commodity price volatility Global hot spots Regional instability and war Water scarcity Increased government regulation
Global interconnectivity and technology	Global events & news available 24/7 Social media Innovation and new solutions Transparency "Internet of things" Ubiquitous cell phones Digital divide Cyber attacks vs. traditional warfare
Increased stakeholder interest in sustainability	Investors' desire for environmental, social, and governance (ESG) information Consumers' desire for green products and services Millennial generation sustainability "hot button" NGO watchdog activities Increasing governmental regulations and standards Shifting global power

are listed in Table 1.2. Growth in population and income in the developing world is creating opportunities for those corporations that are creating value-added sustainable solutions to serve these expanding middle-class populations. Challenges also are increasing with increasing risks from weather-related damage and disruption, raw-material limitations, and growing investor concern about sustainability issues. There are many management teams like Kodak's, who misread megatrends such as the impact of innovation and technology, making their products and services obsolete. Leading management teams are embracing sustainable strategy as a framework for long-term planning in order to maximize opportunities and mitigate the threats posed by these megatrends.

The balance of the chapter takes a deeper dive into megatrends and the potential impacts for governments, organizations, and society.

1.7 Natural Resource Scarcity

Global conflict continues to arise over use of international commons such as oceans, especially concerning access to natural resources. Because of growth in its population and economy, China's demand for energy and food has grown. It has become increasingly aggressive in claiming rights to shared resources in the South China Seas. In 2012, Chinese naval ships fenced off one of the most abundant fishing grounds near the Philippines.[31] In the spring of 2014, China towed an oil drilling rig accompanied by military aircraft and ships into a disputed area of the South China Seas. Vietnam disputed China's right to drill for natural gas in the area, which is close to their coast, and tensions escalated between the two countries. Reaction moved beyond the normal governmental diplomatic discourse. The Vietnamese people responded violently to China's actions by rioting against factories they thought were owned by Chinese nationals. The violence spread to other factories. To control the rioting, the government of Vietnam had to send in police and troops.[32]

Shared access to limited resources remains a difficult global challenge, but one to which we must find a solution in order to facilitate a future of economic prosperity and global stability. International conflict, trade disputes, embargoes all impact an organization's ability to conduct business on a global scale. Potential risks arising from conflict over natural resource scarcity need to be incorporated into an organization's risk management plan and risk response strategy.

Because of acts of aggression stemming from natural resource limitations, we are experiencing an increase in governmental intervention and regulation. The Democratic Republic of the Congo (DRC) has been an area of conflict and abuse of human rights. Militias have used forced labor for extraction of rare earths to fund their military actions. In order to stem the flow of funds to these groups, the U.S. Congress drafted the Conflict Mineral Declaration required under Section 1502 of the Dodd-Frank Act. As part of the legislation signed into law by President Obama in 2010, public companies must report to the U.S. Securities & Exchange Commission (SEC) on the source and usage of conflict minerals such as tantalum, tin, gold, or tungsten originating from the Democratic Republic of Congo (DRC) or adjoining countries in their supply chain.[33] The purpose of the act is to reduce funding of violence in this region and to encourage responsible sourcing by U.S. firms. Going forward, additional governmental regulation is anticipated in order to provide frameworks to allocate natural resources more justly and equitably. Many firms have been challenged by this legislation because of their complex, global supply chains. Projects have been undertaken to identify and track the source of these materials through a complex supply chain. Intel has emerged as a leader on conflict mineral identification. It has gained a competitive advantage by being able to produce the first conflict mineral–free microprocessor. By strategically addressing this regulation and its supply chain complexity in a transparent manner, Intel has gained this competitive advantage in its marketplace. A video about Intel's assurance of conflict-free minerals in their microprocessors can be accessed at https://www.youtube.com/watch?v=HZDsNXtM-rk.

As natural resources become scarcer, prices become more volatile and supply becomes increasingly uncertain for companies. Remaining reserves are in locations that are more difficult and costly to access. As a result, organizations are becoming increasingly interested in raw material and component reclamation, recycling, and reuse, as the source of supply is often less expensive and more stable. As a result, product life cycles are moving from a linear to a circular plan, whereby resource utilization is maximized and waste creation is minimized. As future programs and projects are proposed, natural resource utilization will need to be considered from the design, sourcing, reclamation, and end-of-life disposition perspective. In order to address resource limitations, the process of new product and service development, manufacture, and distribution will need to change.

1.8 Climate Change

According to the U.S. Environmental Protection Agency (EPA), the Earth's average temperature has rise 1.4 degrees Fahrenheit over the past century and is expected to rise another 2 to 11.5 degrees over the next century. Global warming is primarily attributable to higher levels of greenhouse gases (GHG) trapped in the atmosphere. Human activity has been the major source of GHG emissions. Although we may not all agree on the severity of the issue of climate change, it is important to understand how researchers view its impact on our air quality, ecosystems, and biodiversity. Rising global temperatures have been tied to changes in rainfall, which have caused flooding, droughts, and severe storms. Climate change includes not only changes in temperature but also changes in precipitation and wind patterns. Ice caps are melting, sea levels are rising, and oceans are becoming more acidic.[34]

> **What Are Greenhouse Gases?**
>
> Greenhouse gases are gases that trap heat in the atmosphere. Scientists believe that they are a major cause of global warming. Major components include:
>
> - Carbon dioxide (CO_2)—A by-product from burning fossil fuels
> - Methane (CH_4)—Created by livestock and other agricultural processes and the decaying of organic matter
> - Nitrous oxide (N_2O)—Emitted from agricultural and industrial activities and the burning of fossil fuels and waste
> - Fluorinated gases (SF_6, HFC, PFC)—Synthetic gases created in industrial processes[35]

Changing temperatures and rising sea levels are impacting ecosystems and the biodiversity of plant and animal species. According to a recent study by the World Wildlife Foundation (WWF) in conjunction with the Zoological Society of London and several other groups, Earth has lost half of its wildlife in the past four decades. The decline has been seen across habitats and has been attributed to climate change, habitat destruction, commercial fishing, and hunting. The fastest declines have been in rivers and other freshwater ecosystems. From a geographic perspective, Latin America has seen the most significant decline. It experienced an 83% decline in overall population of wildlife.[36] Based on this research, human consumption of natural capital is outstripping the natural world's ability to regenerate itself. Current estimates are that we would need the resources of 1.5 Earths to supply the annual usage of natural goods and services. According to Carter Roberts, CEO of the WWF, "As we lose natural capital, people lose the ability to feed themselves and to provide for their families-it increases instability exponentially."[37] Without these species, we no longer have them in the mixture of the Earth's interconnected life system. A less diverse biosphere provides less opportunity to discover new solutions to our global health and food security challenges.

The most common measure of greenhouse gas (GHG) emissions is CO_2e, carbon dioxide equivalent. Total GHG emissions for the United States in 2012 were 6525 metric tons of CO_2e. As indicated by Figure 1.3, carbon dioxide represents the largest component of GHG emission at 82%. Figure 1.4 identifies the largest producers of GHG emissions by source. Generation of electricity creates the most significant impact. Gaining an understanding of the source of emissions is the first step toward addressing the issue.

Burning of fossil fuels is the largest contributor of carbon emissions in the United States, followed by the combustion of gasoline and diesel to power cars and trucks. The United States experienced a 5% rise in CO_2 emissions from 1990 to 2012, due primarily to fossil fuel consumption. In 2012, the International Energy Agency (IEA) predicted that average temperatures would rise 6 degrees Celsius by the end of the twenty-first century given the current trajectory.[40] Based on a report released by DARA, a humanitarian organization, the cost of air pollution and global warming is currently 1.6% of global Gross Domestic Product (GDP) and is anticipated to rise to 3.2% by 2030.[41] The most significant drain is on the economies of the developing nations. According to former Bangladesh Prime Minister Sheikh Hasina, a 1 degree Celsius rise in temperate is associated with a 10% loss in farming

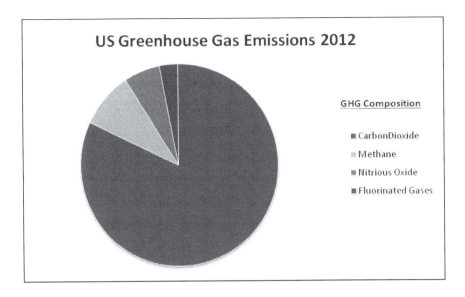

Figure 1.3 U.S. 2012 Greenhouse Gas Emissions by Type[38]

productivity, which equates to about 2.5% of the 2012 Bangladesh GDP.[42] The impacts to developing economies can be significant, driving up prices and reducing availability of life essentials such as food and drinking water.

From an organizational perspective, climate change increases physical, operational, and investment risk. In 2014, the Carbon Disclosure Project (CDP), an organization that gathers climate change–related information from public companies on behalf of their institutional investors, reported that S&P 500 firms were reporting higher actual and projected costs associated with facility damage, operation disruption, lower product demand, loss of productivity, and increased operating costs, owing to storms, flooding, and drought related to climate change.[43] Risk related to climate change is being considered as part of the risk management plan for these leading firms.

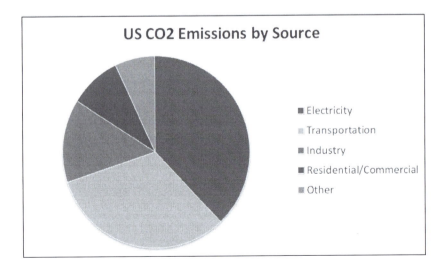

Figure 1.4 U.S. 2012 CO_2 Emissions by Source[39]

1.9 Population Growth, Urbanization, and Demographic Change

By 2050, the world's population is expected to grow 38%, from 6.9 billion in 2010 to 9.6 billion in 2050, with most of that growth occurring in developing nations.[44] The challenge is to provide an environment in which all people are living well but within the constraints of Earth's resources. Approximately half of the world's population is currently living in urban environments, and that number is predicted to grow to approximately 66% of the population living in an urban setting by 2050.[45] For fast-growing cities such as Mumbai, India, challenges will include adequate infrastructure, affordable and ecologically sound housing, energy, waste management, education, and social services. By 2030, the global middle class in low- and middle-income countries is expected to triple, from 400 million to 1.2 billion.[46] As people move up the economic ladder, they consume more resources on a per-capita basis. Demand for labor-saving devices such as washing machines and cars increases. Food preferences change to include more meat in the diet, and energy usage rises. All of this consumption leads to more demand for scarce natural resources and higher GHG emissions. While the growth in population will create many resource allocation challenges, it also creates opportunities for businesses to develop new products and services to address these challenges.

Infrastructure development will be a major growth area in the developing world. Concrete is crucial to constructing roads, bridges, and buildings, but its manufacture and distribution have a significant environmental footprint resulting from raw material mining and high natural resource usage rates. As we discuss in Chapter 4, the Canadian concrete industry has undertaken an analysis of its environmental footprint to reduce its impact while continuing to meet growing demand for concrete from the developing world. Business has tremendous opportunity to rethink its products and services and to deliver value while protecting natural and social capital.

Most interesting is the changing demographics of the population. Across the globe, the population is aging. By 2050, people age 65 and over are expected to reach 1.5 billion compared to 532 million in 2010.[47] The average age in the developed countries is anticipated to rise because of lower birth rates. The majority of the population in Japan, South Korea, and Germany will be older than 50.[48] Also, the average population of Latin America is anticipated to age more rapidly than that of the United States by mid-century.[49] As a result, the need for goods and services in these countries will skew toward those sought by an aging population, such as health care. Also, the working-age population will experience increased pressure to allocate more earnings to support a larger percentage of dependents than in developing countries such as India. Africa's population is expected to grow the most and will ultimately make up the largest portion of the global population, offering great opportunity to the business community to deliver products and services to this expanding market. European and Asian populations are expected to decline, and populations in the Americas are projected to remain constant. Projections suggest that India will become the most populated country, surpassing China, and further increasing its needs for goods and services.[50]

> **Malthusian Catastrophe**
>
> In the eighteenth century, Thomas Malthus wrote *An Essay on Population,* in which he predicted that population would grow to the level supported by food supply limitations. And then the population would be kept in check by the natural world's limitations. His premise was that the population grows at an exponential rate and the food supply increases at an arithmetic rate. Once the population exceeded the sustainable level, civilization would collapse into famine, war, and disease. Of additional concern to Malthus was a society that approached the unsustainable point in terms of the marginal lifestyle for the vast majority and their inability to improve their lifestyle. This point of unsustainable human expansion is known as the "Malthusian catastrophe."[51]

We are several centuries beyond the eighteenth century, and we have experienced continued improvements in life expectancy, food security, and infant mortality in both the developed and developing worlds. So what happened? We innovated and created new technologies to increase food production per acre through the use of fertilizer, labor-saving machinery, irrigation, and improved seed. While we are facing a world of real limitations, we do have the power to make the Earth a place to provide a quality standard of living for both today and into the future. It requires us to think differently about our limited resource pool and how we can reduce, reuse, recycle, and innovate to preserve our Earth for future generations.

1.10 Global Connectivity and Information Transparency

Information moves around the globe instantaneously. By the end of 2014, the number of cells phones was predicted to reach 7.3 billion globally, which means there are now more cells phones than people.[52] Many users have more than one device in order to maximize their connectivity options. Governments, not-for-profits, and businesses can no longer confine information internally. If something happens in a factory in Bangladesh, the world is aware almost immediately via text, video, and photos. For companies with supply chains around the world, this means being better informed about the practices of your suppliers and setting safety and labor standards through the use of a supplier code of conduct. Transparency and full disclosure are the new norm. From a brand perspective, a company can be accused of "greenwashing" if its environmental and social marketing claims are not supported through actual operational performance and transparent reporting. Using social media, watchdog groups can quickly disseminate information on greenwashing, causing damage to the brand of the targeted organization.

As demonstrated in the SAP joint venture in Ghana, the use of mobile technology and an innovative financing structure creates new opportunities for businesses to bridge the challenges of the digital divide. Global connectivity has the potential to improve health and living standards and to deliver solutions to remote locations. The "internet of things" provides new ways for devices and infrastructure systems to communicate, thereby optimizing operational efficiency. One example is having smart garbage cans that are tagged to send data on the fullness of the can. Logistic software can then design a pick-up route to maximize efficiency and reduce fuel usage and GHG emissions by the garbage truck fleet. Governments, institutions, and companies are seeking to innovate to zero in terms of emissions, worker health and safety incidents, and product safety issues. Innovation continues to create adaptive technologies such as wastewater reclamation, water desalination, and on-demand floodgates to meet the challenges of a changing world. Organizations that develop solutions to global challenges create long-term opportunities for value creation.

While technology and innovation can help address many of our resource limitations, they introduce additional risk. Along with greater connectivity comes the increased likelihood of cyber attacks. Some predictions are that cyber attacks will be the next generation of warfare. While big data, technology, and interconnectivity provide solutions to complex problems, they also introduce another layer of both positive and negative risk into the project management process.

After a series of high-profile data breaches such as the one that involved Target having 40 million credit and debit card numbers compromised by hackers, corporate boards are taking the risk of cyber attacks seriously. Approximately, 1500 companies listed on the New York Stock Exchange (NYSE) and the NASDAQ exchanges include cyber-threats as a component of risk management. Boards have enacted a variety of actions, ranging from adding new board members with technology backgrounds to creating new roles such as Information Security Officer. Data security changes include additional governance, policies, protocols such as data encryption, and training of employees.[53] While the "internet of things" can provide innovative solutions to sustainability problems, it also introduces new risks and challenges. The risk of cyber attack needs to be part of the risk analysis, and project managers must ensure that their resources have been educated on the issues and well trained to follow protective protocols.

1.11 Stakeholder Interest in Sustainability

Stakeholders including investors, customers, regulators, communities, and NGOs see the value of sustainable strategy to an organization. From an investor viewpoint, management teams that take this holistic view are thought to be better managers, producing better results through opportunity maximization and risk minimization. Consumers are increasingly demanding more sustainable products and services as they seek to achieve their own sustainability and business goals. Increasingly, employees are seeing sustainable organizations as employers of choice. The Millennial generation, in particular, is focused on purpose-driven employment and is looking for opportunities to have an impact on the environment and society through their employment.[54]

1.12 Portfolio, Program, and Project Management Impact

In order to address global megatrends through the adoption of sustainable strategy, an organization must begin its sustainability journey. Adoption of sustainable strategy involves transforming an organization's values, mission, and priorities to realign its strategic focus on sustainable business strategy. Transforming an organization requires a comprehensive sustainability program designed to deliver strategic benefits through portfolio component identification, selection, and implementation. Project management professionals are well positioned to facilitate this change process utilizing project management methodologies. Following a project management approach helps to frame the material issues and provide rigor and structure to the sustainability transformation process. Organizations with mature project management standards and processes have an advantage in adopting sustainable strategy.

From a project management perspective, the relevance of sustainability is pervasive and can impact each step of the project management process. From initiation through closing, sustainable concepts impact all aspects from developing the business case and understanding the stakeholders to planning each phase. Projects, not just green projects, require greater project manager awareness of organizational sustainability principles and best practices around requirements for design, materials, risk, quality, procurement, production, and logistics. In addition, benchmarks such as water usage, waste diversion, energy usage, or percentage of minorities in leadership roles will increasingly be part of project metrics. From a leadership perspective, managing a project in a sustainable organization impacts many important basic aspects of the business, including stakeholder engagement, communications, risk management, and governance requirements for your team and key business partners. By the very nature of their role, project management professionals help embed sustainable strategy into the fiber of the organization.

From a program management viewpoint, sustainable strategy can provide benefits to all projects, not just so-called green projects, through improved resource utilization, stakeholder engagement, and collaboration. As an organization moves forward on its sustainability journey, project requirements, deliverables, and metrics align with sustainable business goals and targets. Areas such as stakeholder engagement, procurement management, benefits management, and risk management are impacted. Program management office (PMO) standards and templates incorporate sustainability requirements to ensure that projects and programs reflect the organizational vision. Through providing best practices and sharing lessons learned, the PMO facilitates the process of supporting programs and projects that further embed sustainability within the organizational culture. Through embedding sustainability, management seeks to improve projects, processes, and outcomes while ensuring that their organization is a better steward of natural and social capital, preserving them for future generations.

Program directors and portfolio managers identify, evaluate, and recommend projects based on criteria that align sustainable and corporate strategy. Allocation of scarce organizational resources is always a challenge, and portfolio managers are integral to determining a project's alignment with corporate sustainability vision. The adoption of sustainable strategy creates opportunity for practitioners of project management methodologies to use their expertise and skill set to drive an organization forward in its sustainability journeys.

1.13 Conclusion

This chapter has provided an overview of sustainability, the concept of the triple bottom line, sustainability standards, and global megatrends. Its purpose is to provide a sustainable subject matter foundation for the reader. Sustainability is a complex subject, with many books written about the topics that were covered at a high level in this chapter. If you wish to read further in the topic of sustainability, *Sustainable Program Management* by Gregory T. Haugan, another book in the *Best Practices and Advances in Program Management Series,* is an excellent resource.

This book seeks to provide insight into the value-creation potential of sustainable strategy for organizations. It also suggests an approach to promoting adoption of sustainable strategy by an organization, including stakeholder engagement, leveraging internal resources, and engaging employees. Building a sustainability program is an iterative process that benefits from engaging cross-functional leaders. Adopting a sustainable strategy requires creating a culture that promotes and adopts sustainability principals. As drivers of change within an organization, project management professionals play an important role in creating a sustainable culture. While project management professionals are experiencing an increase in sustainability projects, they report many challenges to the implementation process. In the coming chapters, I will provide case studies as well as tools and techniques to improve an organization's sustainable strategy adoption. The audience for this book is organizational leaders, sustainability champions, human capital professionals, and project management professionals.

Notes

[1] Nestlé, "Water Efficiency," http://www.nestle.com, accessed July 21, 2014, http://www.nestle.com/csv/water.
[2] John Elkington, *Cannibals with Forks: Triple Bottom Line of 21st Century Business* (Gabriola Isaland, BC: New Society Publishers, 1998).
[3] Timothy F. Slaper and Tanya Hall, "The Triple Bottom Line: What Is It and How Does It Work?," *Indiana Business Review*, Spring 2011, pp. 1–8.
[4] Ibid.
[5] UNGC, "Overview of the UN Global Compact," accessed August 11, 2014, http://www.unglobalcompact.org/AboutTheGC/index.html.
[6] Ibid.
[7] Ibid.
[8] RobecoSAM, "2014 Methodology Update: RobecoSAM Corporate Sustainability Assessment 2014," Robecosam.com, March 2014, https://assessments.robecosam.com/documents/Methodology_Changes_2014.pdf.
[9] Ibid.
[10] Nestlé, "Nestlé in Society, Creating Shared Value and Meeting Our Commitments 2012," 2012, http://www.nestle.com/asset-library/documents/library/documents/corporate_social_responsibility/nestle-csv-full-report-2012-en.pdf.
[11] Nestlé, "What Is Creating Shared Value," http://www.nestle.com, accessed July 18, 2014, http://www.nestle.com/csv/what-is-csv.
[12] Nestlé, "Nestlé's Corporate Business Principles," http://www.nestle.com, accessed February 18, 2015, http://www.nestle.com/investors/corporate-governance/businessprinciples/businessprincipleshome.
[13] World Commission on Environment and Development, "Our Common Future: Report of the World Commission on Environment and Development," A/42/427 (The United Nations, March 20, 1987).
[14] Ibid.
[15] Ibid.
[16] Ibid.

[17] United Nations Conference on Sustainable Development, "The Future We Want-Outcome Document," *Sustainable Development Knowledge Platform*, June 20, 2012, https://sustainabledevelopment.un.org/rio20/futurewewant.

[18] Dow Jones Sustainability Indices and RobecoSAM, "Industry Group Leaders 2013," accessed February 16, 2015, http://www.sustainability-indices.com/review/industry-group-leaders-2014.jsp.

[19] Alcatel-Lucent, "Alcatel-Lucent 2013 Sustainability Report," 2014, www.alcatel-lucent.com., 3.

[20] SAP, "SAP Integrated Report 2013—Vision, Mission, and Strategy," accessed February 16, 2015, http://www.sapintegratedreport.com/2013/en/strategy-and-business-model/vision-mission-and-strategy.html.

[21] Roche, "Sustainability," *Creating Value for All Our Stakeholders*, accessed February 10, 2015, http://www.roche.com/sustainability/approach.htm.

[22] Citi, "Global Citizenship, A Commitment to the Environment," accessed June 9, 2014, http://www.citigroup.com/citi/environment/index.html.

[23] SAP, "SAP Integrated Report 2013—Vision, Mission, and Strategy."

[24] SAP, "StarShea Nonprofit in Ghana Becomes an Independent Social Business with Technology and Services Donated from SAP," SAP News Center, accessed June 2, 2014, http://www.news-sap.com/starshea-nonprofit-in-ghana-becomes-an-independent-social-business-with-technology-and-services-donated-from-sap.

[25] SAP, "SAP Integrated Report 2013—Vision, Mission, and Strategy."

[26] World Commission on Environment and Development, "Our Common Future: Report of the World Commission on Environment and Development."

[27] Garrett Hardin, "The Tragedy of the Commons," *Science* 168, no. 3859 (December 13, 1968): 1243–1248, DOI:10.1126.

[28] National Geographic Society, "Overfishing—National Geographic," 38.90531943278526 and -77 0376992225647 800-647-5463, accessed July 21, 2014, http://ocean.nationalgeographic.com/ocean/critical-issues-overfishing.

[29] S. Singh, "The 10 Social and Tech Trends That Could Shape the Next Decade," *Forbes*, accessed June 17, 2014, http://www.forbes.com/sites/sarwantsingh/2014/05/12/the-top-10-mega-trends-of-the-decade.

[30] David A. Lubin and Daniel C. Esty, "The Sustainability Imperative," *Harvard Business Review*, accessed February 16, 2015, https://hbr.org/2010/05/the-sustainability-imperative.

[31] Andrew Browne, "Beijing Pays a Price for Assertiveness in South China Sea," *Wall Street Journal*, May 14, 2014, sec. Asia, http://online.wsj.com/news/articles/SB10001424052702303627504579558913140862896.

[32] Vu Trong Khanh in Hanoi, Jenny W, and Hsu in Taipei, "Anti-China Rioting Turns Deadly in Vietnam," *Wall Street Journal*, May 16, 2014, sec. Asia, http://online.wsj.com/news/articles/SB10001424052702304908304579562962349248496.

[33] U.S. Securities and Exchange Commission, "Specialized Corporate Disclosure," accessed June 24, 2014, http://www.sec.gov/spotlight/dodd-frank/speccorpdisclosure.shtml.

[34] Climate Change Division U.S. EPA, "Basics," Overviews & Factsheets, accessed May 20, 2014, http://www.epa.gov/climatechange/basics.

[35] Climate Change Division U.S. EPA, "Greenhouse Gas Emissions: Greenhouse Gases Overview," Overviews & Factsheets, accessed May 21, 2014, http://www.epa.gov/climatechange/ghgemissions/gases.html.

[36] Gautam Naik, "Wildlife Numbers Drop by Half Since 1970, Report Says," *Wall Street Journal*, September 30, 2014, sec. World, http://online.wsj.com/articles/report-wildlife-numbers-drop-by-half-since-1970-1412085197?KEYWORDS=Study%3A+Half+of+Wildlife+Lost.

[37] Ibid.

[38] U.S. EPA, "Greenhouse Gas Emissions."

[39] Ibid.

[40] Fiona Harvey and Damian Carrington, "Governments Failing to Avert Catastrophic Climate Change, IEA Warns," *The Guardian*, April 24, 2012, sec. Environment, http://www.theguardian.com/environment/2012/apr/25/governments-catastrophic-climate-change-iea.

[41] DARA, "DARA Impact Matters," February 13, 2013, http://daraint.org/2013/02/13/4388/bloomberg-businessweek-climate-change-reducing-global-gdp-by-1-2-trillion.

[42] Nina Chestney, "100 Million Will Die by 2030 If World Fails to Act on Climate: Report," *Reuters*, September 25, 2012, http://www.reuters.com/article/2012/09/25/us-climate-inaction-idUSBRE88O1HG20120925.

[43] CDP North America, "Major Public Companies Describe Climate-Related Risks and Costs: A Review of Findings from CDP 2011-2013 Disclosures," May 2014, https://www.cdp.net/CDPResults/review-2011-2013-USA-disclosures.pdf.

44 World Business Council for Sustainable Development, "Vision 2050" (WBCSD, February 2010), http://www.wbcsd.org/WEB/PROJECTS/BZROLE/VISION2050-FULLREPORT_FINAL.PDF, 2.
45 Ibid.
46 Ibid.
47 Rakesh Kochhar, "10 Projections for the Global Population in 2050," Pew Research Center, accessed May 23, 2014, http://www.pewresearch.org/fact-tank/2014/02/03/10-projections-for-the-global-population-in-2050.
48 Ibid.
49 Ibid.
50 Ibid.
51 Thomas Malthus, *An Essay on the Principle of Population* (London: Electronic Scholarly Publishing Project, 1998).
52 Joshua Pramis, "Number of Mobile Phones to Exceed World Population by 2014," *Digital Trends*, accessed June 18, 2014, http://www.digitaltrends.com/mobile/mobile-phone-world-population-2014.
53 Danny Yadron, "Corporate Boards Race to Shore up Cybersecurity," *Wall Street Journal*, June 29, 2014, sec. Tech, http://www.wsj.com/articles/boards-race-to-bolster-cybersecurity-1404086146?tesla=y&mg=reno64-wsj.
54 "The Business of Doing Good: How Millennials Are Changing the Corporate Sector," *Forbes*, June 18, 2014, http://www.forbes.com/sites/jeancase/2014/06/18/millennials2014.

Chapter 2
Building the Business Case for Sustainability

Sustainable development is a philosophy based on the three pillars of people, planet, and profit. The fundamental concept is to meet the needs of our global society while ensuring that future generations can meet their needs as well. Sustainable strategy incorporates these principles into a management approach that believes that, through engaging with a broad range of stakeholders, management becomes better informed about opportunities and threats that impact the long-term health and viability of their organization. As a result, management creates a business strategy to generate opportunities, manage risks, provide competitive advantage, and generate a positive financial impact while protecting natural capital and improving social impact. Sustainable strategy provides a framework for a management philosophy that transforms people, process, and practice, creating business value while addressing financial, environmental, and social challenges.

Creating the case for company-wide adoption of sustainable strategy requires a sustainability champion to drive the agenda forward and a well-conceived plan to demonstrate correlation between the agenda and business value. Champions building a case for a sustainable transformation must demonstrate alignment between sustainability goals and business mission. Gaining CEO and C-suite support is crucial to a sustainable transformation. Sustainability champions improve their success rate by supporting their recommendation with research and case studies providing tangible examples of organizations that have adopted sustainable strategy, and become leaders in their industries and outperformed their competition. As senior leadership gains perspective on the benefits of sustainable strategy on business value creation, the sustainability champion is able to propose a full sustainability agenda. The process involves mapping a strategy that addresses natural, social, and economic strategies and their impacts on long-term value creation. In order to effectively engage senior leadership in a meaningful conversation about sustainability, a sustainability champion must make a meaningful business case. Benefits include cost savings, improved compliance, competitive advantage, improved financial returns, and greater access to capital. Considering value creation from a sustainability perspective includes a broader set of factors that are selected to assess the long-term impacts of management decisions. This approach considers not only economic metrics but also social and environmental metrics that, while not normally captured under financial reporting standards, have a significant impact on the financial performance of an organization. As with any project, the first step in recommending a sustainable strategy is demonstrating the business value it creates. Increasingly, investors and other stakeholders are seeking these alternative forms of operational assessment because these factors have a significant impact on the long-term viability of the organization and its return on investment. Ultimately, changes to the definition of organizational value creation will impact program and project selection and approval. Once management accepts the value proposition of sustainability, it changes how business is conducted within the organization.

2.1 Sustainable Strategy Drives Value

Sustainability drives value because it focuses on levers that drive opportunity, mitigate risks, reduce resource utilization, and leverage cooperation and collaboration. In addition, taking a broader holistic approach to organizational management by considering local and global communities, resource scarcity, air quality, and the health and happiness of employees creates a more productive and effective workplace.

This framework, illustrated in Figure 2.1, provides a roadmap for sponsors of sustainability initiatives to demonstrate the strategic benefits that a sustainable approach delivers to the core business mission through revenue generation, resource preservation, cost savings, risk mitigation, and stakeholder engagement. Revenue generation can come through new sustainable solutions and the opening of new markets, risk optimization through managing your supply chain in order to protect your reputation and brand, cost savings through resource conservation, and stakeholder engagement through investor relations, employee engagement, and customer relations. Global companies that embrace sustainable strategy reap the strategic benefits of this approach in terms of generating higher returns for their investors than their peer group. Sustainable strategy creates positive environmental, social, labor, human rights, and governance impacts while mitigating risk and generating long-term profitability. Focusing a sustainable lens on an organization's risk mitigation strategy identifies additional physical and operational risks. Proactive identification and planning for climate change risk, such as facility site impact assessment, alternative sources of raw materials, and assessment of exposure concentrations, has a significant impact on financial performance. Increasingly, mainstream investors are requesting transparency on environmental, social, and governance (ESG) issues. They perceive that management teams that report on ESG information operate better companies and generate more stable returns to investors. Strategic stakeholder engagement has moved beyond the traditional group of investors, employees, and customers to include a broad array of interested stakeholders. Through this processes of engagement, management can identify new opportunities and uncover previously unknown threats. Other areas for business value creation include operational improvements, resource savings, and employee attraction, retention, and engagement. Sustainability champions that build the business case for sustainable strategy using industry-specific examples, market-leader case studies, publically reported data, and metrics to support their proposal provide a compelling case for adoption.

2.2 CEO Perspective

Gaining chief executive officer (CEO) support is crucial to implementing sustainability within an organization. In those organizations that have embraced sustainability, CEOs have bought into the belief that they must be personally and publically committed to their organization's sustainability mission. Obtaining the CEO's support is crucial for a successful sustainability program. The CEO's perspective on sustainability shapes the program commitment, resource allocation, and organizational buy-in.

In 2013, the United Nations Global Compact (UNGC) and Accenture conducted their third survey of CEOs on the topic of sustainability as defined by the 10 principles of the UNGC on human rights,

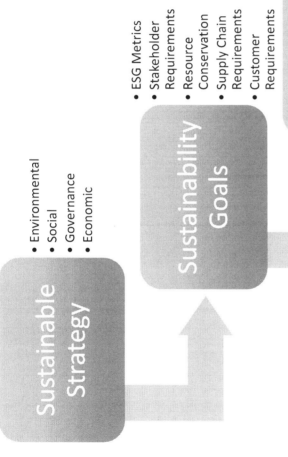

Figure 2.1 Creating Value Through Sustainable Strategy

labor, environment, and conflicts of interest. (See Chapter 1 for a complete listing of the UNGC's Principles.) The response group was the largest yet, including 1000 CEOs from 103 countries in 27 industries. According to the survey, 93% of the CEOs polled viewed sustainability as a key component of their business's success. Further, 76% of the CEOs said that they believed that embedding sustainable strategy into their core operations will results in new opportunities and revenue growth.

One of the key barriers to embedding sustainability has been the ability to tie sustainability to business value. The 2013 survey indicates that 37% of the CEOs cited the difficulty in linking sustainability initiatives and business value creation as a reason for not taking more action on sustainability.[1] Linking sustainability to business value creation remains a challenge for market leaders and is even more of a challenge for the next tier of organizations. In order to move sustainability forward within an organization, it is a crucial link that needs to be demonstrated. As a project sponsor supporting a sustainability project, it is important to identify and quantify the project's business value. In today's business climate, many projects are competing for resources. Projects undertaken for philanthropic or charitable reasons, without sound core business value creation with measureable metrics, are not likely to have long-term success. As organizations travel along their sustainability journey, those firms that have reached a more mature phase in their journey are better able to demonstrate value through creating innovation and competitive advantage, engaging customers and employees, and communicating strategy to communities, governments, and investors. These firms are able to clearly link sustainable strategy and value creation. A key to success is selecting areas that are material to your business. Ask yourself, "Where can we have the most significant impact?" Often, organizations that are beginning their journey select a project that has an environmental impact. While environmental projects create benefits, selecting one that has marginal impact will not garner sufficient managerial support. Encouraging employees to reduce printing through centralizing printers and promoting double-sided printing is a good first step in creating awareness and changing the culture, but it doesn't significantly impact core operational performance or improve financial performance. To embed sustainability requires selecting projects that align with core business values such as addressing customer requirements or reducing manufacturing costs. While sustainability projects incorporate environmental or social benefits, these projects need to be structured to include metrics that measure contribution toward business goals. Sustainability projects must withstand vetting against portfolio criteria to ensure that they deliver value for the organization.

CEOs are interested in sustainability and the impact that it has on organizational performance. Figure 2.2 represents an excerpt of the most frequently cited drivers for sustainability from a CEO's perspective. *The UN Global Compact-Accenture CEO Study on Sustainability* asked the question, "Which factors are currently driving you, as a CEO, to take action of sustainability issues?"

Not surprisingly, market factors such as brand, trust, and corporate reputation as well as the opportunity for revenue growth and expense reduction rank as the top motivations for CEOs to take action on sustainability. Clearly, they value the innovation, new solutions, and operating efficiencies realized through sustainable strategy. With global communication and round-the-clock access to information, protecting corporate reputations and brands is of crucial importance. Leading CEOs understand that sustainable strategy must be transparent and create alignment among environmental, social, and governance goals and corporate actions. Too often, the headlines are filled with stories of companies that have not followed a sustainable approach, and the consequences to their image and reputation are significant. Five years after its massive oil spill in the Gulf of Mexico,, BP's name is still in the headlines on issues ranging from settlement claims to the long-term devastation of marine wildlife in the gulf. As a result of a perception that BP lacked business integrity, the company was banned for 2 years from obtaining new contracts with the U.S. government.[3] In the CEO survey, consumer/customer demand is currently ranked as third in importance and has surpassed personal motivation as a factor from the 2010 survey. Increasingly, both business customers and end consumers are seeking sustainable solutions to support their own sustainability initiatives or to align with their own personal views on the environment and society. Employee engagement also rose in importance to CEOs from the 2010 survey

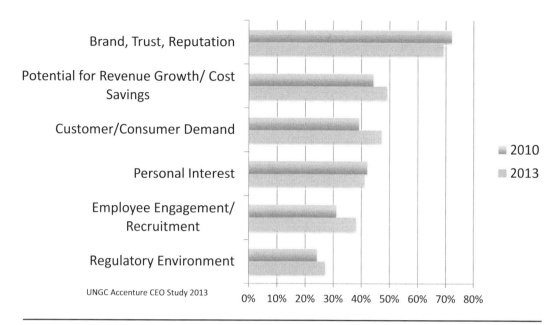

Figure 2.2 CEOs' Top Drivers for Sustainability. (Reproduced with permission)[2]

results as a driver of sustainability. Surveyed business leaders cited consumers and employees as the most influential stakeholder groups over the next 5 years.[4] The impact of sustainable strategy on employee engagement, retention, and attraction is an area of emerging importance for CEOs. They realize that in order to attract and retain the best and the brightest employees and to remain an employer of choice, their corporate sustainability image, message, and actions need to be in line with employee expectations. CEOs are becoming more aware of the need for collaboration and dialogue with consumers, employees, communities, and governments in order to offer new solutions and to broaden the pool of voices defining the role of business in meeting the environmental, social, and governance issues of the world. Those companies that have reported the most success with business value creation have used the triple bottom line (TBL) approach to evaluate internal and external ESG priorities and to create innovative and sustainable solutions for customers.

Mary T. Barra, CEO of General Motors Company, has provided insight into how she perceives the value of a customer-centric sustainability strategy:

> We have a customer-driven sustainability strategy. Customers expect us to help mitigate, if not eliminate, issues like congestion, pollution and traffic accidents, among others. If we expect our industry to continue to thrive, we must provide solutions. This extends to how we build our products and how we engage with the world around us. When it comes to sustainability, we pursue outcomes that create value for both GM and our customers. This has led to expanded use of renewable energy, a "zero waste" mindset and other initiatives that have sharply reduced our energy intensity, resource consumption and greenhouse gas emissions worldwide.[5]

This CEO's viewpoint demonstrates her belief in opportunity for innovation and new market solutions through a sustainability lens to provide alternatives and solutions for customers. Sustainable strategy creates benefits for both the organization and its customers.

While CEOs of large global organizations are often the most vocal on the benefits of sustainable strategy to their organization, leaders of small and mid-size organizations have found benefit as well.

Even though much of the published information is on public companies, it serves as a proxy for the impact that these types of strategies can have on privately held companies. As a starting point, select several market leaders in your industry and use the stories from their CEOs to help build your own organizational business case.

The owners of a mid-size wholesale distributorship became aware of sustainability because it was a strategic initiative for their major suppliers. When the opportunity arose to purchase the headquarter building, management began to think seriously about their own environmental impact and focused on energy efficiency projects. They began by exploring options such as building upgrades, including lighting retrofits, HVAC improvements, fuel cells, and solar power. Several peer businesses had benefited from these types of initiatives, and from speaking with fellow business leaders, management was ready to consider energy efficiency strategies and alternative energy sources.

In order to better understand the implication of these options, a historical analysis of energy usage and intensity relative to similar facilities in the area was prepared. Based on these energy usage patterns, opportunities were identified for education, behavior change, monitoring systems, and retrofits that could generate $100,000 in energy savings. This project also included investigating demand response options to generate conservation savings and to monetize energy assets. While the energy savings through new solutions and efficiencies were significant, the most interesting benefit was the potential to improve the market value of the building by lowering its operating costs. A building with lower operating costs generates higher net profits, resulting in a higher market valuation. Regardless of the size of the organization, there are numerous examples for sustainability champions to use to build the case of sustainable strategy adding value to an organization.

2.3 Financial Performance/Competitive Advantage

Research on stock market performance indicates that, over the long term, adoption of a sustainable strategy provides a competitive advantage generating financial performance that outstrips the competition. A study over a period of 18 years by Eccles, Ioannou, and Serafeim demonstrates that firms that had adopted sustainability policies and were transparent about their usage generated financial results and a corresponding stock price that outperformed their peer group. A $1 investment in 1993 in a portfolio of sustainable companies grew to $22.6 by 2010, representing a 50% higher return than the same $1 invested in a portfolio consisting of low-sustainability firms.[6] As the study found, those companies that have embedded sustainable strategy from the board level through senior management and into operations have reaped a financial benefit. Some of the best practices include tying compensation to sustainability metrics, establishing a process for stakeholder engagement, and measuring and reporting on sustainability results. This research demonstrates that firms that adopt sustainable strategy outperform traditional firms over time.

Another source supporting better financial performance for organizations that embrace sustainable strategy is the Carbon Disclosure Project, recently renamed the CDP. The CDP is an international nonprofit whose mission is to provide a global system for cities and businesses to measure, share, and manage environmental data on impacts to natural capital such as carbon emissions, water usage and quality, policies about forests, and energy. Their goal is transforming the way businesses and governments operate to reduce climate change impact and to protect natural resources. Research by the CDP found that industry leaders in climate change initiatives and disclosure outperformed industry laggards by a return-on-equity (ROE) margin of 5.2%, cash flow stability of 18.1%, and dividend growth of 1.6%.[7] In 2013, the CDP collected environmental disclosure information from organizations on behalf of its 722 investors, including banks, investors, wealth advisors, and pension funds that control $87 trillion in assets.[8] In order to be included in The Carbon Disclosure Project Leadership Index (CDLI), a firm must be in the top 10% of the Global 500 sample. In 2013, 81% of companies in the Global

500 responded to the questionnaire. For those companies that have reached leadership positions within the CDP indexes (CPLI or CDLI) because of their climate change policies and actions, their financial performance has outperformed the broader index. From 2005 to 2013, the CDLI companies generated total returns of 82.8%, outperforming their peer group of the Global 500 (49.6% in total returns) by two-thirds.[9] This analysis supports a correlation between quality sustainability performance and disclosure and organizational financial returns.

The benefits are not limited to large organizations. In order to translate these financial benefits to small to mid-size firms, our consulting practice created the Sustainability Business Model to provide a roadmap of sustainable strategies and their impact on generating new revenues and reducing operating costs. The model provides examples of strategies used by organizations and demonstrates the impact of implementation on an organization's income statement.

As illustrated in Figure 2.3, this model looks at traditional income statement categories such as sales, labor costs, energy, transportation, and packaging and gives examples as well as suggested impacts from these actions that can be traced directly to an organization's bottom-line impact. It provides sustainability champions with a model to begin assessing the value-creating impact that sustainable strategy can have on their own organization. The interactive model can be found on the HRComputes website at http://www.hrcomputes.com/sustainability.html.

Making the business case with financial metrics such as higher net income, improved market valuation, return on equity (ROE) and return on investment (ROI) adds credibility to the argument for sustainable strategy adoption. The economic benefit is clear for creating a sustainable framework that supports adopting policies, processes, programs, and projects that create value for the organization while protecting natural and social capital.

2.4 Managing Risk

Compliance with governmental requirements and regulations is a major driver of policy for all organizations. With growing frequency, organizations are being asked to comply with ever-increasing national, state, and local regulations for environmental, social, and governance standards. While Europe has been the traditional leader on the sustainability forefront, other countries are joining the bandwagon. U.S. President Barack Obama and Chinese President Xi Jinping made a joint announcement in the fourth quarter of 2014 that their two countries have negotiated an agreement to reduce greenhouse gas (GHG) emissions. China has agreed to increase its use of zero-emission energy sources to 20% by 2030 and to cap its emissions by 2030. The United States has pledged to reduce emissions to 26–28% below 2005 levels.[10] This agreement represents a significant increase in commitment to sustainability from two economic powerhouse countries. In order to reach these goals, all types of organizations from businesses to cities are going to be called upon to change their processes significantly in order to reduce emissions. As a starting point, organizations need to be aware of their carbon footprints in order to understand how they generate carbon. In Chapter 1, we included the most significant sources of carbon in the United States. Once management understands how they are generating GHG emissions, then actions can be taken to reduce them.

In 2014, The European Parliament adopted a directive requiring disclosure of nonfinancial and diversity information for large firms that are considered "public-interest" organizations. These are defined as organizations with more than 500 employees, balance sheets of 20 million euros, or net turnover of 40 million euros. Public-interest entities include exchange-listed firms, credit institutions, insurance firms, and others to be defined by the member states.[11] The scope of this directive includes approximately 6000 companies that operate across the European Union (EU). Impacted organizations will need to disclose results on policies, risks, and outcomes as they relate to environmental, social, and employee issues concerning human rights, anticorruption, bribery, and board diversity.[12] A number of

Small Business	Income Statement	ACTION	IMPACT	OUTCOME
Sales	$10,000,000	Green Products New Opportunities Price Premium Joint Ventures	5%	$500,000
Labor Cost	$1,050,000	Green Design Green Teams Workforce Engagement	10%	$105,000
Energy	$55,000	Benchmarking/ Monitoring Energy Usage Lamp Replacement/ Programmable Thermostats	10%	$5,500
Transportation	$45,000	Logistics/ Maintenance/ Telematics	10%	$4,500
Packaging/Waste	$22,000	Repurpose/ Reuse Recycling	25%	$5,500

Click on Action Links for further details

next

Figure 2.3 Sustainability Business Model

foreign-registered companies that are listed on EU-regulated exchanges will need to comply. The directive relies on several existing frameworks for guidance, such as the UN Global Compact, ISO 26000, and the GRI Sustainability Reporting Guidelines. Compliance with these guidelines will become part of an organization's license to operate in the EU.

These are just a few examples of the increased regulatory pressure for organizations that operate both globally and locally. While an organization may not fall directly under these mandates, there is a good chance that a valued client must comply. As larger institutions reach back into their supply chain to meet compliance requirements, small to mid-size companies are increasingly asked for information and evidence of sustainability programs, often with rigorous compliance requirements. Walmart is an example of a firm with supplier sustainability requirements.

2.5 Climate Change Risk

According to a Carbon Disclosure Project (CDP) report issued in May 2014, which reviewed the findings from CDP reporting disclosures from 2011 to 2013, companies are increasingly assessing the risks from climate change, and many have already experienced operational disruptions and the corresponding costs. Overall, the findings indicate that management is concerned and is evaluating physical risk from climate change. Operations are experiencing greater physical disruptions and greater costs associated with climate change. As indicated in Table 2.1, these risks are being tied to actual and potential business impacts that these management teams deem significant.

As indicated by the CDP report, climate change is not something that management of major corporations is taking lightly. They are assessing risks, reporting on them to stakeholders, and developing strategies, plans, and alternatives to address the very real business impact presented by climate change. In order to manage risk within an organization effectively, both compliance and operational climate change risk need to be evaluated as part of the risk assessment equation. Mitigating negative risk and maximizing positive risk adds business value. As risk plans are developed and risk responses created, climate change risk should be incorporated. Projects such as developing emergency preparedness to handle natural disasters at various facilities need to be planned and implemented. Managing these types of risks goes beyond emergency preparedness. Managers of programs and projects need to consider the impact of climate change risk on ongoing and proposed operations and projects. A site selected for a new production facility could become compromised if sea levels rise. Suppliers of cotton could lose their crops because of drought. Increasingly, insurance providers are requesting identification and disclosure of these risks from clients to determine underwriting standards.

While insurance industry leaders are taking climate change seriously, there is room for improvement across the industry. Ceres, a nonprofit organization that advocates for sustainability leadership, received 330 insurer responses to their 2014 Climate Risk Disclosure Survey, which represents approximately 87% of the U.S. insurance market. It should be noted that the response rate to the survey had increased 80% from 2011, reflecting the growing importance of this issue to insurance companies. The assessment was based on climate risk focus in the following categories:

1. Company governance structure
2. Enterprise-wide management programs
3. Computer modeling to manage
4. Stakeholder engagement
5. GHG emissions reduction programs and monitoring[14]

Large market leaders such as The Hartford, Prudential, and Zurich Insurance demonstrated much more robust climate risk management practices than the smaller firms. Property and Casualty (P&C)

Table 2.1 S&P 500 Organizations Description of Climate Change Risk and Business Impact[13]

Industry	Company	Event/Potential Risk	Business Impact
Consumer staple	J. M. Smucker Co.	Rising raw material costs due to weather impact on coffee and peanuts.	Rising operating costs.
Consumer discretionary	GAP	Drought led to lower cotton yields and correspondingly higher cotton prices, impacting gross margins.	Rising operating costs.
Energy	Hess	Superstorm Sandy caused approximately $20M in damage to refining facility, terminal networks, and retail operations.	Disruption in production capacity.
Finance	Allstate Corp	Potential for catastrophic losses from hurricanes in major metropolitan areas in the Eastern and Gulf Coast areas of the U.S.	Inability to do business.
Health care	Humana	Climate change has the ability to influence disease vectors, impacting the spread of infectious disease and the rate of incidences, impacting the types of services offered to members.	Rising operating costs.
Industrials	Union Pacific Corp	Extreme weather can impact customers such as the agricultural business, which can negatively affect the needs for transportation services.	Decreased demand for goods/services.
Information technology	Hewlett-Packard	Flooding created supply chain disruption. The potential impact is $1B loss in annual revenue.	Disruption in production capacity.
Materials	Alcoa	Hydroelectric power in the Southeastern U.S. has been impacted by lower rainfall, increasing costs. Drought conditions in Australia have inhibited bauxite refining.	Increased operating costs. Supply disruption.
Telecommunications	Verizon	Rebuilding of wireline and wireless networks in the NY and NJ area after Superstorm Sandy. Impact of 0.07/per share to 4th Q 2012 earnings.	Disruption in production capacity.
Utilities	Ameren Corporation	Climate change could impact consumer usage patterns, disrupt maintenance, and increase downtime. Demand for electricity and natural gas could be impacted reducing revenues.	Access to capital. Decreased demand for goods/services.

insurers exhibited a greater understanding of the risks of climate change to their business than Life & Annuity (L&A), and Health insurers. This finding is not surprising given that the data in Table 2.1 highlights business interruption and property & casualty risks. P&C insurers have experienced this type of risk from extreme weather, rising sea levels, and changes in weather patterns. However, L&A and Health insurers were found to show less action relative to both their core operations and their investment strategies. Lastly, only 10% of respondents issue a public climate risk management statement explaining their view of climate change risk and its impact on their core business operations and investment policies.[15] While climate change risk is a major challenge for the insurance industry, these data suggest that the majority of the industry needs to play catch-up to the market leaders in terms of their risk management strategy.

Ceres recommended best practices for a climate change risk management plan include:

1. Create a Climate Risk Committee at the Board and C-Suite Level
2. Develop and Publish a Public Climate Risk Policy
3. Integrate Climate Risk into Risk Assessment Methodologies and Modeling Systems
4. Assess Climate Risks and Opportunities in Investment Portfolios
5. Engage with Key Stakeholders on Climate Risk
6. Provide Climate Risk Disclosures to Regulators
7. Participate in Industry Initiatives around Climate Risk

The Ceres best practice findings for the insurance industry are applicable to other industry sectors as well. A climate change risk management plan should include senior management oversight, internal and external stakeholder engagement, risk assessment, technology selection, and implementation considerations. Engaging in this planning process allows for a risk mitigation strategy and an emergency preparedness plan in the event of a major storm or natural disaster. In order to identify and manage risk for an organization effectively, climate change risk needs to be part of the risk management plan.

2.6 Creating New Opportunities and Products

Unilever has developed its business platform to maximize the impact of its sustainable strategy. Their philosophy considers Earth as a planet with its resources under strain to meet the needs of 7 billion people. From a global perspective, Earth's population faces many challenges from resource scarcity, water quality and shortages, living standard in equities, biodiversity threats, and climate change. Through its Unilever Sustainable Living Plan (USLP), CEO Paul Polman explains how this blueprint for sustainability will lead to doubling their business while reducing their environmental footprint and enhancing their social impact. Three years after the introduction of the USLP, he attributes their sustainability focus with helping to drive profitable growth of their brands, fueling innovation, generating cost savings, and reducing risks. In 2014, they are focusing on three major initiatives to support environmental and social change. These include reducing threats resulting in climate change through assisting in ending the practice of deforestation, improving food security through promoting and supporting agriculture and small farmers, and ensuring water quality and hygiene by facilitating better access to safe drinking water and soap.[16] Unilever's leadership team is committed to the concept that sustainability and profitability are not mutually exclusive, and that driving social and environmental change and creating long-term value are both possible.

Here is an example of Unilever's success in reinvigorating their largest brand, Knorr, which includes soups, bouillon cubes, sauces, seasonings, and snacks, based on a sustainably sourced ingredients strategy. They created a new market opportunity by focusing on customer requirements for sustainably sourced products.

Based on research in 11 countries, they found that 75% of consumers would be more likely to purchase a product if they were aware that it was sourced from sustainable ingredients. In response to this stakeholder input, Knorr focused its marketing message around its strategic commitment to sustainably source 100% of its agricultural ingredients. In order to communicate this strategic benefit to consumers effectively, the company created a package logo for the Knorr Sustainability Partnership, which helps consumers understand and make choices for sustainably sourced products. In Germany, which is Knorr's largest market, these efforts have further strengthened its brand value.[17] As a result,

Knorr is growing in consumers' product preference and resulting in improved financial performance for the brand. Unilever is an example of a transformational company that has fully integrated sustainability into its culture. As a result of Unilever's strategic focus, it has improved stakeholder engagement with both customers and investors by offering a product better aligned with consumers' values and generating higher returns for investors.

2.7 Access to Capital

Increasingly, investors are asking for information about an organization's sustainability programs. They view sustainability initiatives as a means for companies to maximize returns by generating new opportunities, reducing costs, and reducing risks. The UN Principles for Responsible Investment (PRI) has two UN partners, the UNEP Finance Initiative and the UN Global Compact and 1325 signatories from international asset owners, investment managers, and service providers with $45 trillion in assets under management.[18] The goal of the PRI is to better understand the impact of sustainability for investors and to encourage signatories to include sustainable criteria on environmental, social, and human capital in their investment decisions. UNPRI is the largest network for investors around the globe to demonstrate a commitment to responsible investment.[19]

> **UNPRI Six Principles for Responsible Investing[20]**
>
> Principle 1: We will incorporate environmental, social, and governance (ESG) issues into investment analysis and decision-making processes.
> Principle 2: We will be active owners and incorporate ESG issues into our ownership policies and practices.
> Principle 3: We will seek appropriate disclosure on ESG issues by the entities in which we invest.
> Principle 4: We will promote acceptance and implementation of the Principles within the investment industry.
> Principle 5: We will work together to enhance our effectiveness in implementing the Principles.
> Principle 6: We will each report on our activities and progress towards implementing the Principles.

Asset owners that are signatories include major corporations, unions, and pension funds, such as Lloyds Banking Group, the AFL-CIO, and CalPERS.[21] In order to facilitate reporting on their progress, PRI created a framework to provide a common structure, language, and metrics for its institutional investors. It better informs the market on their process for embedding ESG criteria into their investment decisions.

As the PRI data indicate, investors are increasingly considering ESG data to better understand the future risks and opportunities of their investments. In order to have the broadest possible access to capital, chief financial officers should be aware of these investor trends. As the senior management team evaluates acquisitions, major projects, and new lines of business, they need to be aware that investors will increasingly be asking for transparency on sustainability performance and future projects. A dedicated investor group known as Socially Responsible Investors (SRI) aligns their values with their investment strategy considering environmental, social, and governance performance as well as financial returns. SRI investments have grown by 22% to $3.74 trillion in assets under management. In the

United States, $1 of every $9 under management can be classified as an SRI investment.[22] This trend has not gone unnoticed by some of the most significant investment management firms.

Under the leadership of Chairman James Gorman, ,Morgan Stanley launched in 2013 the Morgan Stanley Institute for Sustainable Investing to provide financial market–based solutions to economic, social, and environmental challenges. They have a three-prong approach, which includes providing products and solutions for clients to invest in sustainability-oriented strategies, thought leadership on increasing the flow of capital toward sustainable investments, and strategic partnerships to develop best practices and build capacity in the area of scalable sustainable investment. Morgan Stanley has created the "Investing with Impact Platform" with a goal of generating $10 billion in client assets under management with the next five years. They have a focus on creating new products in which environmental and social impact is a crucial component of the investment strategy. To support thought leadership and innovation, they have established an annual Sustainable Investing Fellowship program at Columbia Business School. This fellowship enables graduate students to pursue sustainable investing thought leadership, internships, and product innovation. In addition, they are investing $1 billion in a sustainable communities initiative and working with strategic community partners to support affordable housing for low- and moderate-income households, with access to other key services such as health care and nutrition. Morgan Stanly believes that the Investing with Impact Platform provides products and services to investors that both meet their return objectives and address global social and environmental challenges. According to Mr. Gorman, "Our clients are increasingly turning their attention to what it takes to secure lasting and safe supplies of food, energy, water, and shelter necessary for sustainable prosperity."[23]

While Morgan Stanley is offering this type of investment platform to support its own core sustainability mission, it is also entering this marketplace with a significant investment and resource commitment because the number of investors interested in responsible investing has grown to a critical mass. Offering socially responsible investing has become a viable business model for the firm, with a long-term revenue potential.

In making the case for sustainable strategy, CEOs and CFOs are paying attention to the growing numbers of investors and investment managers interested in investing in companies that incorporate sustainable strategy into operations and report on results using environmental, social, and governance standards. For those who are sponsoring or managing large multiyear projects that require funding from investors or financiers, sustainability standards are increasingly relevant for accessing this type of funding.

2.8 Engaging Stakeholders

Stakeholders represent both internal and external groups that have an interest in the organization's mission, function, and operation. Traditionally, stakeholders were shareholders, customers, and employees. Today's definition includes community groups, environmental groups, and human rights groups who are seeking an avenue to voice their concerns to the company. Increasingly, stakeholders are becoming shareholders in order to have their voices heard by management. They are using shareholder proposals as a means of asking tough questions about corporate sustainability policies and programs. A group of investors, known as Socially Responsible Investors (SRIs), limit their investments to companies that meet their political, environmental, social, and humanitarian philosophies. Traditionally, SRIs were foundations, religious groups, and tax-exempt organizations. However, environmental, social, and governance (ESG) issues are increasingly becoming more of a concern for mainstream investors such as pension funds, unions, and institutional investors. In 2014, there were 417 environmental and social shareholder proposals filed, which is an increase of 50 over 2013. Major issues were political activity (30%), environment issues focused on climate and energy (19%), sustainability (12%) and diversity

(11%), and human rights (9%).[24] Although in 2014, traditional SRIs continued to file the most shareholder resolutions (31%), other groups are growing in their activity. Investment managers such as Calvert Investment or Trillium Asset Management often file shareholder proposals on behalf of their foundation clients. Mainstream investment managers such as pension funds have become increasingly significant, filing 24% of ESG shareholder resolutions. Both New York and California public pensions are active in filing resolutions. The New York State Common Retirement Fund and the New York City Pension Fund each filed 12 proposals. Other significant environmental and social shareholder proposal sponsors include religious groups (17%), unions (9%), and foundations (7%). A significant trend has been the increasing support that these shareholder proposals have received from mainstream shareholders. Overall shareholder support almost doubled during the 10 years from 2004 to 2013, rising to 21.3%. As shareholder support approaches meaningful threshold levels (usually 25%), boards of directors are taking these proposals more seriously. It should be noted that these numbers are somewhat understated because they exclude proposals that were withdrawn because the issue was resolved through meeting with management.

> **Ceres**
>
> Ceres is nonprofit organization that promotes sustainable leadership. As part of its mission, it tracks shareholder resolutions on ESG issues such as climate change, energy, water scarcity, and sustainability reporting. The resolutions are filed by its investor network and include some of the largest public pension funds, foundations, and religious organizations in the United States. In 2014, SRIs filed the largest number of shareholder resolutions on corporate political activity so far. Their concern is that management is making political contributions without shareholders' knowledge, approval, or monitoring. The issue is one of transparency and corporate governance. From a shareholder perspective, what controls are in place to keep executives from using corporate funds to promote personal interests? In response to continued concern by investors, U.S. Senator Robert Menendez (D-NJ) has introduced the Shareholder Protection Act into the U.S. Senate. The act requires greater transparency and disclosure on money spent for political purposes and advocates that shareholders be given an opportunity to vote on political budgets.[25]
>
> For those who think that investor activism is something that can be easily managed and dismissed, think about the 2010 Dodd-Frank law, which requires that CEO pay for public companies be voted on by shareholders. CEO pay had been a top concern for investors and appeared frequently in shareholder proposals.[26] As a result, these concerns moved beyond the boardroom and were addressed via legislation.

Shareholders remain key stakeholders for sustainable organizations, and shareholder activists are an influential group. As a result, corporate leaders are taking notice and action on these types of environmental and social shareholder proposals. They are establishing programs to engage with shareholders to better understand their concerns and to address issues proactively.

2.9 Competitive Ratings and Rankings

Investors and other stakeholders are demanding increased transparency of information on ESG risks and opportunities. As a result, leading corporations are making sustainability reporting part of their reporting process. Because of increased demand from investors, customers, governments, and NGOs for a way to make sense of this information by comparing different companies' performance, there has been a significant increase in the number of ratings and rankings indices over the past decade. The number of sustainability ranking indices grew from 21 in 2000 to 108 in 2010.[27] Competitive industry rankings have garnered attention from the boardroom and the C-suite because investors and consumers are increasingly turning to them to evaluate investments and purchasing decisions. Because of the proliferation of indices, the original goal of offering consumers and investors a simple way to compare companies on sustainability performance has become clouded. With so many, which ones are the best to consider? According to a joint survey from SustainAbility and GlobeScan of 850 sustainability professionals across 70 countries spanning business, government, NGOs, and academia, the most widely recognized and credible indexes are

1. Dow Jones Sustainability Index
2. Carbon Disclosure Project Leadership Index
3. FSTE4-Good Index Series 55[28]

These are useful indices not only for investors, consumers, and competitive rankings, but also business leaders who have been tasked with driving sustainable transformation within their organization. Even though an organization may not meet the eligibility requirements to be considered, these indices' questionnaires provide a framework to create a sustainability business case and offer guidance on establishing policy, managing the process, and reporting on sustainability outcomes.

The Dow Jones Sustainability Indices (DJSI) are a joint product of S&P Dow Jones Indices LLC and RobecoSAM AG, a sustainability investment specialist founded on the premise that integrating ESG criteria into financial analysis provides better informed investment decisions. The DJSI is developed based on the Corporate Sustainability Assessment, which is an annual ESG analysis of over 2500 companies. While the rankings are important for competitive assessment, taking a deeper look at the questionnaire for annual assessment gives an idea of the depth and breadth of value-add variables considered material by sustainability industry leaders. The questions include economic, environmental, and social criteria, including items such as board structure and diversity, governance, tax strategy, and anticorruption policies. The criteria reach into both the strategy and operations of the organization to measure the degree to which sustainability has been embedded. A copy of the questionnaire is available at http://www.robecosam.com/en/sustainability-insights/about-sustainability/robecosam-corporate-sustainability-assessment.jsp.

Reviewing this questionnaire provides a great framework for sustainability champions to develop a case for creating business value by adopting policies, programs, and processes in support of environmental, social, and governance agendas. Organizations that are listed on the DJSI are recognized as sustainable business market leaders, benefitting from preferred access to capital, brand loyalty and awareness, new opportunities and partnerships, employee engagement, and talent pool preference, to name a few of the business benefits of being listed.

The CDP develops the CDP Leadership Index (CDPLI) from the top performers in its annual climate change questionnaire issued to over 5000 global public companies. In 2014, the CDPLI reflected 187 businesses chosen from a pool of 2000 companies that had their climate disclosure independently assessed and ranked by the CDP's scoring methodology. Companies that are listed on the CDPLI are doing the most to combat climate change. As part of their mission to include a wider measurement of impact on the earth's natural capital, CDP has broadened the scope of its questionnaire to include water and forests along with carbon, energy, and climate. These results are reported to the world's largest investors in order to facilitate their investment decision process.

The 2014 CDP report on S&P 500 industry leaders indicated that the metrics from respondents to the CDP questionnaire had 67% higher ROE than nonresponders. Top performers demonstrated

36 Becoming a Sustainable Organization

Table 2.2 Largest CDP S&P 500 Top Performers[29]

Top CDP Performers from S&P 500 Q1 Ranking 2014	Market Capitalization (billion)
Apple Inc.	$612
Microsoft Corporation	$370
Johnson & Johnson	$290
Wells Fargo Na Co.	$267
WalMart Stores, Inc.	$245
Chevron Corporation	$244
JP Morgan Chase & Co	$223

lower earnings volatility and better dividends to shareholders. Some of the largest and most successful companies in the United States are on this list, as excerpted in Table 2.2.

The CDP questionnaire focuses on three major areas: governance, risk & opportunity, and emissions (see Figure 2.4). The purpose is to understand how climate change is managed within the business, how climate change regulation and the physical impact of climate change such as storms and flooding impact operation, and an organization's scope and tracking of emissions. Assessing an organization on these categories provides insight into opportunities to create or protect value in an organization.

Major global investors view adoption, disclosure, and reporting on sustainability performance as a business imperative and a competitive benchmark for investments. Benefits of disclosure include greater transparency to stakeholders, demonstrating preparedness for climate change challenges, highlighting new business opportunities, and improved operating efficiencies and reduced costs.

Drilling down into the reports by industry-specific data is useful for identifying commonly reported opportunities and risks related to climate change. This level of detail is useful to sustainability

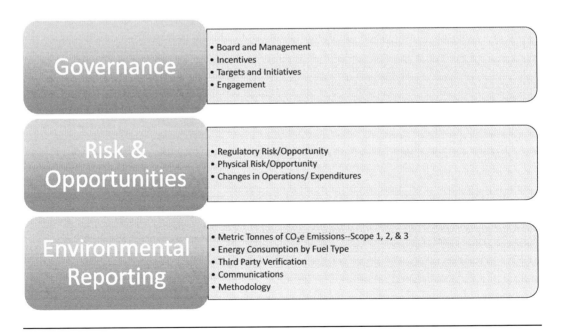

Figure 2.4 CDP Climate Change Questionnaire Highlights[30]

Table 2.3 Healthcare Industry Sector Responses[31]

Opportunities	Percentage of Respondents
Reputation	63%
Changing consumer behavior	40%
Cap-and-trade schemes	37%
Other physical climate opportunities	27%
Emission reporting obligation	23%

Greatest Risks	Percentage of Respondents
Cap-and-trade schemes	55%
Changes in precipitation and droughts	50%
Reputation	40%
Fuel/energy and regulations	35%
Tropical cyclones	35%

champions in preparing their business cases as well as to program and project managers in identifying risks that require action in the risk response plan. Table 2.3 highlights some of the opportunities and risks as reported by the healthcare industry.

Depending on the vantage point from which management considers a challenge, it can represent both an opportunity and a threat. While physical climate change presents threats such as operational disruption from storm damage and flooding, it also provides opportunities for growth through new markets and solutions. Changes in mean temperatures may lead to new opportunities for healthcare providers to expand their market, as the location of disease may change with climate changes.[32] While this information is specifically for the healthcare industry, these CDP industry surveys provide a reference point in terms of risks and opportunities for a variety of industries. For sustainability champions, they can facilitate developing a case for sustainable strategy through identifying specific industry risks and market-based opportunities as well as highlighting actions of market leaders concerning sustainability.

Similar to the other indices, the FTSE4Good Index is designed to meet the needs of responsible investors. Criteria for inclusion requires a firm to be working toward environmental sustainability, developing positive stakeholder relationships, supporting human rights, supporting good labor standards in their supply chain, and ensuring antibribery policies. Certain industry sectors, such as tobacco producers, weapons manufactures, and nuclear power stations, are excluded from eligibility. Reviewing the criteria and standards for your industry provides a proxy for reporting standards, threats, and opportunities to consider in creating your organization's sustainability plan.

The leading organizations on these indices are not only the leaders in sustainability, but also some of the most successful organizations within their industries. As sustainable market leaders, they benefit from opportunities such as invitations to participate in joint ventures, entry into new markets, and a seat at the table to discuss legislative changes. They have meaningful and documented sustainability stories to share with customers, investors, and employees, making them organizations of choice for this group of stakeholders. Leadership's belief that alignment between sustainable and business strategy adds value to their organization creates a culture in which meaningful TBL value is created.

For organizations that have the opportunity to create rather than re-create their organizational structure, incorporating sustainability as part of the foundational process allows for more rapid adoption and scaling for global growth. For privately held companies, there is extensive information on standards and requirements to become a B or a Benefit Corporation, which promotes environmental, social, and governance standards as part of its corporate charter. B Corporations are founded on the principle of "using business as a force for good."[33] Information sources include B Corp member annual reports and the Impact Assessment, which rates organizations based on alignment with metrics and standards on issues such as transparency, environment, community, and governance. Companies that have been identified as creating the most impact for a better world are listed in a ranking index known as the B Corp Best for the World.

In addition, some U.S. states, such as New Jersey, are offering sustainable business registries that provide a marketplace for consumers and other businesses to locate products and services from "green" businesses. They help promote sustainable businesses, provide a competitive advantage, create opportunity for new revenue, reduce cost, and reduce risk. The purpose of the New Jersey Sustainable Business Registry is to promote sustainable practices by providing information, tools, and support to help organizations begin their sustainability journey. In addition, the website tracks the aggregate impact these sustainable businesses are having on the state in areas such as reduced carbon emissions, water savings, and overall cost savings. For those who wish to engage in the sustainability process, there is a series of questions to help business owners identify opportunities to incorporate sustainability into their own operations. Organizations meeting the selection criteria are listed as a sustainable organization in a searchable database. The competitive advantage comes through being identified as a sustainable business for other like-minded organizations, businesses, and consumers. Other benefits include technical assistance, information, webinars and conferences, contests, and the ability to promote your status as a sustainable business. The New Jersey Small Business Development Centers (NJSBDC) launched the New Jersey Sustainable Business Registry in 2014, and it can be found at http://registry.njsbdc.com.

> According to Edward Kurocka, the NJSBDC's sustainability program manager, even though the registry is relatively new, some of its members are leveraging their status and accomplishments to create new opportunities or further existing initiatives. More often than not, members who are willing to share their stories are often already committed to sustainability, and have a strong desire to see it spread into all sectors and the community. Working in conjunction with the NJSBDC, these members are approaching and encouraging their chambers of commerce and business/industry associations to promote this initiative to their memberships (articles about members, the registry, and the NJSBDC's no-cost. sustainability consulting are appearing in chamber publications throughout the state). These efforts to enlist support from business organizations are valuable in themselves, but when a community sustainability initiative engages with the business community—such as is the case with Sustainable Morristown piloting programs with the local chamber of commerce—it creates the kind of synergy necessary to broaden sustainability's reach and impact.[34]

Impact of the NJ Sustainability Business Registry on Small Business

Talia Manning is the founder and Chief Creative Officer of Chameleon Studios—a marketing and graphic design firm based in Princeton, NJ. Chameleon Studios works with a variety of industry sectors, and creatively adapts their services to achieve the specific vision of each of their clients. From the beginning, sustainability was integrated as one of four core pillars in the Chameleon Studios company mission: commitment to good design, outstanding customer service, making a positive impact, and having fun along the way.

Chameleon Studios began by implementing sustainability initiatives focused on their small facility: adjusting behaviors to conserve water, gradually replacing lights with LED bulbs, and double-siding prints to reduce paper usage. After establishing energy benchmarks, the firm introduced additional changes, such as opting to use fans and open

> windows whenever possible, rather than continuously air conditioning—which helped to realize a 25% decrease in energy usage during the peak summer months in 2014 over the previous year.
>
> Another critical aspect of Chameleon Studios' commitment to sustainability is that they encourage clients to establish good marketing practices. Some changes are relatively simple, such as trying not to overprint (and when you do print, work with printers that use vegetable-based inks and nontoxic processes). Other changes require more help, such as learning to leverage electronic marketing opportunities and designing multiuse materials that are longer-lasting and reduce waste. And some changes require big shifts in approach, such as implementing inbound marketing techniques that inform buyers about how to make purchase decisions based on "green" practices in your industry that set you apart from your competitors.
>
> Consumer polls in a variety of industries show that people will spend more money and invest greater loyalty in businesses that are more sustainable than in their competitors. So how did Chameleon Studios, as a firm that has already invested in their commitment to sustainability, demonstrate this competitive edge in a way that will capture the attention of prospective clients and stakeholders?
>
> That is where the Sustainable Business Registry comes in. By offering vetted recognition of the firm's sustainable activities with measurable outcomes, it gives Chameleon Studios credibility as a sustainable business. The firm's marketing statements are now backed by the Sustainable Business Registry seal, which sets them apart as not merely "greenwashing" but substantiating a meaningful sustainability commitment. As a marketing firm, Chameleon Studios intuitively understands the value this has to business development—particularly among clients who share the firm's passion for sustainability, which is an important niche market in which sustainability-minded businesses have the opportunity to grow actively. The registry affords a fantastic opportunity for all businesses that understand the importance of sustainability to achieve greater recognition, value, and growth.[35]

Even for privately held companies, competitive ranking and ratings of sustainable organizations are becoming more widely available for customers, communities, investors, bankers, and insurance companies to review and evaluate. Perceptions around sustainability are driving decisions about where and with whom to conduct business.

2.10 Operational Improvements and Energy Savings

In conjunction with the World Wildlife Fund (WWF), in 2013 the CDP published the 3% Solution study, which outlines a plan for the U.S. corporate sector to reduce carbon emissions by 3% per year between 2010 and 2020 while creating $780 billion in savings after covering program costs. Savings will come from improved energy efficiency as a result of behavior and management changes, introduction of technological improvements, and further deployment of low-carbon energy solutions.[36]

Operational improvement is an area where portfolio, program, and project managers traditionally have a significant impact. Opportunities abound for energy savings, design improvements, new production processes, waste reduction, and low-carbon solutions. Energy savings are often an entry point for organizations just beginning their sustainability journey. Engaging employees to generate new ideas and innovative solutions as well as advocating and promoting behavior change furthers the impact of these programs.

2.11 Project and Portfolio Management Impact

Creating the link between sustainability and long-term value creation in an organization lays the foundation for senior management engagement. The role of the sustainability champion is to develop a plan that demonstrates value creation through adoption of sustainable strategy. There are many drivers of sustainability including higher financial returns, brand and reputation protection, new markets, customer demands for sustainable solution, improved access to investors, employee engagement, cost savings, and risk mitigation to highlight. Reviewing competitive ranking criteria and industry-specific data facilitates identifying material issues and areas of impact for the organization. There is a wealth of information including case studies to demonstrate the link between adopting sustainable strategy and creating business value. The Sustainability-Driven Business Value Framework is designed for sustainability champions to facilitate a conversation about the ways in which sustainable strategy adds value to an organization.

This framework, as illustrated in Figure 2.5, identifies how taking a sustainable approach drives value through opportunity creation, risk mitigation, competitive advantage, and financial impacts. It further breaks these areas into specific benefits. Sustainability champions can use this framework to help identify and communicate with senior management about opportunities and threats to the organizations as well as benefits associated with harnessing the power of a sustainable approach. Lay out the framework to highlight sustainable strategy's impact in areas such as improving competitive ranking, developing new sustainable products or services, reducing supply chain risk, engaging employees, and broadening the investor pool.

Once senior management understands the business value proposition and adopts a sustainable strategy, the definition of creating business value within the organization reflects the broader definition of a triple bottom line approach. Adopting a sustainable strategy reframes the conversation around portfolio component selection and the requirements for programs and projects selection, implementation,

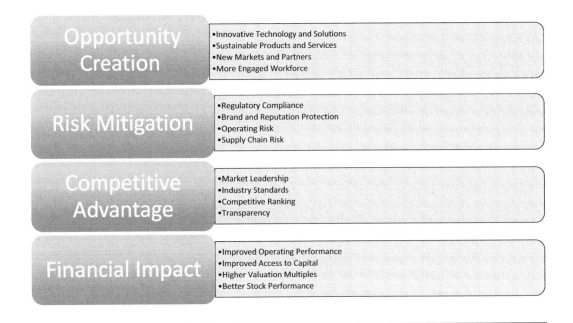

Figure 2.5 Sustainability-Driven Business Value Framework

and success. Sustainability projects become integrated into core business functionality rather than being one-off projects.

According to the UNGC-Accenture survey, CEOs are increasingly concerned that their organizations have reached a plateau in terms of their ability to further integrate sustainability into operations. They are reporting that their companies are not advancing in their journey in sustainability in terms of reaching new levels and scaling sustainability at a pace to meet global challenges and to achieve business objectives.[37] CEOs are seeking tools to close the sustainability performance gap, moving toward systemic and structural changes that fully integrate sustainable strategy into people, policies, and processes that fundamentally change the way in which the organization operates. This performance gap creates opportunity for sustainable portfolio management to create better alignment between strategic sustainability objectives and selection of programs and projects. Moving sustainable strategy from a management principle to an operational action is an important role for portfolio, program, and project managers.

2.12 Conclusion

This chapter has focused on the value creation benefits of sustainable strategy. We have confirmed that CEOs are interested in creating sustainable organizations because in the long term this type of organization provides longevity and economic benefits for all stakeholders. Several studies and surveys have verified that sustainable organizations outperform their less sustainable peers. Benefits come from improved investor relations, greater access to capital, new opportunities and markets, lower risks, operational improvements, and competitive advantage. Creating a business case for a sustainable strategy is the first step in creating a sustainable organization. In order to begin the journey, it is crucial to have C-suite support.

Notes

[1] Accenture and United Nations Global Compact, "The Accenture-UN Global Compact, CEO Study on Sustainability (2013)," 2013, 12, www.unglobalcompact.org.
[2] Ibid.
[3] "U.S. Lifts Ban Blocking BP from New Government Contracts," *Reuters*, March 13, 2014, http://www.reuters.com/article/2014/03/13/us-bp-usa-contracts-idUSBREA2C24E20140313.
[4] Accenture and United Nations Global Compact, "The Accenture-UN Global Compact, CEO Study on Sustainability (2013)," 37, 38.
[5] CDP North America, "Climate Action and Profitability, CDP S&P 500 Climate Change Report 2014," January 14, 2015, 26, https://www.cdp.net/CDPResults/CDP-SP500-leaders-report-2014.pdf.
[6] Robert Eccles, Ioannis Ioannou, and George Serafeim, "The Impact of a Corporate Culture of Sustainability on Corporate Behavior and Performance," *Working Papers—Harvard Business School Division of Research*, November 2011, 21.
[7] Bruce Kahn and Marc Fox, "Linking Climate Engagement to Financial Performance: An Investor's Perspective" (Sustainable Insight Capital Management and CDP, September 2013), https://www.cdp.net/CDPResults/linking-climate-engagement-to-financial-performance.pdf.
[8] CDP, "Global 500 Climate Change Report 2013," September 12, 2013, https://www.cdp.net/CDPResults/CDP-Global-500-Climate-Change-Report-2013.pdf.
[9] Ibid., 17.
[10] Lenore Taylor et al., "US and China Strike Deal on Carbon Cuts in Push for Global Climate Change Pact," *The Guardian*, accessed January 15, 2015, http://www.theguardian.com/environment/2014/nov/12/china-and-us-make-carbon-pledge.

11. Governance & Accountability Institute, Inc., "New Corporate Sustainability Reporting Disclosure Mandate in the European Union," accessed October 20, 2014, http://grifocalpointblog.org/usa/wp-content/uploads/sites/12/2014/09/EU-Directive-for-Non-Financial-Reporting-US-Companies-FACT-SHEET-FINAL.pdf?dm_i=1VZV,2WG3V,HC55H0,AHQDQ,1.
12. European Commission, "The EU Single Market, Non-Financial Reporting," accessed October 20, 2014, http://ec.europa.eu/internal_market/accounting/non-financial_reporting/index_en.htm.
13. CDP North America, "Major Public Companies Describe Climate-Related Risks and Costs: A Review of Findings from CDP 2011-2013 Disclosures," May 2014, https://www.cdp.net/CDPResults/review-2011-2013-USA-disclosures.pdf.
14. Ceres Insurance Program, "Insurer Climate Risk Disclosure Survey Report & Scorecard: 2014 Findings & Recommendations," October 2014, http://www.ceres.org/resources/reports/insurer-climate-risk-disclosure-survey-report-scorecard-2014-findings-recommendations/view.
15. Ibid.
16. Unilever, "Making Progress in 2013," accessed January 15, 2015, http://www.unilever.com/sustainable-living-2014/our-approach-to-sustainability.
17. Unilever, "Embedding Sustainability: Sustainability-Led Growth," accessed August 18, 2014, http://www.unileverusa.com/sustainable-living-2014/embedding-sustainability/index.aspx.
18. PRI, "About PRI Initiative," *Principles for Responsible Investment*, accessed August 11, 2014, http://www.unpri.org/about-pri/about-pri.
19. Ibid.
20. PRI, "The Six Principles," *Principles for Responsible Investment*, accessed August 11, 2014, http://www.unpri.org/about-pri/the-six-principles.
21. PRI, "About PRI Initiative."
22. Michael Chamberlain, "Socially Responsible Investing: What You Need to Know," *Forbes*, accessed June 16, 2014, http://www.forbes.com/sites/feeonlyplanner/2013/04/24/socially-responsible-investing-what-you-need-to-know.
23. Morgan Stanley, "Morgan Stanley Establishes Institute for Sustainable Investing," *Morgan Stanley*, November 1, 2013, http://www.morganstanley.com/about/press/articles/a2ea84d4-931a-4ae3-8dbd-c42f3a50cce0.html.
24. Heidi Welsh and Michael Passoff, "Proxy Preview," 2014, http://www.asyousow.org/wp-content/uploads/2014/03/ProxyPreview2014.pdf.
25. Ibid.
26. Ibid.
27. Christopher Thomas and Sarah Corrigan, "Ratings and Rankings: How Competition Promotes Corporate Sustainability," *GreenBiz.com*, accessed June 2, 2014, http://www.greenbiz.com/blog/2013/07/03/ratings-and-rankings-how-competition-promotes-corporate-sustainability.
28. GlobeScan and SustainAbility, "Rate the Raters 2012 Polling the Experts, A GlobeScan/SustainAbility Survey," *SustainAbility.com*, June 2012, http://www.sustainability.com/projects/rate-the-raters#projtab-9.
29. CDP North America, "Climate Action and Profitability, CDP S&P 500 Climate Change Report 2014," 22.
30. CDP, "CDP's 2014 Climate Change Information Request," accessed January 16, 2015, https://www.cdp.net/CDP%20Questionaire%20Documents/CDP-climate-change-information-request-2014.pdf.
31. CDP, "Global 500 Climate Change Report 2013."
32. Ibid.
33. "What Are B Corps?," accessed October 15, 2014, http://www.bcorporation.net/what-are-b-corps.
34. Edward Kurocka, "Chapter 2 Becoming a Sustainable Organization," personal correspondence, March 3, 2015.
35. Talia Manning, "NJ Sustainable Businesses, Chameleon Studios Presentation" (Creating Sustainable Businesses in New Jersey, NJDEP Headquarters, Public Hearing Room, Trenton, NJ, March 3, 2015).
36. CDP and WWF-US, "The 3% Solution: Driving Profits Through Carbon Reduction," 2013, http://www.worldwildlife.org/projects/the-3-solution.
37. Accenture and United Nations Global Compact, "The Accenture-UN Global Compact, CEO Study on Sustainability (2013)."

Chapter 3

Gaining Stakeholder and C-Suite Support and Sponsorship

In order to engage the C-suite and garner their support and sponsorship, the sustainability champion must create a program proposal that aligns sustainable strategy with the core business mission and demonstrates value creation for the organization. Making the case for sustainability involves demonstrating the impact of sustainable strategy on key areas such as revenue growth, brand value, risk mitigation, operating performance, talent management, investor relations, finance, and technology. Sustainable strategy is a business methodology that engages both external and internal stakeholders on a variety of environmental, social, and economic issues in order to establish strategic priorities and better manage threats and opportunities to the organization. The key to a successful sustainable strategy is to identify issues that are material for the organization from an industry, management, and stakeholder perspective and then to assess their impact on the organization. A successful sustainable strategy then focuses on how best to support and implement a strategy that addresses these material issues through C-suite sponsorship, cross-functional support, change management, and process and program changes. For an effective sustainable strategy, the C-suite needs to embrace the triple bottom line (TBL) philosophy and agree to systemic changes and resource reallocations to support adoption of a sustainable approach. Many barriers stand in the way of adopting sustainable strategy, but utilizing the tools and techniques in this chapter will help to facilitate moving a sustainability agenda forward within an organization. As management's experience with sustainability grows and the organization moves forward successfully on its sustainability journey, the process becomes more embedded into the culture and operations. As a result, sustainability becomes part of the management processes and cascades throughout the organization. The change management process, especially in terms of program and project management, begins to consider sustainability in terms of defining value creation, cost calculations, stakeholder engagement, risk management, timelines, and operational process. Ultimately, the portfolio management process for portfolio component selection, approval, and funding adopts sustainability criteria as a key part of the process. However, the first step in becoming a sustainable organization is garnering board of directors, CEO, and C-suite support. For long-term strategic organizational change, the sustainability champion must begin at the top of an organization to make the business case for sustainability.

3.1 Creating Value Through Sustainable Strategy

Sustainable development is a philosophy focused on natural capital preservation and positive social impact. The objective is to meet the needs of our global society while ensuring that future generations can meet their needs as well. Sustainable strategy incorporates these principles into a management approach that believes that, through engaging with a broad range of stakeholders, management becomes better informed about issues that impact the long-term health and viability of their organization. Through sustainable strategy, management creates a viable business strategy to generate opportunities, manage risks, provide competitive advantage, and generate a positive financial impact while protecting natural capital and human rights. This approach leads to management creating business value while helping to solve environmental and social challenges. As a result, the approach is more inclusive of stakeholders, which means a change to the traditional stakeholder engagement strategy.

3.2 Planning for Sustainability Success

At its core, sustainability is about protecting natural, social, and organizational assets overtime. An important part of the function for a sustainability champion includes focusing senior leadership on long-term business impacts of sustainable strategic initiatives. As a result, long-range visioning with senior leadership should be part of the sustainable planning process. Encourage leadership to think about the impact of their current business decisions on the organization, community, and environment in 10 years. Ask them to consider what their world might look like 10 years from now. How will the decisions made today impact the environment, society, and the health of the company in the future? Taking the time to consider long-term rather than just quarterly performance metrics begins to give leadership the perspective to think about sustainability issues and their role in securing the long-term health and viability of the organizations that they are steering.

Once leadership understands the value of sustainability to the organization, consider widening the pool of influence by including external stakeholders. Bringing external stakeholder input into the visioning process is a technique used by leading organizations to help frame sustainability objectives. Engaging a diverse group of stakeholders in the process of planning for the future is a best practice used by firms such as Unilever and Dow Chemical. They have established advisory boards comprised of non-governmental organizations (NGOs), academia, government, and business members to provide diverse viewpoints on organizational priorities.

While planning is crucial to any program development, it is especially important for sustainability because many C-suite members are not familiar with the concept. Part of the planning process is developing a meaningful story around industry best practices and relevant case studies to highlight the benefits of sustainable strategy within your organization. Figure 3.1 lays out a series of steps to facilitate the process of creating a sustainable strategy for an organization.

Begin by identifying and understanding organizational values, mission, and goals. A successful sustainable strategy must align with these organizational foundations. A key tenet of sustainable strategy is linking sustainability programs to core business objectives. Assess the culture of the organization and its readiness for change. A sustainability champion must understand how change happens within

Figure 3.1 Planning for Sustainable Strategy Success

an organization. Who are the key influencers and drivers of change? Gather information on your industry, operating environment, and value chain. Look at both internal operations and the impact of suppliers and customers on the product life cycle. Neil Hawkins, Corporate Vice President and Chief Sustainability Officer, The Dow Chemical Co., advises, "It takes a life cycle view of products and services up and down the value chain to achieve sustainable development—so wherever that's not happening is the real opportunity"[1] Understand your industry and its sustainability issues. Who are the market leaders, and what are they doing about sustainability? Are there industry sustainability standards? Is there an opportunity for collaboration with business partners on mutual sustainability challenges? Are there particular operating environmental challenges such as regulatory or compliance issues? Countries around the globe have very different policies relative to sustainability requirements. Understand where your organization operates and the current and future legislation that will impact your operation. While nationally the United States has been slow to adopt broad-based regulatory requirements for environmental sustainability, certain states such as California and cities such as New York City have enacted legislation. In addition, European Union (EU) countries have adopted new legislation that requires certain organizations to report on their sustainability efforts. The impact on affected organizations that conduct business within the EU is expected to be significant, with the number of companies required to report on sustainability rising exponentially. Understanding the areas in which an organization operates goes beyond the national level to the regional and community levels. In order to really understand how an organization is perceived, you must engage with local community stakeholders. Stakeholder engagement is a fundamental part of planning for success. Sustainability is at its core with regard to collaboration. Build a base of stakeholders that includes customers, suppliers, academics, NGOs, trade groups, and community groups. Stakeholders are an invaluable resource for ideas, information, and collaboration. Engaging with a diverse group of internal and external stakeholders provides different perspectives on organizational opportunities and threats. This approach to stakeholder engagement is much broader than the traditional grouping of stakeholders such as investors, customers, and regulators. As an organization moves forward on its sustainability journey, the practice of diverse stakeholder engagement becomes embedded into program and project management methodologies designed to manage the sustainability change management process. In Chapters 8 and 9, stakeholder management is discussed in greater detail.

3.3 The Sustainability Journey

Adopting and embedding sustainable strategy is a journey: It doesn't happen overnight. Senior leadership support of sustainable strategy is a function of their understanding and experience with sustainability programs and projects. As they see the relationship between business value creation and sustainable strategy in practice, their confidence level and support will grow. Educating senior management, building foundational trust, and creating understanding are key to getting approval for a sustainable strategy roadmap.

As part of the planning process, assess your organization's maturity on its sustainability journey. The Exposure Phase begins the journey. Drivers for sustainability at this phase are environmental stewardship, health and safety, compliance, and resource conservation. Companies often enter this phase to meet regulatory requirements or customer sustainability requirements. One of the most significant drivers at this stage is a vendor sustainability survey from a customer. Companies are reaching back into their supply chain to gain better data on the environmental and social impacts created by their suppliers on their products. Usually these initial forays focus on measuring carbon footprints, checking labor practices, and verifying human rights policies.

Projects undertaken in the Exposure Phase are often considered "low-hanging fruit" because they are relatively low in cost and easy to implement. The focus is often on energy conservation, with

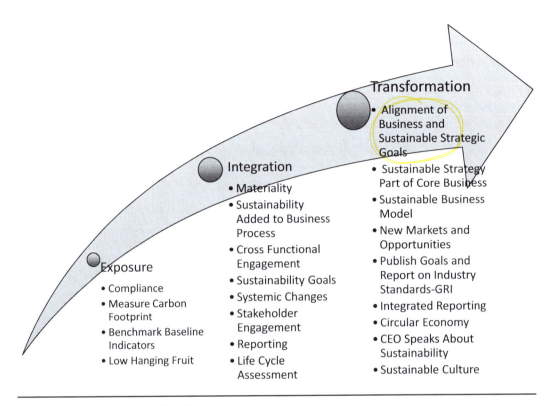

Figure 3.2 The Sustainability Journey

common projects including lighting retrofits or installation of a building management system. Projects in the Exposure Phase create quick wins that generate positive impact with relatively small budgets. They give senior management experience with the strategic benefits of sustainable strategy through risk management and operational efficiency. These early wins build a confidence level about sustainability projects that sustainability champions can leverage further to embed sustainability into the organization. Once senior management experiences the strategic benefits, they become more interested in moving forward with sustainability initiatives on a broader scale. Figure 3.2 depicts the stages and key indicators of the sustainability journey.

Moving forward into the Integration Phase, sustainable strategy adds value to the business process. Management begins to understand that sustainability can add value across the business spectrum from revenue generation, cost savings, employee engagement, and risk mitigation. Sustainability programs consider the full life-cycle impact of products and reach backward to suppliers and forward to consumers to effect changes in design, raw material usage, production, product usage, and disposal. Management engages with stakeholders to understand their concerns and then evaluates the impacts these issues have on their operation. They begin to identify opportunities created through sustainable strategy to enter new markets and offer new sustainable solutions to customers.

As an organization moves through the Integration Phase, senior leadership takes a much more holistic view of sustainability and agrees to targets and reporting on environmental, social, and governance issues. They understand the relationship between sustainable strategy and business value creation. While senior management is engaged at this point, sustainability often remains a separate strategy rather than being fully integrated and aligned with business strategy. According to a Society for Human Resource Management survey, approximately 90% of the responding organizations are equally divided between the Exposure and the Integration Phases.[2]

The game-changing phase is Transformation. A sustainability champion who guides her organization in this phase has achieved strategic alignment of sustainable and business strategies. Sustainability is embedded into core operations, transforming the organization's business model and approaching challenges and opportunities through a sustainability lens. Both Unilever's Sustainable Living Plan and GE's Ecomagination are examples of this type of transformation. Each organization has developed branded sustainability platforms designed to transform their organizations holistically.[3] These platforms engage both internal and external stakeholders on sustainability issues as part of core strategy to create opportunities and mitigate risks.

In 2010, Unilever launched the Unilever Sustainable Living Plan, which is their blueprint for success. Their CEO, Paul Polman, is an avid sustainability advocate, which is a sign that an organization has reached the transformational stage. As management moves to make sustainability part of every function and every job, it becomes part of the organizational culture. At GE, management has focused resources on its fast-growing Ecomagination products and solutions, which reduce environmental impacts while saving customers billions of dollars.[4] This approach has created a new revenue stream for GE with significant economic impact. Both Unilever and GE have systemically changed their business to make sustainability part of all decisions, functions, and processes.

GlobeScan and SustainAbility, leaders in the field of sustainable strategy, have coined a term, "extended leadership," to explain the type of holistic change at organizations such as Unilever and GE. This approach encourages organizations to focus on advocacy and action aimed at system change. The "extended leadership" concept is based on the sustainable developments issues raised at the UN summits such as the Rio+20 summit in Brazil centering around the role of business in helping to accelerate and scale progress on the most pressing sustainability issues.[5]

Senior leadership needs to drive an organization forward in its sustainability journey to effect transformational change. One of the issues highlighted in the Accenture–UNGC CEO survey is that sustainability challenges are significant and growing. Business leaders are frustrated by the pace and results of their own organizations' efforts to create meaningful change.[6] As organizations mature in their sustainability life cycle, their senior leaders become more aware of the importance and the global impact of sustainability to both their organizations and to society as a whole.

Senior management's experience, trust, and confidence depend up where the organization sits relative to its own sustainability journey. Understanding your organization's sustainability history, experience, and outlook will inform a sustainability champion's program recommendations. As an organization moves deeper into the process, sustainability champions need to consider including sustainability requirements in portfolio component selection criteria. This recommendation builds a foundation for embedding sustainability into the decision-making process. Programs and projects will need to include standards and reporting requirements in terms of sustainability targets and goals. As the organization moves further along in its sustainability journey, the more embedded sustainability should become in policies, processes, and standards. As a result, project management professionals have ever-more impact in the transformation process.

3.4 Building the Case for Sustainability

Identifying opportunities for sustainable strategy to support and enhance the C-suite's vision, mission, and goals lays the foundation for creating a business case for sustainability. *Shared Value* is a concept that will help in educating senior management on the benefits of adopting a sustainable approach. Shared Value is a business model designed to incorporate stakeholders' agendas into business decisions. An example is Unilever's Compass strategy, which details their plan for their brands and services to reach and inspire people across the globe while doubling their revenues, reducing their environmental footprint, and improving their social impact.

Table 3.1 Reconceiving Products and Markets

Organization	Wells Fargo	Intel	GE	Thomson Reuters
Solution	Financial products, services, and tools to help clients budget, manage credit, and reduce debt.[8]	Founded Digital Energy Solutions Campaign to minimize life-cycle impact of products, providing better solutions for clients.[9]	Ecomagination products and services designed to reduce own and clients environmental footprint, generating $160 billion in revenues.[10]	Reuters Market Light Provides weather, market data, and financial information to remote farmers in India; subscriber income increased up to 25%.[11]

Through developing creative solutions designed to help address social and environmental challenges, management teams that have adopted Shared Value have significantly improved their organization's value. Industry leaders have embraced the transformative power of Shared Value, creating new products and services, modifying life-cycle impacts, and engaging with stakeholders in a more collaborative manner. As part of this value-creating process, organizations have reconceived their offerings to develop innovative products or services to meet an unmet need or to solve a social problem.[7] The end result has been a significant contribution to overall performance. Some examples of organizations that have reconceived their offerings are given in Table 3.1.

Another method in Shared Value is to evaluate an organization's value chain for efficient use of natural capital, developing talent or growing supplier capabilities. For example, Johnson & Johnson saved $250 million on healthcare costs by investing proactively in employee wellness programs.[12] By providing resources to improve employee health, they not only saved healthcare costs but also boosted productivity and reduced absenteeism.

Spending money on employees improves the organization's financial return and social impact. Walmart reduced carbon emissions by reducing its packaging and reconfiguring truck routes to reduce mileage. While the financial savings were $200 million, the environmental impact both global and locally were significant as well.[13]

Another approach to Shared Value creation is a strategic investment in support of developing solutions that align with your business or region by supporting local cluster development. Clusters such as the Silicon Valley or the North Carolina Research Triangle have played a significant role in creating successful and growing regional economies by driving innovation, productivity, and competition. Clusters may include businesses, academic institutions, trade associations, local governments, and NGOs. They help provide solutions to social issues and environmental challenges facing a common group.

From a leadership engagement perspective, the concept of Shared Value is that business needs a healthy and sound community in which to operate in order to have sufficient demand for its products and positive support through public assets. Communities need businesses to create jobs and generate wealth. Both groups depend on the other for future growth and prosperity as well as protection of shared natural and social capital.

From the sustainability champion's perspective, the Shared Value model offers a holistic approach to business value creation. As an organization moves forward on its sustainability journey, the concepts of Shared Value enter the portfolio components selection process. New product and service ideas need to be evaluated under a full product life-cycle assessment to better understand internal and external impacts as well as the roles of key stakeholders. As a result, projects and programs that had been accepted under traditional business impact assessment might be rejected, and others that were deemed too expensive in the past might be accepted based on this broader set of criteria. From a portfolio, program, and project viewpoint, Shared Value changes the definition and drivers of business value creation, resulting in a new approach to management's definition of business value creation.

3.5 Engaging the C-Suite

To engage the C-suite, demonstrate the ways that sustainable strategy facilitates attaining organizational mission and business goals. Focus on the value created by improving stakeholder engagement, new business opportunities, operational efficiencies, investor relations, access to capital, and risk management. Figure 3.3 details the benefits of sustainable strategy in order to promote C-suite engagement.

On a macro level, sustainability champions must align sustainable and business strategies in order to create greater long-term organizational value. (In Chapter 2, research was presented that demonstrated that organizations that have fully adopted sustainable practices outperform their peers.) Demonstrate the impact that adoption of sustainable strategy has on revenue growth by opening access to new markets, creating new partnership collaborations, and developing products and services that solve customers' sustainability challenges. Highlight operational efficiencies including cost savings from resource conservation, energy efficiency, and design modification. Discuss the impacts on human capital strategy including culture, engagement, retention, and talent acquisition. Culture is a hot topic for the C-suite: They realize that creating an adaptive culture that focuses on core values is crucial for organizational success. In addition, a culture of sustainability is a competitive advantage in terms of attracting and retaining top talent. Other benefits include the risk management impact in terms of operational and physical plant risk mitigation as well as protection of an organization's license to operate and its brand and reputation.

Sustainability champions should create their sustainability proposal using organizational resources and templates so that the formats are familiar to the reader. If the organization has a Program Management Office (PMO), it will be useful to access these types of resources. In order to measure the impact of sustainable strategy, define metrics to measure success, track data, and report on results. If your organization does not have a PMO or proposal standards, the following is a list of points to include in the proposal.

Sustainability Program Proposal Outline

1. Executive Summary
2. Clear Proposal Statement
3. Project Sponsor
4. Type of Program/Project
5. Alignment with Mission, Goals, Targets
6. Material Issue(s) Addressed
7. Stakeholder Issue(s) Addressed
8. Business Value Creation
9. Partners and Functional Areas Involved
10. Stakeholders Included
11. Benefits and Risks
12. Timeline
13. Resource Requirement
14. Budget and Projection
15. Clear Identification of Operating and Capital Expenses
16. Net Present Value of Cash Flows (NPV), Payback Period, Return on Investment (ROI)
17. Cost Breakdown
    a. Internal
    b. External
    c. Hardware Software
    d. Capital
    e. Implementation
    f. Training

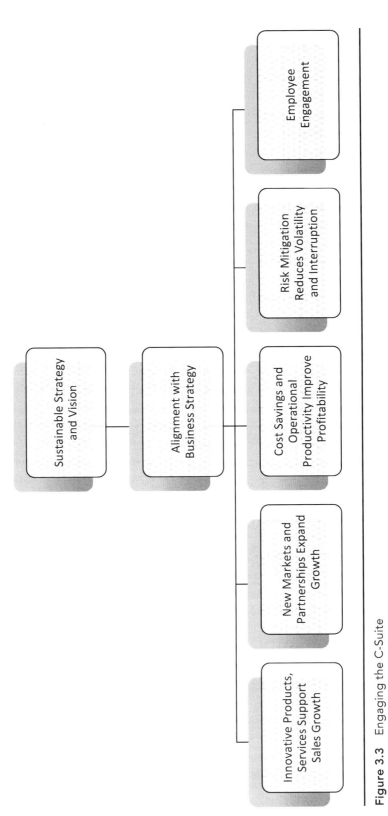

Figure 3.3 Engaging the C-Suite

While the proposal is for a sustainability program, be careful not to get caught up in the language of sustainability while drafting. Use the language of business and the specific terminology used by the C-suite to make the business case for the proposal. Utilize financial metrics such as NPV, ROI, and payback period to support the justification for resource allocation. The sustainability champion needs to be the translator between sustainability and business issues for senior management. Develop the sustainability roadmap around missions and goals that are relevant to the future success of the organization. Ensure that the proposal creates clear alignment between sustainability and business benefits.

To facilitate the process, utilize the following questions to gain perspective from a cross section of the organizational leadership. Ask them of different levels of management within the organizational hierarchy from the C-suite to line managers. Their responses will help identify where the organization is in terms of its sustainability experience and its readiness and capacity to adopt sustainable strategy. In addition, the responses will guide the sustainability champion in creating the organization's sustainability roadmap by clarifying management's perception of the organizational mission and material issues as well as outline tangible benefits and concrete actions.

Questions to Create a Sustainability Roadmap

1. What does sustainable strategy mean?
2. What are the most impactful or material areas in our industry?
3. In what way does sustainable strategy support these material issues?
4. How does sustainable strategy support our current corporate mission? How could it?
5. Are there any current drivers of sustainability within our organization?
6. If so, what are the current drivers of sustainability?
7. Which industry material issues are most impactful to our business?
8. How are these material issues impactful to our business?
9. Who are the stakeholders with whom we should engage?
10. How can sustainable strategy be combined with core business goals to create shared value for all of our stakeholders?
11. How ready is our organization for change?
12. What new opportunities or markets may be created through a sustainable approach?
13. What risks can be mitigated or maximized?
14. How can sustainable strategy impact our competitive advantage?
15. What is our organizational reputation in the communities in which we operate?
16. What cost savings can be recognized through energy, water, waste, or raw material reductions?
17. What messages are we receiving from our stakeholders such as investors, employees, and customers in regard to environmental, social, and governance issues?
18. How does our sustainability plan impact value chain partners?
19. What departments in our organization can be most significantly impacted by sustainable strategy?
20. With whom does it make most sense to build strategic alliances to create the case for sustainability?
21. How well informed is our board, CEO, and leadership team on the issues of sustainability?
22. What relevant industry examples are pertinent in order to build a meaningful business case using ROI and key performance indicators (KPIs)?
23. How will this program impact our corporate brand and reputation?
24. How will it impact our corporate image with current and future employees, specifically our recruitment and retention efforts?

The answers to these questions will provide insights and information from multiple levels of management, including their understanding of sustainable strategy, the organization's sustainability agenda, and how it currently as well as prospectively impacts key business decisions and relationships. In addition, the responses will provide perspective on perceived readiness to either adopt or expand sustainable strategy. Additional perspective is provided on what actions are currently being taken and what opportunities exist for further improvements and impacts. Use these answers to highlight opportunities and potential internal partners to support sustainable strategy adoption By identifying material issues and the most impactful areas on which to focus, the sustainability champion is able to create a proposal that is more meaningful to the C-suite, thereby improving the likelihood of approval. Combining this survey data with the initial plan for success provides a roadmap for advocating for the adoption of sustainable strategy.

3.6 Finding a Champion in the C-Suite

Getting the CEO's commitment is crucial to successful sustainable strategy adoption and integration. Without it, the sustainability champion will not be able to align other members of senior management or garner budget and resources for the sustainability program. While some advocate for grass-roots efforts, obtaining CEO support is the most effective way to make the type of systemic and cultural changes that are required to become a sustainable organization. As we saw from the Accenture–UNGC CEO survey discussed in Chapter 2, there are a variety of drivers for adoption of sustainable strategy. In the early stages of the sustainability journey, there is usually a trigger event that helps to get leadership's attention. It might be the threat of a lawsuit that could be damaging to an organization's reputation or brand. With 24-hour access to news from around the globe, a value chain event such as a supplier human rights violation could be another trigger. Supply chain problems can no longer be contained. In 2013, the Savar building collapse and loss of human life that occurred in Bangladesh created a significant public relations nightmare for GAP.[14] A photo of a GAP T-shirt in the building rubble went viral. While GAP uses contractors in Bangladesh, they were not producing goods in this factory. Because of the photo, they were targeted by a number of human rights groups and had to spend resources protecting their brand and image. The news has a global reach through the proliferation of cell phones. While one of the goals of sustainable strategy is to address this type of situation proactively, a crisis can be a leverage point for a sustainability champion. Out of a supplier crisis may come an opportunity for a discussion with the VP of Operations about life-cycle assessment, including supply chain standards, supplier scorecards, and third-party assessment and compliance verification. Often a crisis provides the platform for a conversation with senior management about taking a sustainable approach to management and operations of an organization.

Customer compliance is also a significant driver for adoption of sustainable strategy. Increasingly, customers are seeking information from all size organizations about their own sustainability compliance. In addition they are seeking sustainable solutions from suppliers to help solve their own sustainability challenges. Companies that can offer sustainable solutions to clients create a competitive advantage. Identifying gaps in the marketplace between competitive offerings and customer requirements for sustainable solutions offers real opportunity for a new revenue stream for organizations that provide these solutions.

Some CEOs are personally driven in their interest in sustainability. They have been enlightened about environmental and social concerns from peers, industry events, and personal experience. Sometimes their interest stems from the legacy that they will leave behind for their organizations, communities, and families. Legacy can be a powerful driver for sustainable strategy adoption.

Obtaining the CEO's sponsorship for sustainable strategy is vital to introducing sustainability programs and projects effectively within the organization and to obtaining cross-functional resources and support. According to Mark Weick, Director, Sustainability Programs and Enterprise Risk Management, at Dow Chemical Company, "CEO support is critical. In the 90s, our CEO, Frank Popoff, championed sustainability and supported transformational changes within our organization. It has become part of the Dow culture. Our current leadership team has grown up in an environment where sustainability is part of the core mission of Dow." (The full interview with Mark Weick on "Engaging the C-Suite" appears in Section 3.10 at the end of the chapter.)

Even with CEO support, garnering support from the balance of the C-suite can remain a challenge. Business leaders are focused on their own area of responsibility, with operations often performed in a silo with very little cross-functional engagement. Their budgets, goals, scorecards, and compensation are aligned to promote focusing on their own function rather than the organization as a whole. Convincing other members of the C-suite will require not only CEO support, but also evidence that sustainability positively impacts their functional area. Identify those areas of the organization that will be most beneficially impacted by the adoption of sustainable strategy and partner with those senior leaders. The greater the impact, the greater the level of interest will be for supporting sustainability programs and projects in their area. The VP of Human Resources can be engaged through evidence supporting the links between sustainable organizations, employee engagement, and rising productivity levels. The CFO might be engaged through risk mitigation and lower cost of capital. The VP of Sales can be engaged through new markets, increased sales, and innovative solutions. The VP of Operations by improved process, resource conservation, and risk mitigation. (In Chapter 9, we take a much deeper look at this topic.)

Before walking into a C-suite steering committee meeting, a sustainability champion must ensure that she knows the audience and their major concerns and issues. Plan to meet with key members of the committee individually prior to the presentation in order to present your proposal, understand their issues, and solicit their support. If the steering committee meeting will be the first time that members of the C-suite will hear about your sustainable strategy proposal, there is a great likelihood that the decision will be deferred or declined. More often than not, the committee members don't fully understand the concepts and, as a result, decline them. As leadership, they need time to evaluate the impact of your recommendations on their own function as well as on the organization as a whole. Meeting with key decision makers ahead of time provides an opportunity to address individual concerns and gain allies within senior leadership. Know who is a supporter, a bystander, a dissenter before you enter the room. Having several members of the C-suite lined up to support your sustainability program will go a long way toward convincing other members to join the bandwagon. Be prepared for objections, political infighting, and wavering alliances.

Focus on demonstrating how your sustainability program not only helps the organization reach its goals, but also how it can help senior management achieve their own performance scorecard metrics. Collaborating with senior management to meet their personal performance objectives is a path to success for your sustainability program.

While technical knowledge of environmental impacts, product stewardship, and social impacts is important for managing sustainability projects, sustainability practitioners report that the soft skills are most vital when seeking senior management support. The following comment from Kathrin Winkler, Vice President of Corporate Sustainability at EMC, highlights the importance of relating sustainability issues to core business drivers: "A significant part of my role is translating those social and environmental issues that intersect with our business in ways senior management will appreciate. Showing how they can either provide a competitive edge or respond to a customer demand is critical for success."[15]

Utilizing established business protocols and building alliances greatly enhances the probability of success for sustainable strategy. Following are lessons learned from 32 sustainability practitioners from Fortune 100 companies:

1. Interpersonal skill is crucial to moving your organization forward on its sustainability journey.
2. Use the language of business to make your case to the C-suite, rather than the language of sustainability.

3. Support your request for investment in sustainability strategy with quantifiable data and metrics that support your organizational mission.
4. A sustainability champion is much more impactful through collaboration with other business leaders.
5. Traditional business drivers such as pressure from customers and competitors are also key drivers for sustainable strategy adoption.[16]

As we can see from the above list, making the business case for sustainability is very much like making the business case for other programs or projects. What problem or set of issues does the adoption of the sustainability initiative help solve? What does cost–benefit analysis show? Sustainability programs and projects need to be presented and evaluated like other business decision. Many of the most compelling reasons are the same business-case reasons outlined in the practice of project management, such as customer demand, market demand, business requirement, regulatory or legal requirement, technology improvement, economic or social impact. In terms of getting the sustainability agenda moved forward, it is imperative that the sustainability champion understand which business issues are the most pressing for the organization. Go to a C-suite steering committee with a proposal that supports the organization's key business mission, discuss the proposal in business terms, and support the initiative with data and metrics.

3.7 Materiality

Materiality means focusing on issues that are most important to management and stakeholders in terms of economic, social, and environmental performance. What is material to an organization varies depending on its industry, mission, structure, scope of operations, and strategy. Sustainable strategy focuses on engaging both internal and external stakeholders to help inform senior management on major issues facing the organization and to inform direction of the company strategy. In addition to traditional stakeholders, this group may include employees, unions, trade groups, local governments, NGOs, and community groups. By gathering a diverse set of opinions, management can better align stakeholders' concerns with business strategy in order to prioritize issues. Stakeholders make decisions such as purchasing products, investing, supporting local operations, and seeking employment. Stakeholders' issues might include topics such as sustainably sourced ingredients, packaging design, rural development, women empowerment and equality, human rights, facility expansion, employee health and wellness. A materiality matrix lays out the attractiveness of issues to stakeholders relative to their impact on the business. It provides a tool to help management decide where to focus and invest in its programs and projects. The matrix in Figure 3.4 lays out issues and their importance to stakeholders relative to issues and their perceived importance to the business by management.

This approach allows for stakeholders to provide feedback on issues that are important to them and then have management rank them on importance to the organization. The most important factors to both stakeholders and the company appear in the top right section. These issues are the most material issues for an organization and represent priorities for senior management action. Focusing sustainable strategy on these issues helps to better engage stakeholders while focusing on business imperatives that create value. The following example from Ball Corporation illustrates the materiality matrix.

Ball Corporation, founded in 1880, is a provider of metal beverage packaging, metal food and household products packaging, and of aerospace and other technologies, to commercial and government clients. They are listed in top sustainability indices such as the DJSI and FTSE4Good. Worldwide, they employ 14,500 employees. They have adopted the "Drive for 10" vision, which highlights their corporate sustainability priorities which were developed using their materiality matrix. Approximately 200 global stakeholders, including customers, shareholders, suppliers, trade associations, academics,

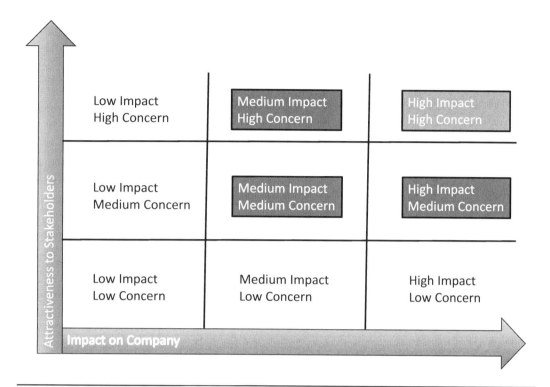

Figure 3.4 Materiality Matrix

and employees participated in the development of their materiality matrix.[17] The responses are classified based on the impact category for creating long-term value at Ball. As can be seen in Figure 3.5, customer satisfaction rates as the most important issue, followed by packaging recycling, and innovation and design. Ranking high on the list is health and safety, with other employee-centered issues such as employee engagement and talent management falling within the top third of concerns.

Ball's objective is to create value for customers by being a sustainable supplier of choice through meeting client needs and providing solutions to help customers meet their own sustainability goals. Their top materiality issues are incorporated into their corporate priorities of innovation, operations, talent management, recycling, supply chain, and community. Figure 3.6 depicts Ball's priority materiality issues in a triple bottom line format considering each in terms of social, environmental, and economic impacts. As the chart indicates, management is focusing on issues that fall into each of these categories to drive long-term organizational success. Categories include ethics, risk management, employee engagement, talent management, and health & safety, as well as packaging, resource management, and customer satisfaction. Both management and stakeholders find these issues relevant to the success of the organization.

In order to address these priority materiality issues, operational priorities are further broken down into safety, electricity, gas, water, waste, and VOCs (volatile organic compounds).[19] Each of the priorities plays an important role in the Drive for 10 vision, which provides a clear roadmap for leadership and employees on the foundation and direction of the organization. Within the Drive for 10 vision, they have identified five key drivers to the long-term success of their business:

1. Maximize their existing business
2. Expand through new products and capabilities

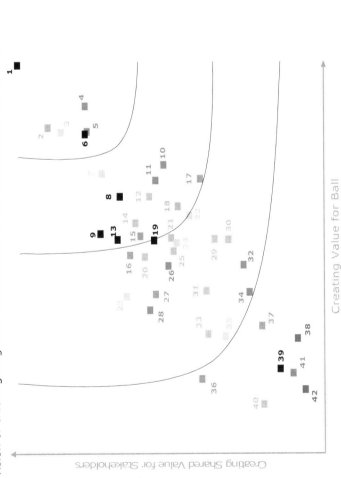

Figure 3.5 Ball Corporation Materiality Matrix. (Reproduced with permission)[18]

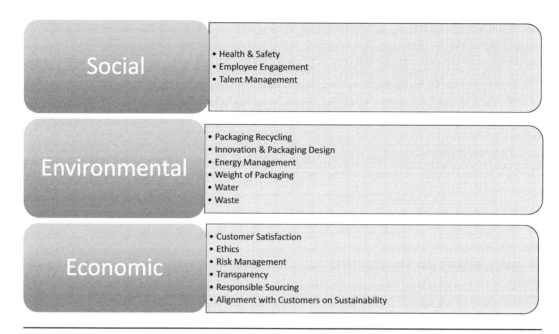

Figure 3.6 Ball's Priority Materiality Issues

3. Align with key customers and markets
4. Continue to expand global reach
5. Leverage knowledge and technology to provide a competitive advantage.

In order to reach their goals, management knows that they need to focus on their customers, operational excellence, people, culture, sustainability, and innovation. Each is a driver of success. According to John Hayes, Chairman and CEO of Ball, "Our commitment to sustainability has never been stronger due to the value it creates, including greater customer satisfaction, enhanced employee engagement, and improved financial results."[20] From Ball's CEO perspective, sustainability is a key driver of current and future success in their organization.

Talent management is a material issue for stakeholders and management. Ball is committed to developing their talent pool in order to continue to grow their business. Each employee is tasked with creating an "individual development plan," which includes specific learning objectives for growth within their current position and to prepare for the next steps in their career progression. This program is supported by the adoption of the "70-20-10" principle, by which management acknowledges that 70% of learning happens on the job through solution of real-life challenges. Dialogue and feedback from supervisors and peers provides another 20%. Formal training and eLearning represent the final 10%. Their performance management framework focuses on setting and reviewing goals and core competencies, ongoing coaching and feedback, support for individual development through the "individual development plan," and recognizing and rewarding performance.[21]

The following is an example of Ball's talent management practices in action to support growth, meet client objectives, and embed sustainable practices across the organization.

Ball and Thai Beverage Can Ltd (TBC) joined forces to seize opportunities in the attractive Vietnam market and officially broke ground on a new plant in June 2011. The recruitment and development of strong team members was critical to the joint venture's success. Seventy-two new hires were sent to Ball's can plants across China as well as to the TBC plant in Thailand to learn everything about beverage can making. Administrative employees in finance, logistics, sales and sourcing also visited partner locations to receive training in their respective areas. In early 2012, 12 top performing TBC-Ball trainees were rewarded with a visit to Ball's Beijing plant, where they gained additional insights into efficient operation of high-speed can making equipment.

Hands-on training and experiential learning were instrumental in developing a talented and high-performing workforce, which guaranteed a successful start of commercial beverage can production in March 2012. Our team continues to fine-tune its talent management program to ensure we have highly talented, dedicated and engaged employees who can succeed in an excellent work environment.[22]

Understanding the material issues impacting an organization and developing internal and external partnerships to develop solutions to either maximize or minimize these issues is important for sustainability champions in developing an organization's sustainable strategy.

3.8 Barriers to Adoption

Despite the benefits of adopting sustainable strategy, many sustainability champions struggle to move their programs forward. The following reasons were offered at a roundtable of sustainability practitioners. Often the C-suite reaction is to delay giving a response, hoping that the request will go away and be taken off the long list of issues being presented. Another related objection is competing priorities. Leadership has too many projects with limited resources; they can't accept all of the proposals. Sometimes sustainability is seen as a compliance exercise and not as adding business value. When sustainability is considered as a truly transformative management approach, the magnitude of change can be daunting.[23] Transforming into a sustainable organization requires change that reaches deep into the organizational processes and procedures. It requires rethinking the business model and sometimes breaking down and rebuilding current systems. Sometimes C-suite members feel that this level of change is too much to consider given other pressing challenges. They don't feel that the organization has the capacity or capability to implement this type of change.

While all of these objections have been raised in C-suites, the biggest obstacle to adoption may come from the approach the sustainability champion uses to present sustainability programs. Often, he presents sustainability programs highlighting environmental and social benefits of sustainable development, rather than business value created through sustainable strategy. Sell the sustainability proposal like any other business strategy, focusing on its ability to drive the organization toward its mission and to attain its goals. Adopting a sustainable strategy provides an organization with tools to outperform the competition. By pitching sustainability as a means to better achieve existing targets and goals, the sustainability champion heads off a common barrier to adoption.

While education on the advantages of sustainability is key to engaging the C-suite, using the pitch time as an educational time is not going to be productive. Take the time to provide education to members of the C-suite prior to pitching the proposal. If possible, enlist C-suite champions who are prepared to support your proposal by offering their group as a pilot site. If the sustainability champion is seeking approval for an energy-reduction pilot, having the COO on board with running a pilot before the pitch adds credibility to the proposal and a defined next-step action.

In developing the proposal, ensure sure that the audience and their concerns are addressed. Demonstrate the linkage between the sustainability proposal and the most pressing challenges facing the C-suite. If your proposal doesn't solve any of these key challenges, you may need to rethink your proposal. A "nice to have" sustainability project should not get preference for limited resources over other high-priority projects that impact the ongoing success of the operation. Utilize project management office (PMO) templates and request-for-funding protocols to ensure that the proposal meets funding criteria. Include tangible metrics to track progress. Find out the preferred presentation format for C-suite proposals and use that style. Just because sustainable strategy is a different approach to creating business value, it doesn't mean that the approval process is different. Ensure that your proposal has the rigor of competing projects.

Sustainable strategy is a broad management approach and potentially encompasses all aspects of organizational operation. Ensure that the sustainability proposal is clear on what is being requested. Are you proposing a pilot for an energy efficiency retrofit, modifications to a product based on a life-cycle assessment, a new wind turbine–powered manufacturing facility, development of a sustainability platform as part of core business offering? While all of these proposals require a sustainability commitment, the magnitude of investment and the scope of the commitment vary widely. Sustainable strategy is a journey, and a sustainability champion must begin with manageable first steps and then grow the program in scope and complexity. Senior management will become more comfortable supporting sustainable strategy as results are demonstrated. The ultimate objective of the sustainability champion is to embed sustainability into the organization and have it become so ingrained in the culture that it fundamentally changes the business model.

3.9 Project, Program, and Portfolio Management Impact

In order to unleash the power of sustainable strategy, the sustainability champion must obtain CEO and C-suite support. In the early stages of the sustainability journey, it is common for the sustainability champion to take on the role of program or project manager. Once the C-suite is convinced of the benefits of sustainable strategy to the long-term prosperity of the organization, the process of strategic visioning begins. As the organizational vision incorporates sustainability, goals and targets begin to cascade downward within the organization, changing organizational programs, projects, and processes. The portfolio component selection process is modified to include sustainability criteria, bringing programs and projects into alignment with the organizational vision. As sustainability cascades throughout the organization, policy structures such as governance around ethics, anticorruption, supplier codes of conduct, and diversity and inclusion are reviewed and modified. Communication plans are developed to ensure that information on sustainability vision and supporting policies are disseminated throughout the workforce. Compensations structures are aligned to reward actions and behaviors that are consistent with sustainability goals and objectives. Adopting a transformational strategy is a complex program management endeavor, and the tools and methodologies of project management are well suited to support this type of organizational change. While there are many barriers to adoption, developing a sustainable strategy ultimately provides an organization with a competitive advantage.

As sustainability becomes more embedded, the benefits from project and program management increase with maturation of an organization's sustainability strategy. The definition of benefit creation includes the impact on people, planet, and profit. New project charters reflect how the project helps to achieve sustainability goals and targets. Performance metrics focus less on short-term results and more on long-term value creation. The definition of product and service costs incorporates the costs of natural and social capital. As a result, program and project management considerations and requirements evolve to reflect senior management's vision of sustainable strategy. Projects are selected based on their ability to help the organization meet both business and sustainability targets and goals. Standards and requirements become part of the PMO templates, protocols, and organizational assets. Beginning the sustainability journey is the first step on pathway to fundamentally changing an organization's approach to conducting business and creating value, and program and project managers are an important link in driving sustainable transformation.

3.10 A Conversation About Engaging the C-Suite

Mark Weick, Director, Sustainability Programs and Enterprise Risk Management, the Dow Chemical Company, shares his thoughts on engaging the C-suite on issues of sustainability.[24]

I had the great fortune to be able to interview Mark, an industry leader in the area of sustainability, on the topic of engaging the C-suite. Many of the topics discussed in this chapter are reinforced by his comments. While the sustainability community often speaks about a grass-roots approach, for long-term systemic change within an organization, CEO and C-suite support is crucial. Creating the framework for sustainability at the top focuses on materiality and strategic alignment that then can be cascaded throughout the organization, impacting people, process, and culture. While implementing sustainability is a journey and an iterative process, C-suite support and buy-in is a foundational building block. Providing a wider perspective through broad-based stakeholder engagement is a useful resource to establish long-term vision and mission for the organization. Hearing a variety of viewpoints and opinions opens the door to new possibilities and identifies unseen risks. Collaboration and partnership are tools to be leveraged to create Shared Value. Creating a sustainable culture drives not only business mission but also enhances corporate image and brand valuation, including becoming an employer of choice for employees—especially those in the Millennial generation.

About Mark Weick:

> In his role, Mark directs the coordinated planning and implementation of the 2015 Sustainability Goals as well as sustainability integration across the company and business units. He is also responsible for directing Dow's future sustainability strategy, as well as the company's enterprise risk management strategy. In addition, he leads Dow's global collaboration with The Nature Conservancy on valuing ecosystem services and biodiversity.

About Dow:

> Dow (NYSE: DOW) combines the power of science and technology to passionately innovate what is essential to human progress. The Company is driving innovations that extract value from the intersection of chemical, physical and biological sciences to help address many of the world's most challenging problems such as the need for clean water, clean energy generation and conservation, and increasing agricultural productivity. Dow's integrated, market-driven, industry-leading portfolio of specialty chemical, advanced materials, agrosciences and plastics businesses delivers a broad range of technology-based products and solutions to customers in approximately 180 countries and in high-growth sectors such as packaging, electronics, water, coatings and agriculture. In 2014, Dow had annual sales of more than $58 billion and employed approximately 53,000 people worldwide. The Company's more than 6,000 products are manufactured at 201 sites in 35 countries across the globe.[25]

1. In your opinion, how important is CEO support for adoption of sustainable strategy?

> CEO support is critical. I was fortunate to begin my tenure at Dow under then-CEO Frank Popoff. In the 1990s, he made the case for sustainable development at Dow and helped initiate the World Business Council for Sustainable Development (WBSCD), a CEO-led sustainability council. Along with Livio DeSimone, he co-authored *Eco-Efficiency,* a book on implementing sustainable strategy. Subsequent CEOs have continued our culture of sustainability that he fostered. Our current CEO, Andrew Liveris, considers sustainable development a core mission at Dow.

2. How do you engage the C-suite in developing sustainable strategy?

At Dow, we are fortunate to have a long history with sustainable development. Our C-suite has grown up in a culture of sustainability. In 2006, we announced 2015 Sustainability Goals centered on Sustainable Chemistry, and then developed the Sustainable Chemistry Index (SCI). The SCI has focused our efforts on creating more sustainable products and solutions for clients and has helped build awareness of our sustainability mission throughout our organization. This program has been a decade-long journey and it has been highly effective; but we are always looking to move forward and are thinking about what is next. Our planning process is to engage in regular visioning sessions with senior leadership in order to develop the strategy for our next long- term goals.

We began the process of developing our next approach to sustainability within a few years of launching our 2015 Sustainability Goals. In 2009, we invited our Sustainability External Advisory Council (SEAC) to spend the afternoon with the CEO and his direct reports to engage in a vision casting exercise. SEAC is an external advisory group made up of 8–10 people from academia, NGOs, government, business, environmental, and sustainability communities. We gathered these two groups together to facilitate a discussion about what sustainable human society and sustainable chemistry would look like in 2097, Dow's 200th year. Visioning a world this far into the future was really unique. We asked John Elkington to help with the effort, and initially he thought we were overly optimistic to think it would generate meaningful results. Looking back over the past 100 years, who could have envisioned WWII or the creation of the Internet? After thinking about it for a day, he called to say that it was such a ground-breaking idea that he had to be part of it, and developed a video that helped us introduce a public dialogue on this long-range vision at the 2009 Business for the Environment conference in Seoul, Korea. From this process, we began developing our Next Generation goals, which will be publically announced later this year.

3. Are there any best practices that you can share for obtaining C-suite approval and sponsorship?

Commissioning the C-suite to hear from outsiders. We use the SEAC to give senior leadership a different perspective on issues, challenges, and alternative solutions. Engaging with external stakeholders, some of whom might appear to be more likely to protest Dow rather than attend a meeting in our Boardroom, and giving them a credible voice through the external advisory council, provides an invaluable resource. It is an expensive resource to maintain, but it is worth it. Our position on an issue may be different from members of our advisory team, but through the council we have created a forum for open dialogue and listening. Ultimately, this process has created opportunities and offered solutions that would not have been possible otherwise.

4. Do you have any lessons learned that would help other organizations in the early stages of their sustainability journey?

In the initial phase of sustainability, the two driving factors are crisis and legacy. For Dow, the 1970s were an unsustainable period, especially in terms of our impact on the environment. We were under attack from environmental groups. Initially, leadership was defensive as leaders from NGOs, governments, and academia raised concerns around some of our practices. In response to this crisis, senior leadership began the process of listening to others and engaging with stakeholders, who had varying perspectives and ideas. Out of this crisis rose the foundations of our sustainability program.

Use crisis to your advantage. In the early 2000s, senior management presented a plan, of which they were very proud, to address mainly environmental concerns in the ten-year plan leading to 2015. The reception of the plan by the advisory committee was not as expected. They were advised by one SEAC member to consider how their plan affected the poorest woman on the planet, a 7th generation child, and a porpoise. Out of that meeting came a senior management team that was determined to develop a much better plan. This crisis led to the development of a better sustainability strategy that focused both internally and externally, and led to collaborations with NGOs such as The Nature Conservancy, governments, and business to offer new sustainable solutions. Joint research with customers included the development of bonding materials that enabled lighter-weight materials for Ford's F150 truck, and teaming with Unilever to develop products that enabled soap to last longer and build awareness about the importance of hand washing to combat disease. Through this process, we have fostered a culture of sustainability at Dow, and our C-suite champions that culture.

Another example of crisis giving rise to a better program comes from our early years of setting targets. In 1992, we established annual goals around safety, emissions, and energy efficiency. Because the goals had to be achieved within one year, management was having difficulty meeting them. At Dow, we have a culture of focusing long-term which is partially a function of the long-term nature of investments in the chemical industry. Once we make an investment, it is there for a long time. Management became aware that the types of operational changes and capital investments that were required to make a significant impact on sustainability goals were long-term, multi-year projects. As a result, we were able to focus attention on sustainability as a long-term strategic goal rather than a short-term objective. It became much more embedded in our culture.

Legacy is the other trigger that I mentioned. Our C-suite members are concerned about the world that they are leaving to their grandchildren. In addition, they care deeply about the company that they are leaving behind. How will their tenure with the company be remembered? Legacy is not often discussed publically, but in private conversations it is a driver for sustainability in the C-suite.

5. How important is employee engagement as a factor to drive sustainability?

It is vital. Our recruiting campaign in China is totally built around sustainability. Dow's reputation as a Green company in China attracts and helps retain top talent, especially in highly competitive markets such as Shanghai. We have built sustainability into the entire process beginning with the initial contact via our web portal, through the interview process, and into the daily functions of our people. Not only does this approach help recruit top Millennial talent in the engineering and technical sectors, it supports presenteeism. Our employees are more engaged and productive because they understand their role and its impact on sustainable development within the organization.

Latin America is another market in which we focus our talent acquisition and retention strategies around sustainability. We have found cultural factors, resource depletion concerns, and awareness of sustainability issues because of the Rio+20 meeting in Brazil have made this workforce more aware of sustainability. Our sustainability program is an important component of our talent acquisition strategy. In my experience, the Millennial generation expects to find sustainability in the workplace.

3.11 Conclusion

Engaging the C-suite and garnering CEO support is crucial for a sustainability champion to create an impactful sustainability program. While planning is crucial to the process, one approach does not fit every organizational situation. A sustainability champion needs to understand where the organization is on its sustainability journey to better understand the drivers and appetite for sustainable transformation. Building the case for sustainability requires focusing on organizational mission and specific goals and targets and creating a sustainable approach that facilitates alignment.

Focusing on materiality creates a compelling business case as the sustainability program addresses issues that are important to stakeholders and management and ultimately drive organizational success. This strategic focus then drives the portfolio component selection process to better align program and project selection to support sustainable strategy. Obtaining C-suite support is a starting point for creating a sustainable strategy, but in order to create organizational change, the sustainability champion must reach far into the organization in order to impact organizational culture. Project management professionals are a valuable resource to begin this change management process.

Notes

[1] Neil Hawkins, "Six 'Know Thys' of Sustainability," *Ensia*, accessed January 26, 2015, http://ensia.com/voices/six-know-thys-of-sustainability.

[2] SHRM, BSR, and Aurosoorya, *Advancing Sustainability: HR's Role,* Society for Human Resource Management, 2011.

[3] Chris Coulter and Chris Guenther, "The Expert View: Top Corporate Sustainability Leaders of 2014," *The Guardian*, sec. Guardian Sustainable Business, accessed May 14, 2014, http://www.theguardian.com/sustainable-business/blog/sustainability-leaders-2014-unilever-patagonia-interface-marks-spencer.

[4] GE, "Ecomagination," accessed January 21, 2015, http://www.ge.com/about-us/ecomagination.

[5] Coulter and Guenther, "The Expert View."

[6] Accenture and United Nations Global Compact, *The Accenture-UN Global Compact, CEO Study on Sustainability (2013)*, 2013, www.unglobalcompact.org.

[7] Michael E. Porter and Mark R. Kramer, "Creating Shared Value," *Harvard Business Review*, accessed January 20, 2015, https://hbr.org/2011/01/the-big-idea-creating-shared-value.

[8] Ibid.

[9] "Intel Sustainability Initiatives and Policies," *Intel*, accessed January 21, 2015, http://www.intel.com/content/www/us/en/corporate-responsibility/sustainability-initiatives-and-policies.html.

[10] GE, "Ecomagination."

[11] Reuters, "Reuters Market Light," accessed January 21, 2015, http://www.reutersmarketlight.com/products_services.html.

[12] Porter and Kramer, "Creating Shared Value."

[13] Ibid.

[14] Jim Yardley, "Report on Bangladesh Building Collapse Finds Widespread Blame," *The New York Times*, May 22, 2013, http://www.nytimes.com/2013/05/23/world/asia/report-on-bangladesh-building-collapse-finds-widespread-blame.html.

[15] Vox Global, Weinreb Group Sustainability Reporting, and Net Impact, "Making the Pitch: Selling Sustainability from Inside Corporate America," 2012, http://www.aiacc.org/wp-content/uploads/2013/05/VOX-Global-2012-Sustainability-Leaders-Survey-Full-Report.pdf.

[16] Ibid.

[17] Ball Corporation, "Creating Long-Term Shared Value; Stakeholder Engagement," accessed January 21, 2015, http://www.ball.com/stakeholder-engagement.

[18] "Ball Corporation Sustainability Materiality Matrix 2014," accessed October 20, 2014, http://www.ball.com/images/ball_com/uploads/Ball_Sustainability_Materiality_Matrix_2014.pdf.

[19] "Ball Sustainability," accessed October 20, 2014, http://www.ball.com/sustainability.

[20] Ball Corporation, "Ball Corporation 2014 Sustainability Report," Ball Corporation, 2014.

[21] Ball Corporation, "Developing and Inspiring Great Talent," accessed January 28, 2015, http://www.ball.com/talent-management.

[22] Ibid.

[23] Nelson Switzer, "How to Engage and Convince the C-suite about Corporate Sustainability," *Environmental*

Management & Sustainable Development News, accessed October 20, 2014, http://www.environmentalleader.com/2011/09/27/how-to-engage-and-convince-the-c-suite-about-corporate-sustainability.

[24]Mark Weick, A Conversation About Engaging the C-suite, January 30, 2015.

[25]Dow, "Dow Reports Fourth Quarter and Full-Year Results," accessed February 2, 2015, http://www.dow.com/investors/earnings/2014/14q4earn.htm.

Chapter 4

Alignment of Business and Sustainability Strategy Creates Sustainable Strategy

As an organization moves through its sustainability journey, reaching the transformative stage is the "sweet spot" where business and sustainability strategies are aligned to create sustainable strategy. Adoption and integration of sustainability strategy transforms an organization's strategic focus to include environmental and social issues in order to maximize long-term business value. It is at this point that sustainability moves from playing a supporting role in achievement of business strategy to becoming the foundation of business strategy. Sustainable strategy is a multiyear, cross-functional approach to creating business value through addressing material business, social, and environmental challenges. The approach focuses on identifying material industry, community, and business issues and creating programs and projects to address these challenges. Adoption of sustainable strategy focuses on developing policies, changing people and process, creating opportunities, and addressing global sustainability challenges. At the point of sustainable strategy adoption, organizational leaders understand that sustainability and business value creation are not mutually exclusive. Rather, they address challenges and opportunities through a sustainability lens, creating new options, solutions, and benefits. Adopting sustainable strategy has preserved industries and allowed individual firms to flourish in the face of rising regulatory and customer concerns about their operations' impact on natural and social capital. Developing a sustainable strategy requires approaching business operations with a more holistic view, including looking both internally and externally for solutions. Adopting sustainable strategy requires changing organizational perceptions, priorities, and behaviors. This change management process benefits from utilizing project management methodologies, techniques, standards, and protocols. Organizations with mature project management practices, mature benefits realization processes, high organizational alignment, and mature program and portfolio practices have an advantage in adopting and executing sustainable strategy.

4.1 Alignment of Business and Sustainable Strategy

The benefits of aligning business and sustainability strategy include risk mitigation, operational efficiencies, and cost savings (see Figure 4.1). However, they also include impacts to employees in terms of improved engagement and productivity, because they understand the goals and values of the organization and they feel that they are part of the solution. This leads to employees acting as ambassadors for the organization, touting accomplishments and good works. Adopting a sustainable strategy benefits the environment through preservation of resources such as water, air, biodiversity, and other natural resources. Sustainable strategy considers both internal and external stakeholders and benefits society through community engagement, human rights, and fair labor standards. For leadership teams that have adopted sustainable strategy, alignment between sustainability and business strategies has been the key to effectively transforming into sustainable organizations. By this process, management utilizes sustainability levers to drive business value.

Figure 4.1 Benefits of Sustainable Strategy

Aligning business and sustainability strategy considers the impact of sustainability in achieving business mission. As an example, reformulating product packaging to reduce packaging and to utilize recycled materials benefits an organization through reduced costs and the environment through a reduction of content in the waste stream. As an organization moves into fully adopting sustainable strategy, sustainability moves from becoming a tool to facilitate business outcome to a whole new platform for business operations based on the core values of sustainability. The concept of Shared Value discussed in Chapter 3 is based on this tenet. Organizational examples include Unilever's Sustainable Living Plan and GE's ecomagination platforms, on which whole new lines of business have been developed to address customers' needs for sustainable solutions. Sustainable strategy focuses on creating environmental, social, and economic value by fundamentally changing operations, people, and process. How this change is accomplished is of interest to program and project managers as they provide the crucial link by managing programs and projects designed to facilitate the process.

In order to create a sustainable culture, business and sustainable strategy need to be aligned, forming a transformative sustainable business strategy that fundamentally changes how work is performed in an organization. This approach aligns people and process around platforms, methodologies, programs, and projects to create economic, environmental, and social value (see Figure 4.2). Sustainable strategy transforms leadership's perspective on the benefits of a triple bottom line—people, planet, profit—approach. Senior management views strategy through a sustainability lens in order to create long-term business value, minimize business risks, and maximize opportunities. Long-term value is created by aligning organizational policies, processes, and goals to support transformative changes that better align with sustainable strategy. Alignment of business and sustainability strategies is crucial to the value creation process in terms of developing new sustainable products, new markets, redesigning existing products, managing risks, and engaging internal and external stakeholders.

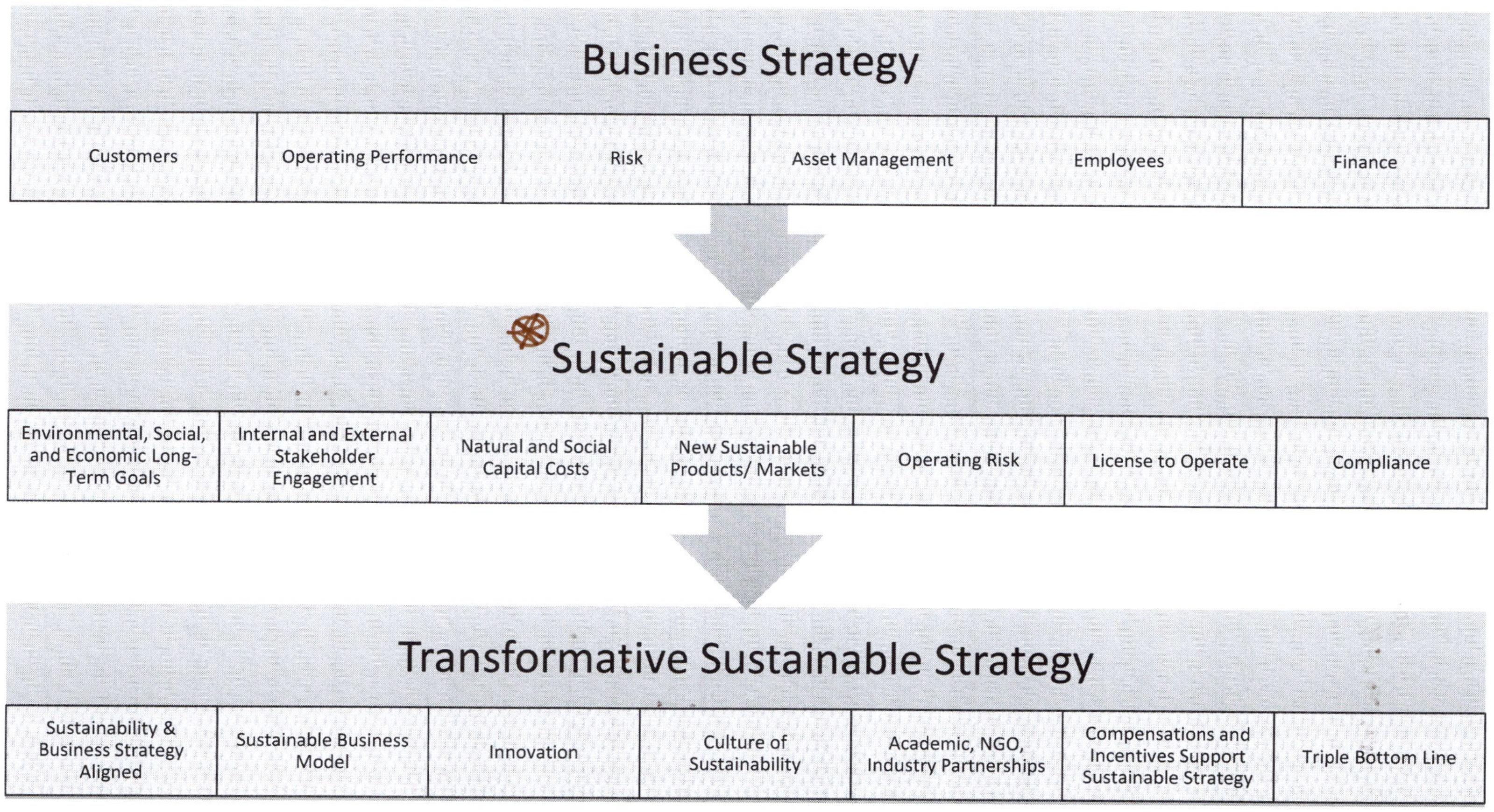

Business Strategy

| Customers | Operating Performance | Risk | Asset Management | Employees | Finance |

Sustainable Strategy

| Environmental, Social, and Economic Long-Term Goals | Internal and External Stakeholder Engagement | Natural and Social Capital Costs | New Sustainable Products/Markets | Operating Risk | License to Operate | Compliance |

Transformative Sustainable Strategy

| Sustainability & Business Strategy Aligned | Sustainable Business Model | Innovation | Culture of Sustainability | Academic, NGO, Industry Partnerships | Compensations and Incentives Support Sustainable Strategy | Triple Bottom Line |

Figure 4.2 Creating a Transformative Sustainable Strategy

Table 4.1 Sustainable Strategy 5-Year Plan

Year 1	Years 2–3	Years 4–5
Benchmark Performance. Engage C-Suite. Engage Business Function Leaders. Engage Employees. Identify Material Issues. Engage Stakeholders. Identify Goals.	Select Goals & Targets. Select Reporting Protocol. Assess Product Life Cycle. Identify Internal & External Impacts. Identify Investments. Train & Develop Employees. Include Subject-Matter Experts. Align Incentives.	Develop Long-Term Sustainability Goals. Create Academic, NGO, & Industry Partnerships. Participate in Global Initiatives. Verify Results with Third-Party Audit. Benchmark Against Industry Peers. Align Management Performance Scorecards. Modify Job Descriptions.

Table 4.1 lays out a general framework and timeline for developing a 5-year sustainability program. An organization's sustainability strategy will reflect particular industry concerns, regulatory, and competitive issues. It will also reflect the organization's maturity with sustainability. The more experience management has had with sustainability, the greater the detail included and the more specific the plan will be. The purpose of this framework is to provide perspective on the many layers, players, and considerations that must be included in the plan.

4.2 Creating and Implementing Sustainable Strategy

In order to implement sustainable strategy and reap the organizational benefit, sustainability must be integrated into the organizational culture. Engaging in this process requires a change in the way that strategy is developed and evaluated as well as the way that work is performed and rewarded. Program and projects are an important link in effecting change within an organization. Undertaking programs and projects to promote and incorporate sustainability into the project management process is an important step. The sustainability champion or, if you are fortunate to have one, the chief sustainability officer (CSO) acts as program manager, promoting the strategic value of sustainability programs and projects, coordinating with project sponsors, and communicating the intended benefits to stakeholders. The sustainability champion or CSO acts as the chief sustainable strategy officer and provides a link among the C-suite, the sustainability steering committee, and project sponsors and managers implementing sustainability projects within the organization. The sustainability champion serves multiple roles, from advocating for sustainability with senior management to managing sustainability projects. Over time this role will evolve, but in early stages, the sustainability champion is often the project manager.

Undertaking sustainability projects that align with core business value improves the likelihood of success. "Projects and programs that are highly aligned to an organization's strategy are completed more successfully more often than projects that are misaligned (48% versus 71%)."[1] The lack of alignment of projects to strategy is the most frequently cited issue when projects are reported as unsuccessful. Other causes of project failure include lack of C-suite involvement, inappropriate C-suite micromanaging, lack of appropriate skills, or insufficient human resource allocation.[2] All of these hold true for sustainability projects.

Sustainability champions, project sponsors, and managers need to plan for these challenges. Strategic alignment and clear benefits identification are crucial for maximizing scarce corporate resources. According to PMI Pulse research, organizations that have high alignment between strategy and function are more agile and better able to adapt quickly to changing market conditions, maximizing operating efficiency, and meeting customer demands.³ Adoption of sustainable strategy allows an organization to be more adaptive to challenges and opportunities through sustainability visioning, stakeholder input, and alignment of resources around core mission.

Aligning business and sustainability goals requires a holistic management strategy that looks at both internal operations and external impacts along the organization's value chain. Undertaking a program to minimize greenhouse gas (GHG) emissions is often selected as an early-stage sustainability project. If a project is undertaken without understanding how the organization's operation creates GHG emissions, it is usually unsuccessful. The sustainability champion may discover that the firm's internal impact on product GHG emissions is 30% of the total, with the majority of the GHG emissions coming from supplier and customer activities. Taking the time to understand the operations of an organization and the relevant impact points makes projects significantly more successful.

While taking a look at internal impacts is a good starting point, broadening that assessment to include the full value chain is optimal. Looking up and down the value chain reveals challenges and solutions from suppliers, customers, and other stakeholders. Early-stage value chain assessments tend to be linear, looking from product sourcing through manufacture to distribution and end use. While this assessment will provide insight into GHG emission creation, it doesn't really offer a complete range of solutions to mitigate GHG emissions. Taking a step further and moving the organizational view of the value chain from linear to circular gives perspective on ways a product or service can be designed, created, shipped, utilized, recaptured, and reused, thereby reducing the environmental footprint across its full life cycle. Sustainability champions who focus on a life-cycle assessment gain insight into current practices and challenges as well as potential solutions and alternatives that form the foundation for a sustainable strategy roadmap. Benefits include a more sustainable approach to product design, sourcing, and development. Evaluating the life-cycle process provides opportunities to conserve resources, protect employees' health and safety, reduce risk, and improve quality. Gathering life-cycle information provides management with the data and information to communicate effectively with stakeholder groups such as investors, customers, and regulators about the environmental and social impacts of the organization's products and services. It also provides insight into new product and service delivery mechanism designed to reduce environmental and social impacts.

In order to benefit from a life-cycle assessment, senior management needs to be guided in thinking about organizational systems and their interrelationships within their value chain with customers, communities, investors, and suppliers (see Figure 4.3). This approach identifies opportunities within the value chain to improve product design, reuse materials, utilize renewable energy, optimize production, remanufacture products or components, and explore sustainable products and solutions. Once the core life cycle is understood, this approach can be expanded to look for solutions and opportunities from outside the current value chain—from other industries, new partners, and other stakeholders.

A core tenant of sustainable strategy revolves around leveraging the relationship between an organization and its stakeholders to develop better business solutions. Rather than looking solely internally for solutions, engage with external stakeholders in order to gain a variety of opinions on issues as well as solutions and opportunities. Forging this network of diverse relationships provides the foundation for a sustainable stakeholder management strategy. (Chapters 8 and 9 discuss stakeholder management in greater detail.) Embedding sustainability into an organization benefits from approaching challenges from a more inclusive perspective in order to uncover a wide array of potential options and solutions.

As an organization's sustainability journey evolves, business leaders need to adopt a broad view of challenges and consider the interconnectivity of the three components of sustainability—economic, social, and environment.

Understanding the relationships among natural, social, and economic systems and how decisions are made in terms of operating practices, policies, technologies, facilities, products, and services impact these systems guides development of sustainable strategy. The U.S. Environmental Protection Agency (EPA) refers to this concept as "systems-thinking." Figure 4.4 highlights interactions within and

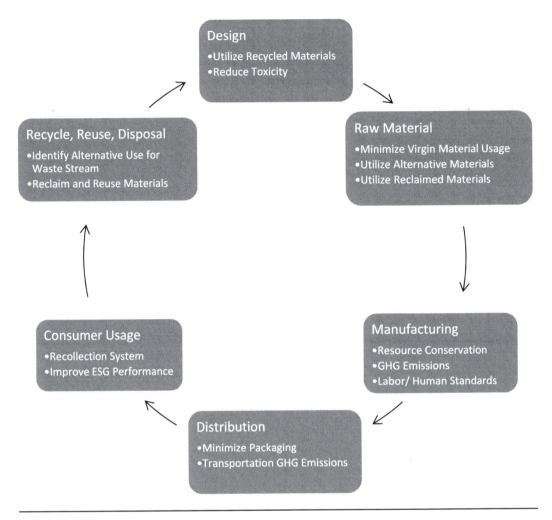

Figure 4.3 Product Life-Cycle Assessment (LCA)

between these systems. Like a product life-cycle analysis, these systems tend to be circular rather than linear cause-and-effect relationships.[5] Following a triple bottom line approach highlights these interdependencies and demonstrates how social and economic value rely on the planet's ecosystems for food, fuel, raw materials, water, energy, biodiversity, air, and other essential services. Engaging in a life-cycle assessment sheds light on management's decisions and how they impact these system interdependencies.

Once the product life cycle of an organization's operation is understood, management can further improve the alignment between business and sustainable strategy by adopting a systemic view of the organization's operating environment, focusing on the interconnectivity among various functions within the industry as well as externalities from government, regulators, and even other industries.

As a starting point, sustainability champions need to address internal organizational issues that limit collaboration. Many businesses are organized linearly by business function, with each business leader immersed in his or her operation. Internal cross-functional collaboration is limited. As a result, complex cross-functional programs and projects require a change in organizational decision making. (This topic is discussed in greater detail in Chapter 11.) The scope of collaboration needs to expand to the value chain and beyond to include stakeholders such as customers, suppliers, and communities as well as industry partners and academic institutions. The Shared Value discussion of Silicon Valley in Chapter 3 outlines some of the benefits of industry collaboration. Sustainability focuses on interconnectivities not only within organization but also between organizations, such as partners, academia, NGOs, and governments. As management searches for strategic solutions to improve their organization's economic, environmental, and social impacts, they begin to discover the complex web of interconnectivity both within their industry and between industries and organizations. From this holistic systemic approach come new strategic options, platforms, and solutions to address environmental, social, and economic challenges and opportunities.

As an example the Canadian concrete industry was looking for new methodologies to "green" the process of manufacturing concrete. Portland cement, a basic ingredient in concrete, is a highly desirable material used to repair and build infrastructure. The developed world relies on it to repair aging infrastructure and the developing world needs it for new construction to support population growth. While Portland cement is vital to economic development, its manufacture has a significant environmental impact, degrading natural capital and generating GHG emissions. The manufacturing process uses a large amount of natural resources. Annual global use of raw materials is approximately 9–10 billion tonnes of crushed rock, gravel, and sand, and 1 billion cubic meters of water.[6] The process is also energy-intensive, with the manufacture of 1 tonne of cement producing an equivalent amount of CO_2 emissions.[7] Mining of raw materials impacts biodiversity and the local ecosystem, as the mining process releases heavy metals that had been embedded into the rock formations, further polluting the

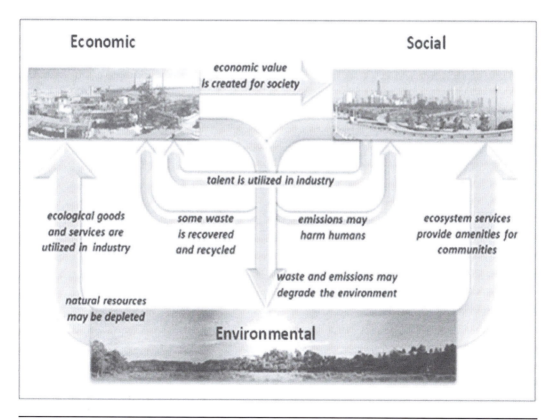

Figure 4.4 Sustainability Systems[4] → SYSTEMS THINKING "CIRCULAR SYSTEM"

environment. Processing and transportation adds to the industry's carbon footprint. The waste stream from construction and demolition activities is about 900 million tonnes per year.[8]

Gathering this industry product life cycle detailing the environmental impact facilitated a conversation among industry leaders to develop greener solutions. This approach facilitated improved environmental performance and preemptively addressed potential regulation from the Canadian government. The industry leaders realized that addressing sustainable development in the concrete industry required looking at the interconnectivity among the social, economic, and environmental impacts both within and outside their value chain.. In order to address these challenges, leadership proposed a "system-thinking" solution approach to evaluate collaborative and innovative concepts both within and beyond its own industry. The model that was developed outlines a complex web of causal loops that evaluate systems interrelationships and opportunities across the concrete industry, other industries, and other stakeholders. Rather than looking at discrete actions of a single firm, they looked at the interrelationships among actions and the array of impacts within the industry value chain and across other industries as part of the solution.

As they outlined relationships between the industry and its stakeholders, these leaders found numerous interrelationships, which acted as either enablers or barriers to the greening process. The model looks both inside and outside the concrete industry for solutions. While many opportunities were identified throughout the industry value chain, a raw-material solution emerged from outside the industry—fly ash, which is a by-product of the utility industry. Using the waste product, fly ash, diverts it from landfills and reduces demand for mining virgin raw materials in the concrete production process. By incorporating fly ash, the environmental footprint of concrete manufacturing could be significantly reduced.

However, there were also potential barriers to adoption of this approach. As utilities adopt improved smoke stack scrubbing technology, there was concern about the quantity and quality of fly ash that will readily be available for use in the concrete manufacturing process. In addition, fly ash creation, disposal, and reuse correlate with the perceived environmental friendliness of both the concrete and utilities industries. While concrete production is able to utilize this by-product of the utility industry, the utility industry is facing current and pending government regulation to limit fly ash creation. Other related system concerns include production capacity and costs, and product quality and performance. The model calculates the impact of concrete manufacture on the environment in different sectors such as consumption, natural resources, waste, emissions, carbonation, fly ash use, market and technology, and the electric power industry, and then combines them in a proportional ratio to calculate the cumulative impact. After running simulations based on varying assumptions, the researchers found that there was an opportunity for the cement industry to reduce CO_2 emissions and conserve natural resources.[9]

Leaders in the concrete industry discovered that through focusing on a system approach, they were able to identify options for addressing environmental, economic, and social challenges facing the concrete industry as a whole. From a market perspective, industry leaders realized that energy efficiency and resource conservation were required for their businesses to remain competitive. Also, emerging national CO_2 emission standards required new technology and business models to meet targets and retain their "license to operate." Also, customers were favoring environmentally friendly products and services. Through taking a more holistic systemic approach, industry leadership was able to preemptively address significant environmental, regulatory, economic, and social challenges.

The Canadian concrete case highlights the interconnectivity of aligning business and sustainability goals in order to remain globally competitive while reducing environmental and social impacts. In order to effect major change, industry leaders looked beyond traditional industry boundaries to incorporate a broader range of stakeholders, providing an innovative solution that addressed customer demands and pending environmental regulation. "System thinking" solutions require engagement of a broad base of stakeholders, creation of innovative processes, and adoption of new technologies to create impactful sustainable solutions.

4.3 Organizational Alignment

The board of directors is charged with the responsibility for providing oversight of the long-term viability and success of an organization. Fully integrated sustainability strategies include board oversight and support. Boards that have adopted a sustainable vision to create long-term value signal an organization's commitment to sustainability principles. The challenge of realizing the sustainable vision is given to the C-suite to translate the vision into organizational mission, goals, and specific performance targets. Establishing an organizational mission statement provides a means to communicate the desired outcome both internally and externally. Selecting long-term goals gives management and employees a clear message about desired organizational outcome. Theses long-term goals frame an organization's sustainability strategy, focusing management's attention on programs and projects that move the organization toward achieving these goals. In the organization's published sustainability report, annual progress toward these goals is measured and assessed. While an organization's sustainability journey does not always move in a straight line, it is important to identify impactful long-term goals and to keep the momentum moving forward.

> **Example Long-Term Strategic Goals**
>
> 1. Reduce greenhouse gas (GHG) emission to net zero by 2030
> 2. Reduce water usage to net zero by 2030
> 3. Eliminate all waste to landfill by 2020
> 4. Achieve 100% renewable sources of energy by 2025
> 5. Grow percentage of sales from sustainable products to 20% by 2020
> 6. Increase percentage of products with sustainably sourced ingredients to 50% by 2020
> 7. All new construction to meet Gold LEED certification standards
> 8. Increase the percentage of women/minorities in senior management to 50% by 2020
> 9. Increase the percentage of women on the board of directors to 50% by 2020

In order to reach any of these strategic long-term goals, the goal needs to be broken down into more manageable, shorter-term targets that measure progress along the way. Suppose, for example, that one goal is to increase the organization's diversity and inclusion performance as defined by increasing the percentage of women and minorities in senior management to 50% by 2025. This goal needs to be broken down into interim targets in order to garner management's attention and to allow for creation of programs and projects to promote changes within the organization to realign resources, policies, procedures, and people to reach the goal. A plan with a project target of moving from a 15% to 25% ratio by year 5 keeps the needle moving in the right direction in terms of the organization's diversity and inclusion, yet allows sufficient time to address organizational barriers and to create a pathway for new opportunities for women and minorities.

Table 4.2 shows one example of such a 5-year plan. This process of breaking long-term goals into more manageable targets sends a message to both employees and external stakeholders that management is serious about achieving its sustainability goals.

Aligning sustainability and business strategies creates a unified organizational mission that incorporates environmental, social, and governance (ESG) criteria into the strategic decision-making process. In order to successfully execute strategies to support this mission, the organizational structure must support cross-functional collaboration. Many organizations establish a Sustainability Office led by a Chief Sustainability Officer (CSO), who is tasked with developing and coordinating sustainability projects across the organization. Developing a dedicated position signifies the importance of

Table 4.2 Diversity and Inclusion (D&I) 5-Year Plan

1 Year	2–4 Years	5 Years
Benchmark the Ratio of Minorities and Women in Senior Management	Engage C-suite to Promote D&I Program	Achieve Industry Recognition for D&I Program
Segment Results by Business Function	Engage Business Functional Leaders in D&I Program Implementation	Achieve 25% Minorities and Women in Senior Management
Identify Stakeholders	Select Annual Targets	Offer Consistent Opportunities Across the Organization
Create Focus Groups to Identify Barriers	Identify Reporting Standards and Metrics	Seek Feedback for Continuous Program Improvement to Reach 50% Goal
Create D&I Program to Address Barriers and Promote Women and Minorities	Provide Leadership Training	
	Create Mentoring Programs	
	Tie Performance Metrics to D&I Targets	
	Develop Strategic Partnerships	
	Develop Employee Resources	
	Invest in Training & Development	
	Develop Flexible Work Arrangements	

sustainability to the organization and provides cross-functional support for adopting a culture of sustainability. Functional silos need to be broken down in order to integrate sustainability and to promote cross-functional collaboration. Best practice suggests creating a dedicated role to focus on the sustainability program and to facilitate cross-functional collaboration. While the CSO is a facilitator, all organizational functions and levels need to be involved with sustainability in order to effect transformational change. Figure 4.5 lays out the process for creating organizational alignment between business and sustainable strategies and identifies some common barriers to the adoption process.

As management's experience with sustainability grows, creating a sustainability steering committee with senior leaders from the business functions facilitates further adoption of sustainable strategy. The CSO acts as the program manager, keeping the steering committee informed about organizational sustainability programs and projects. This structure engages leaders from across the organization on sustainability and begins the process of educating them on the benefits of sustainability. It helps in identifying new opportunities and establishing cross-functional relationships needed to address more complex projects. As business leaders better understand the impact of the sustainability vision, they will become more supportive of programs and projects within their own functional areas. This acceptance process facilitates cascading the mission of sustainability throughout the organization. Implementing sustainability programs and projects reaches down within the organization to change people, process, and technology, further driving the organization toward its sustainability vision.

As an organization engages in this process, sustainability project sponsors, CSOs, and sustainability champions must anticipate barriers to acceptance even with senior management support. One of the most common barriers is that the business units are not engaged in the process. As a result, the sustainability champion or office functions in a silo. Business-line leaders view sustainability as the role of the sustainability champion or CSO and not within their own scope of responsibilities and duties. Engaging the line leaders in the process by creating a collaborative organizational structure encourages business leaders to become involved in cross-functional solutions. Highlighting the positive business impacts of engaging in sustainable solutions and encouraging the sharing of stories and experiences helps to create supporters out of would-be blockers.

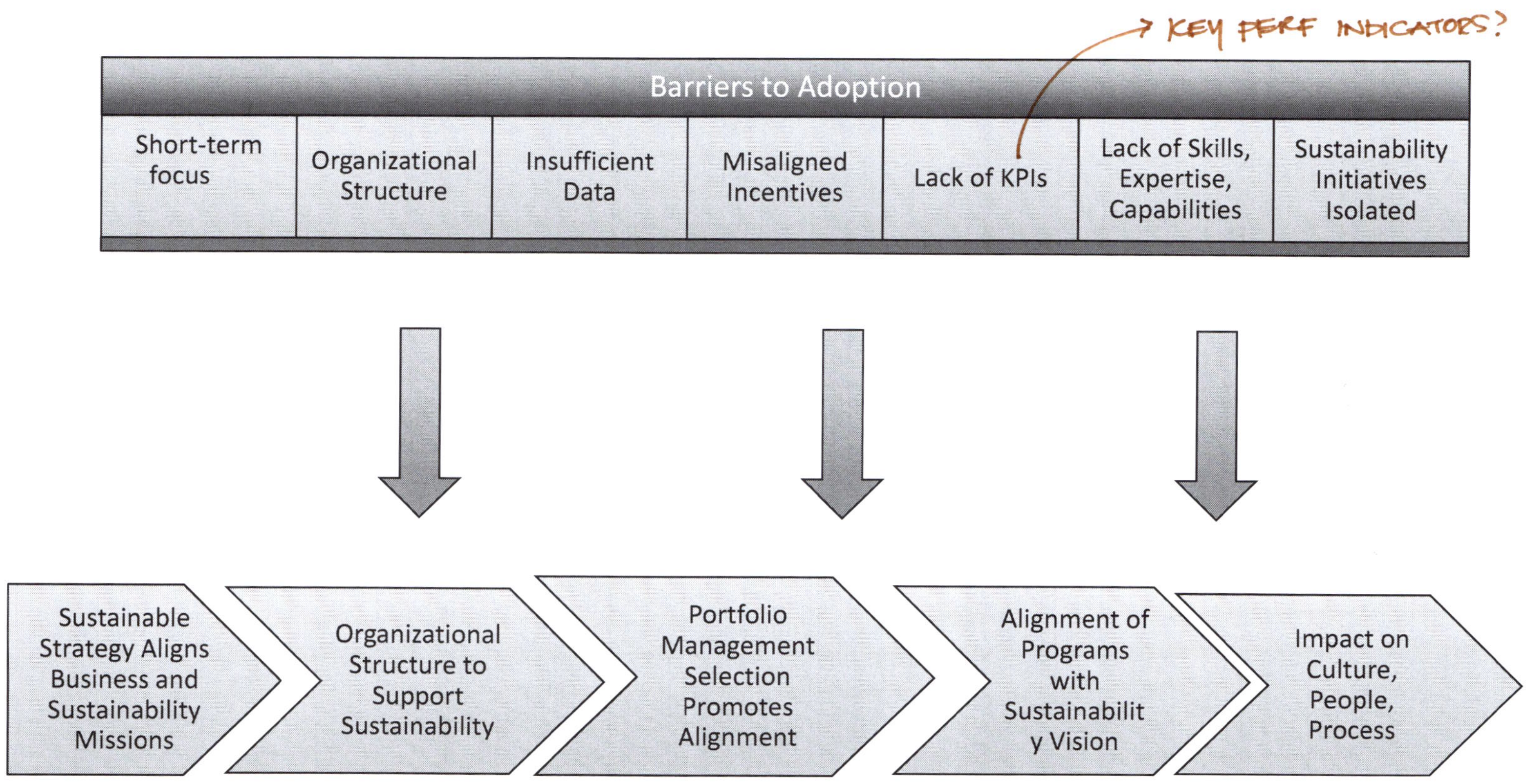

Barriers to Adoption

| Short-term focus | Organizational Structure | Insufficient Data | Misaligned Incentives | Lack of KPIs | Lack of Skills, Expertise, Capabilities | Sustainability Initiatives Isolated |

(handwritten annotation: → KEY PERF INDICATORS, pointing to "Lack of KPIs")

- Sustainable Strategy Aligns Business and Sustainability Missions
- Organizational Structure to Support Sustainability
- Portfolio Management Selection Promotes Alignment
- Alignment of Programs with Sustainability Vision
- Impact on Culture, People, Process

Figure 4.5 Creating Organizational Alignment

Getting business leaders to focus on long-term goals can be challenging. So often, leadership's focus is on quarterly earnings and other short-term targets. Management compensation is often tied to meetings these short-term targets. Successfully implementing a sustainable strategy requires measuring success over a much longer time horizon, as environmental, social, and governance challenges are long-term rather than short-term in nature. Tying incentives and performance scorecards to sustainability goals can significantly change management's perception. Developing a sustainable culture is a journey, and best practice suggests establishing a structure that keeps business leaders engaged and involved. Sustainability champions must learn from successes and failures. Engaging with both internal and external stakeholders to gather feedback and advocate for changes to policies and procedures based on lessons learned ensures the continued relevancy of the sustainability program.

Lack of sufficient data and a definition of success is another barrier. Demonstrating the impact of sustainability requires establishing targets and goals, gathering data, and defining and reporting on metrics. Take the time to select and implement technology that has tools to track data and provide meaningful and timely reports to management. (Chapter 14 further discusses metrics and the role of technology.) Lack of resources is another common barrier. If a sustainability champion cannot demonstrate a program's business value, it is unlikely to receive long-term support. Through identification of programs and projects that truly align with the transformative business mission, business leaders can begin to realign people, budget, and resources to produce results in line with the organization's sustainability vision.

4.4 Portfolio Management Selection to Promote Alignment

In order to manage sustainability effectively for the long-term, projects need to meet both sustainability and business targets. Portfolio management methodologies and techniques facilitate the selection process in order to achieve the long-term goals set by senior management. Embedding sustainability into an organization requires that time, money, and resources be allocated toward projects that align with sustainability targets. Establishing project portfolio criteria to facilitate program and project selection fosters alignment with long-term strategic goals. (Chapter 7 includes a detailed discussion of portfolio management.)

Engaging senior management in the strategic portfolio alignment process creates alignment between business and sustainability goals by clearly defining program and project requirements. Through the selection of portfolio components that align with strategic vision, management lays the building blocks to ultimately achieve long-term goals. Establishing program and project business and sustainability criteria provides meaningful guidance to project sponsors to facilitate project selection. Mapping out where projects fall relative to business and sustainability criteria provides a visual tool to facilitate the process of allocating resources to projects that create the most impact; see Figure 4.6. Nonconforming projects can be eliminated or designated for further assessment before receiving funding approval. Incorporating this process as part of the standard portfolio review encourages the elimination of nonconforming projects, thereby freeing resources to be allocated to projects with higher strategic sustainability alignment.

A variety of factors can cause misalignment, including lack of information, insufficient evaluation, or even conflicting standards and processes within the organization. Projects that are in high alignment with business goals but not sustainability goals might not fully consider long-term risk factors such as supplier integrity, raw material supply disruption, stakeholders' adverse reaction, and brand and reputational damage. Projects that deliver on long-term sustainability goals but not on business hurdles might reflect insufficient business value consideration, or there may be organizational standards that are inherently in conflict. For example, large, capital-intensive sustainability projects often do not meet corporate thresholds for return on investment (ROI) and payback period. As a result, there is a conflict between realizing long-term sustainability goals and meeting short-term financial targets.

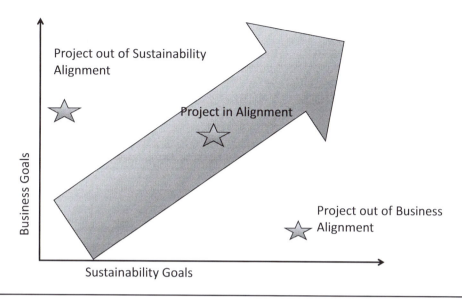

Figure 4.6 Project Portfolio Alignment Mapping

Sustainability projects such as renewable energy projects have to compete internally for limited capital. Because these projects require chief financial officer (CFO) approval, financial threshold metrics such as internal rate of return (IRR) and ROI must be met for selection. Creating alignment between an organization's programs toward promoting renewable energy as part of an overall carbon-reduction strategic vision with financial return thresholds is often difficult. Sustainability is a long-term initiative, and typical financial metrics focus on short-term payback periods. Reconciling these two opposing factors may require "out of the box" thinking.

To address this type of challenge, Johnson & Johnson (J&J) launched a $40 million annual capital relief fund to support renewable energy projects globally. Management's goal was to refocus the renewable energy program, shifting away from purchasing Renewable Energy Certificates (RECs) and focusing on making on-site direct investments in renewable energy infrastructure at their facilities around the world. The capital relief fund provided an option for local facilities to request funding for renewable energy projects so that they didn't have to compete with other facility-level projects. In addition, Finance reduced the threshold IRR requirement from 20% to 15% for projects that have a "carbon reduction impact." The first-year impact was that J&J spent $48.2 million on GHG emission reduction projects, including the installation of 12 solar photovoltaic (PV) systems.[10]

Undertaking a portfolio review process assists in aligning corporate resources with both sustainability and business goals. It also provides a framework in which to identify challenges and to highlight processes and systems within the organizational structure that need to be altered in order to reach the organization's sustainability vision. Engaging in the portfolio review process cascades sustainability requirements within the organization, aligning program and project methodologies such as benefits management, stakeholder engagement, risk management, and vendor and supply management with sustainability vision.

4.5 Alignment of Programs and Projects with Sustainability Vision

While the C-suite is responsible for creating sustainable strategy, it usually falls to the CSO or sustainability champion to lead the sustainability program within the organization. An important aspect of a sustainability program is developing a plan to translate and cascade the sustainability vision through the organization. The CSO is tasked with demonstrating alignment between sustainable strategy and beneficial outcomes for various business units. Once a program or project has been selected, the CSO is tasked with providing strategic updates to the C-suite on sustainability program and project status, lessons learned, and outcomes. Regular program and project status updates keep senior management informed of progress relative to plan to ensure that sustainability programs and projects are in alignment with strategic vision.

Utilizing the benefits management process of project management facilitates adoption of sustainable strategy as business function leaders clearly see the relationship between sustainability and business outcomes. Project metrics are tracked against desired sustainability and business outcomes rather than just activity milestones. This approach keeps all stakeholders focused on the benefits of the project. Moving further down within the project process, project results, deliverables, and tasks must also be aligned within benefits management. Focusing attention on benefit delivery, rather than a series of tasks, promotes alignment between sustainable and business strategy. The process provides clear messaging of the desired outcomes and tracking of metrics that measure core business impact. Figure 4.7 depicts translating sustainable strategy and vision into organizational action.

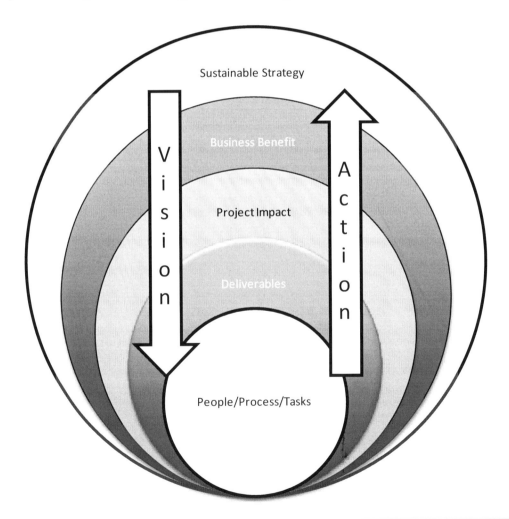

Figure 4.7 Cascading Sustainable Strategy into an Organization

Translating a project's sustainability goals through the layers of an organization is challenging, but using an effective communication plan, common language, and metrics to measure impact helps deliver the strategic vision down to the people, process, and task level. Tracking data and gathering feedback informs sustainability champions about adjustments needed to sustainability programs and projects to improve alignment with the organizational mission.

One of the challenges is communicating the vision through many layers of an organization. While the sustainability vision may be clear to senior management, the message can be lost in translation. A strong communication and change management plan must be created to convey the vision, roles, and responsibilities throughout the organization. In order for sustainable strategy to be impactful, the messaging needs to be clear all the way down to the employee who is performing a specific task. That individual must understand how her behavior and actions affect achievement of the organization's sustainable strategy. (Chapter 11 discusses communication and change management techniques.)

In the early 2000s, academic institutions faced challenges in developing, implementing, and communicating their sustainability programs in order to promote meaningful dialogue with prospective students and college ranking agencies such as Princeton Review. Challenges included aligning strategy to operations, moving sustainability beyond facilities management, clearly communicating with internal and external stakeholders, and fluctuating sustainability rankings based on the vagaries of third-party ranking protocols. At the University of Vermont (UVM), the Sustainability Office addressed these challenges by adopting the Academic Sustainability Tracking, Assessment & Rating System (STARS), providing a framework to align sustainable strategy and core operations while significantly improving internal and external communication and stakeholder engagement. The STARS protocol uses a common language and standards so that both internal and external stakeholders are speaking about the same goals, impacts, and issues. Adopting STARS facilitates effective communication about the institution's sustainability agenda, allowing people and processes to be adapted in support of sustainability programs. In addition, it provides the university with a set of standards that allows prospective students, employees, and ranking agencies to understand clearly the institution's sustainability vision and the desired outcomes. The framework allows academic leadership to focus on material issues and to support meaningful recommendations for program change and improvement.[11] It also provides an industry standard for meaningful benchmarking of academic institutions rather than the alternative of a number of different protocols that didn't allow for meaningful cross-institutional comparison. As a result of these changes, communications were clearer both within the UVM community and with outside rating agencies and prospective students. (For more details, see the discussion of the University of Vermont STARS Implementation in Chapter 14.)

4.6 Materiality Supports Alignment

The clearest way to demonstrate alignment between sustainability and business value creation is to create a strategy that addresses environmental, social, and economic issues that impact operations. This concept is known as materiality. It was introduced in Chapter 3, where the focus was on comparing stakeholder requirements and operational impacts to determine the most material issues for the organization. In this chapter, we are drilling further into the concept of materiality to identify relevant sustainability issues, recommend standards for measurement, and determine industry standards to benchmark performance. In order for a sustainability program to succeed, it must be relevant to management, employees, and other stakeholders.

The Sustainability Accounting Standards Board (SASB), a not-for-profit organization established to develop industry-based sustainability standards for identification and reporting of material environmental, social, and governance issues by publically traded U.S. companies, has developed a set of materiality standards for organizations with specific industry metrics. Their belief is that corporate sustainability reporting standards provide a tool to be used in conjunction with financial accounting standards for companies, investors, regulators, and the public to evaluate corporate performance. Materiality issues fall into five broad categories of environmental, social capital, human capital, business model and innovation, and leadership and governance.

The SASB selected their standards based on the following set of criteria:

1. The standard must be relevant to a material issue and applicable to most companies within an industry.
2. The information provided must be useful for investors and management to make decisions.
3. The data need to be complete, auditable, and comparable to facilitate industry peer-to-peer benchmarking.
4. The metrics must reflect direction so that performance changes can be clearly signaled.[12]

Figure 4.8 outlines the key areas of materiality defined by the SASB.

In order to gain a more in-depth perspective of this universe of ESG issues, the SASB provides identifiers to qualify and quantify responses (see Table 4.3). The Environment qualifiers focus on emissions generation, resource usage, and energy utilization reduction. Social Capital considers community impacts, marketing and communication, customer relations, and product disclosures. Human Capital looks at diversity and equal opportunity, core HR functionality, and labor practices. Business Model &

Figure 4.8 SASB Materiality Assessment.[13] © 2015. Reprinted with permission from the Sustainability Accounting Standards Board (SASB). All rights reserved.

Table 4.3 SASB Materiality Issues and Identifiers[14]

Environment	Business Model & Innovation
Climate Change Risks	Long-Term Viability of Core Business
Environmental Accidents & Remediation	Accounting for Externalities
Water Use & Management	Research, Development, & innovation
Energy Management	Product Societal Value
Fuel Management & Transportation	Product Life-Cycle Use Impact
GHG Emissions & Air Pollution	Packaging
Waste Management & Effluents	Product Pricing
Biodiversity Impacts	Product Quality & Safety
Social Capital	**Leadership & Governance**
Communications & Engagement	Regulatory & Legal Challenges
Community Development	Policies, Standards, Codes of Conduct
Impact from Facilities	Business Ethics & Competitive Behavior
Customer Satisfaction	Shareholder Engagement
Customer Health & Safety	Board Structure & Independence
Disclosure & Labeling	Executive Compensation
Marketing & Ethical Advertising	Lobbying & Political Contributions
Access to Services	Raw Material Demand
Customer Privacy	Supply Chain Standards & selection
New Markets	Supply Chain Engagement & Transparency
Human Capital	
Diversity & Equal Opportunity	
Training & Development	
Recruitment & Retention	
Compensation & Benefits	
Labor Relations and Union practices	
Employee Health, Safety, & Wellness	
Child and Forced Labor	

© 2015. Reprinted with permission from Sustainability Accounting Standards Board (SASB). All rights reserved.

Innovation considers an organization's long-term viability, the impact of externalities, and the product life-cycle impact. Leadership & Governance considers regulatory and legal compliance, ethics, code of conduct, supply chain standards, and board structure. The approach is to evaluate an organization on a broad cross section of criteria that truly measure the long-term viability of an organization and the factors that can threaten or enhance the organization's long-term performance.

For those who are doubtful that these broad-reaching factors impact the long-term viability of an organization, consider the scandal created at Volkswagen AG (VW). In late 2015, VW made headlines around the world because of governance missteps with the admission of using software to improve diesel engine emission test results. By traditional financial metrics, such as vehicle sales targets, VW had been meeting or exceeding targets. Peeling back the layers and gaining transparency into organizational leadership and governance standards provides a much clearer assessment of long-term risk and performance.

The damage that this revelation has done to the organization's brand and reputation with consumers as well as the impact on the ongoing viability of diesel technology is significant. Volkswagen announced a $7.27 billion charge to earnings. The initial impact was that VW became the subject of an EPA investigation, with the possibility of U.S. criminal charges as well as $18 billion in fines.[15] Volkswagen's shares plunged 35% as a result of the news, and CEO Martin Winterkorn resigned his position.[16] The long-term impact on VW's ongoing reputation and operation are still unfolding.

The VW story reminds sustainability champions that sustainability isn't a check-box exercise. In order to garner the benefits, sustainability has to be embedded into the culture of an organization so that managers and employees throughout the organization are making decisions that are in line with organizational values and vision.

The SASB Materiality Map™ (http://materiality.sasb.org) provides a visual tool to sustainability champions to identify industry-specific material issues to better select areas of strategic focus. Using this information as a foundation, SASB has developed a Sustainability Map to evaluate 30 sustainability issues as they relate to 7 industry groups with materiality issues identified relative to their importance to a particular industry. Currently, industry-based materiality information is available in the categories of Health Care, Financials, Technology and Communications, Non-Renewable Resources, Transportation, Services, Resource Transformation, and Consumption. (Renewable Resources & Alternative Energy and Infrastructure have expected release date of 2015 and 2016.)

The standards that SASB utilizes for metric selection are relevance, usefulness, applicability, cost-effectiveness, comparability, completeness, directional, and verifiable.[17] Each industry is further divided into industry subgroups—health care, for instance, is segmented by pharmaceuticals, medical equipment and supplies, health care delivery and distribution, managed care, and biotechnology.[18] Material issues are listed for each subgroup. The SASB has developed industry-relevant disclosure suggestions for the major categories of environment, social capital, human capital, business model and innovation, and leadership and governance. Focusing on the pharmaceutical sector (see Table 4.4), a material social capital issue in the category of human rights and community relations is the safety of clinical trial participants. While this issue is relevant for pharmaceutical companies, for other organizations in the health care industry, such as the medical equipment and supplies subgroup, this issue is not material. As a result, management of a medical equipment company doesn't need to spend time and resources focusing on this irrelevant issue. However, other issues, such as waste and hazardous materials management, are relevant environmental concerns that need to be addressed by the majority of health care industry participants.

The SASB standards include recommendations not only for material issues on which to focus but also on industry best practices for reporting on these issues based on meaningful metrics. It can be challenging to select metrics to measure sustainability program impacts, so these guidelines are very helpful to sustainability champions as they develop sustainability programs and projects. Also, the standardized metrics facilitate benchmarking an organization's sustainability performance relative to peers. The SASB recommended metrics include both numeric measures and discussion and disclosure measures. Working with these recommended numeric and descriptive metrics enables sustainability champions to more clearly define their programs and to measure alignment with desired outcomes. It also provides third-party support for metric choices and selection, which can be helpful in overcoming challenges from business leaders setting up barriers on the road to sustainability.

Table 4.4 SASB Health Care Pharmaceutical Materiality and Metrics (Excerpt)[19]

Material Issue	Disclosure Topic	Metric
Social Capital—Human Rights & Community Relations	Safety of Clinical Trial Participants	Discussion of the management process for ensuring quality and patient safety during clinical trials.
Business Model & Innovation-Life-Cycle Impact of Products	Drug Safety & Side Effects	Product stewardship of return, redistribution, and disposal of unused product at the end of the product life cycle.
		Details should include the amount spent on the initiative and the amount of product returned, reused, or disposed.
Human Capital	Employee Recruitment, Development, & Retention	Description of talent recruitment and retention processes for scientists and other R&D candidates (mentorship, career development, leadership training, incentive programs).
		Training & development budget per full-time employees (break out of advanced industry or professional development.)
		Employee turnover (voluntary & involuntary) by categories of executives, mid-level managers, professional, and other.

As a sustainability program matures, the SASB industry-based metrics become a useful resource for portfolio managers to establish selection criteria to facilitate program and project alignment with strategic vision. Aligning business and sustainable strategies lays the foundation for making sustainability part of an organization's culture. The Materiality Priority Checklist in Figure 4.9 can facilitate the process of assessing areas of materiality. This checklist provides a framework to identify and prioritize materiality issues and to map them to business functions in order to begin the process of developing an impactful sustainability roadmap. This frame work can help to lay out a plan for the sustainability champion to identify business function leaders who have top-priority opportunities and to engage them on the benefits of undertaking sustainability projects within their departments. Based on the results of this assessment, programs and projects to promote sustainability can be identified and prioritized for selection and implementation.

4.7 Developing a Framework to Promote Sustainability

Putting all of these pieces together facilitates alignment of business and sustainability strategy. The Sustainability Framework provides a structure for articulating vision and business objectives, identifying organizational values, defining the mission, identifying stakeholders and material issues, and laying out organizational commitments. Areas to consider include product stewardship, supply chain management, global challenges such as poverty and human rights, community engagement, employee health and wellness, responsible operations, stakeholder partnership opportunities, and establishment of goals and targets. Engaging with stakeholders to complete this framework facilitates alignment between the strategic vision of sustainability and the operations and processes within an organization. The process lays the foundation for program and projects selection and implementation to effect changes to policies, functions, actions, and behaviors in order to develop a culture of sustainability. It also helps to identify areas of organizational strengths and weaknesses in order to create the best organizational structure and environment to support a sustainable transformation.

Materiality Issues	Tier 1	Tier 2	Tier 3	NA	Business Goal	Functional Area Impact
ENVIRONMENT						
Climate change risks						
Environmental accidents and remediation						
Water use and management						
Energy management						
Fuel management and transportation						
GHG emissions and air pollution						
Waste management and effluents						
Biodiversity impacts						
SOCIAL CAPITAL						
Communications and engagement						
Community development						
Impact from facilities						
Customer satisfaction						
Customer health and safety						
Disclosure and labeling						
Marketing and ethical advertising						
Access to services						
Customer privacy						
New markets						
HUMAN CAPITAL						
Diversity and equal opportunity						
Training and development						
Recruitment and retention						
Compensation and benefits						
Labor relations and union practices						
Employee health, safety and wellness						
Child and forced labor						
BUSINESS MODEL & INNOVATION						
Long term viability of core business						
Accounting for externalities						
Research, development and innovation						
Product societal value						
Product life cycle use impact						
Packaging						
Product pricing						
Product quality and safety						
LEADERSHIP & GOVERNANCE						
Regulatory and legal challenges						
Policies, standards, codes of conduct						
Business ethics and competitive behavior						
Shareholder engagement						
Board structure and independence						
Executive compensation						
Lobbying and political contributions						
Raw material demand						
Supply chain standards and selection						
Supply chain engagement & transparency						

Figure 4.9 Materiality Priority Checklist[20]

"MASTER PLAN"

Sustainability Framework

1. Identify Values and Long-Term Mission
   a. Establish Long-Term Vision
   b. Identify Alignment with Business Strategy
   c. Identify Current Organizational Values
   d. Define Desired Organizational Values
   e. Conduct a Values GAP Analysis
2. Stakeholder Identification and Engagement
   a. Identify and Engage Internal Stakeholders
      i. Board of Directors
      ii. C-Suite
      iii. Business Functional Leaders
      iv. Managers
      v. Portfolio, Program, and Project Managers
      vi. Employees
   b. Identify and Engage External Stakeholders
      i. Community
      ii. Academic Institutions
      iii. NGOs
      iv. Government
      v. Industry
      vi. Trade Association/Unions
3. Identify and Prioritize Material Issues (Refer to Table 4.3)
   a. Environment
   b. Social Capital
   c. Human Capital
   d. Business Platform/Innovation
   e. Governance/Leadership
4. Evaluate Regulatory and Operating Environment
   a. Comply with Mandatory Regulations
   b. Voluntary Industry Sustainability Protocols
   c. Consider Adopting Aspirational Protocols
5. Define Organizational Commitments
   a. Project and Program Identification
      i. Product or Service Life Cycle
      ii. Labor and Human Rights Standards
      iii. Product Safety and Sustainable Ingredients
      iv. Emission Reduction
      v. Diversity and Inclusion
      vi. Renewable Energy
      vii. Community Engagement and Volunteerism
      viii. New Markets
      ix. Innovative Products and Services that Address ESG Challenges
   b. Policy Creations and Adoption
      i. Code of Conduct
      ii. Ethics Policy
      iii. Supplier Policy
      iv. Consumer Codes for Standards
      v. Data Confidentiality

c. Budget and Resource Allocation
   d. Incentive Alignment
   e. Training and Development
   f. C-Suite Engagement and Champion
   g. New Stakeholder Partnerships
   h. Change Management Plan
   i. Communication Plan
6. Define Desired Outcomes
   a. Goals Definition
   b. Target Selection
   c. Metrics Definition and Measurement

Completing this Sustainability Framework identifies organizational sustainability goals based on business goals, organizational values, and stakeholder requirements. This process gives insight into the drivers of sustainability for an organization from both internal and external influencers. Taking the time to complete this framework provides insight into the multitude of stakeholders and begins to shape the scope of the sustainability program. Understanding the operating and regulatory environment facilitates identifying risks and developing an appropriate risk response plan. Selecting material issues for the organization gives the sustainability champion or CSO areas on which to focus programs and projects. Gaining perspective on organizational commitments and resources give the sustainability champion a playing field on which to begin to lay out a long-term sustainability roadmap.

The Sustainability Framework serves as a master plan allowing for conversation with senior management and business-line leaders on the fundamental goals and mission for sustainable strategy and the implementation of this strategy within the organization. As the strategy unfolds and programs and projects are undertaken, this strategic plan will need to be modified to reflect lessons learned and changes in the business climate. However, the overall mission and direction remains focused on long-term programmatic change to embed sustainability within the organization through alignment with core business values.

The Dow Chemical Company's corporate strategy is to "Set the Standard for Sustainability."[21] As part of their strategy, they have focused on sustainable chemistry, which is a product life-cycle approach designed to formulate, manufacture, distribute, and dispose of products more sustainably. "Sustainable chemistry is our 'cradle-to-cradle' concept that drives us to use resources more efficiently, to minimize our footprint, provide value to our customers and stakeholders, deliver solutions for customer needs and enhance quality of life of current and future generation." Over the past decade, Dow has focused its sustainability efforts on developing and embedding the Dow Sustainability Chemistry Index (SCI), a metric devised to assess the sustainability performance of their product portfolio. Their goal was to increase sales from the group that they term "highly advantaged by Sustainable Chemistry" to 10% of sales, which is a significant improvement from the 2007 baseline performance of 1.7%.[22] This classification represents commercially viable products that address sustainability challenges such as climate change, water scarcity, food production, health, and safety. The SCI was developed to quantify the sustainability performance of Dow's products and to serve as a focal point for conversations and decisions throughout the organization. The index rates products on 8 categories with a maximum of 5 points per category. Products can have a maximum score of 40 points. The ratings were based on considerations such as renewable or recycled content, resource management, manufacturing efficiency, environmental life-cycle benefit, manufacturing and transportations risk, social benefit, product and value chain risk, public policy, and end-of-life impact.[23] The purpose of this initiative was to embed sustainability, especially a product life-cycle approach, into the organization's culture. The SCI became a shared focal point and a common language across the organization. With the SCI, sustainability-advantaged product goals became quantifiable, allowing for specific targeted action to improve results by highlighting products with sustainability advantage and identifying those that require improvement or elimination. The index provides a snapshot of the organization's sustainability performance as well as a methodology to track improvement over time. In addition, it provides a quantifiable and transparent means for external stakeholders to track the firm's progress relative to product sustainability.

Through this approach, Dow was able to achieve its 2015 target by 2013. In addition, the overall product portfolio SCI reached 24.4 points, up from 20.4 in 2007.[24] Dow attributes the success of the program to improved sustainability awareness within their culture, which facilitated employees better understanding how to integrate sustainability into their daily functions. These results were evident in business strategy, opportunity identification, solutions development, and communications. The process of scoring products and the resulting discussions have facilitated an ongoing sustainability dialogue and learning process throughout the organization. The SCI has not only improved awareness about sustainability and product life-cycle impact, it has also facilitated a shift in strategic product planning and cross-functional discussions across the product portfolio.

In terms of utilizing stakeholder partnerships, Dow has a variety of approaches, including an advisory council as well as academic, NGO, and governmental collaborations. Since its inception in 1992, the Sustainability External Advisory Council (SEAC) has provided an outside perspective on environmental, health and safety, sustainability, and business opportunities and challenges to senior management. Dow has developed partnerships with academic institutions to support advancement in scientific research and development. In alignment with their sustainability goals, Dow created the Sustainability Innovation Student Challenge Award to support the development of creative solutions around social and environmental responsibility. In addition, they are closely aligned with the University of Michigan (UM) and have a UM-Dow Sustainability Fellows Program to support scholars who are committed to finding sustainable solutions on both a local and global scale. Dow also collaborates with The Nature Conservancy, demonstrating that protecting natural capital and creating corporate value are not mutually exclusive strategies. From this partnership, Dow has discovered the value of integrating the value of natural ecosystem services such as water, land, air, oceans, animals, and plant life into its business model and decision-making process. These programs support Dow's goal of advancing sustainability through collaboration across the business, government, and social sectors.[25]

The Dow case demonstrates the value of aligning sustainability programs and projects with organizational strategy. Using the SCI, a quantifiable metric, management is able to integrate the senior-level concept of sustainable strategy, specifically, a product life-cycle approach and the value of sustainable products into their people, process, and programs. According to PMI research, almost 50% of strategic initiatives are unsuccessful because of lack of project alignment with core business strategy.[26] One of the most effective ways in which to add value to an organization is to clearly align projects with strategic vision. Keys to Dow's success include C-suite commitment and involvement, long-term vision, and a clearly aligned project plan. The 2015 target was established and communicated throughout all layers of the organization. A clear metric was selected to measure a successful outcome. Stakeholders were included in the development and ongoing assessment of the program. Adequate resources were allocated to the project in terms of budget and workforce expertise. One crucial aspect was developing their workforce's understanding of and their competencies about sustainability, in particular, life-cycle methodology of product development and evaluation. Despite a slow start to the program, management took a long-term view and remained committed to its successful completion.

4.8 Value Creation Drivers

Increasingly, business leaders are viewing sustainability as a core business driver. It is becoming part of the tool kit for successful organizational management. MIT calls this point in an organization's sustainability journey the "Sustainability Tipping Point."[27] This point is when significant people within the organization understand the benefits that are derived from engaging in sustainable strategy. Sustainability is growing as a strategic priority, and research has found that once sustainability makes it to the C-suite agenda, it stays. According to survey results, 68% of organizations say that their organizational commitment to sustainability has risen in the past year, a significant increase from 2009 when 25% of respondents answered affirmatively.[28] Response to customer demands is the most reported reason for companies to change their business model to incorporate sustainable strategy. With the exponential growth of social media, customers have a significant and powerful voice in an organization's mission and direction. Increasingly, CEOs are engaging with customers to develop business strategy. According to a 2013 IBM study, 60% of CEOs surveyed expect to see customers influence grow over the next 3–5 years on business strategy; however, only 43% are doing so today.[29] In fact, CEO listed customers as second only to the C-suite in terms of impact on business strategy.[30] Engagement plans include constructing platforms and networks to support complex customer interactions. The C-suite is looking to customers to provide guidance on product and service development, and that includes addressing their concerns about environmental, social, and governance issues. Increasingly, organizations are making connections between innovation and sustainability and identifying those opportunities as key to their organizational growth. It is for this reason that Kimberly-Clark included a metric for environmentally innovative products in its 2015 sales goals.[31] Aligning sustainable strategy with organizational mission about meeting customer demands for sustainable solutions creates business opportunity with bottom-line impact. Further, it supports program and project development that focuses resources on the core business mission, allowing for business agility in response to rapidly changing markets. An organization's sustainability agenda becomes embedded into business processes as the link between sustainable strategy and profitability is demonstrated.

4.9 Aligning Programs and Projects with Sustainable Strategy

In order to move an organization forward on its sustainability journey, the CSO needs to develop a framework for considering program and project impacts on the sustainability mission of the organization. As part of the chartering process, considerations should include both the sustainability and business benefits as well as the project's impact on long-term goals and larger programs. The structure for the project, including C-suite involvement, sponsoring group(s), and level of sponsorship, impacts how the project is championed and managed to meet strategic objectives. Incentives and compensation need to be structured to ensure that sustainability projects and programs get the necessary attention and resources from leadership in order to be successful. Figure 4.10 lays out a series of categories and questions to consider in order to ensure that sustainability projects and programs are aligned with sustainable strategy. For maximum success, structure the project so that the right people are doing the right actions with the right resources.

4.10 Conclusion

Sustainable strategy is most impactful when it has been fully integrated into an organization's culture and fully aligned with business strategy. When an organization reaches the Transformative Stage of the sustainability journey, they reap benefits that drive them ahead of the competitive pack. In order to garner these benefits, organizational leadership reaches the "tipping point" where they are embracing the value creation impact of sustainability and supporting forward momentum on the sustainability journey. Sustainable strategy is a long-term commitment that requires support and commitment from all internal stakeholders—C-suite, business function leaders, managers, and employees. Culture change happens as the C-suite's vision of sustainability cascades throughout the organization, engaging employees and promoting understanding of their roles and responsibilities in the process. Identifying

Sustainability Mission	How does the project support the long-term sustainability mission of the organization?
Business Benefits	How is business value created through this project? How will it be defined and measured?
Program/Project Goals	Is this part of a larger program? What are the short-term and long-term goals? How will they be measured? What are the material issues?
Sponsorship	Is the C-suite engaged? How will communications be managed to keep the C-suite updated and involved?
Structure	Is this a Sustainability Office initiative or a cross-functional initiative? Is a senior-level cross-functional steering committee charged with a successful outcome? How have the reporting and functional responsibilities been aligned to support goal achievement? What sensors are in place to allow for changes to maintain strategic alignment?
Compensation/ Incentives	Are the program goals part of leadership's balanced scorecard? Have personal performance plans been adapted to support alignment with these goals? Have incentive programs been created to support cross-functional actions?
People	Is senior leadership involved? Are functional line leaders involved? Do we have the best mix of internal and external talent to support success? Is the mission clear? Have sufficient resources been provided? Is there a timely feedback loop?
Agility	How adaptable is the program/project to address changes in the competitive landscape? How does it match with portfolio management criteria? Is a management system utilized to track performance?

Figure 4.10 Project/Program Alignment with Sustainable Strategy

and empowering a sustainability champion to act as an advocate for sustainability across the organization facilitates the process of adopting a sustainable culture. In order to maximize the effectiveness of the sustainability program, the sustainability champion identifies the organization's material issues so that resources are channeled toward the most impactful areas. The sustainability champion serves many roles, from enlisting C-suite support to acting as program and project manager for early-stage projects. Project management methodologies are useful in guiding the process of translating sustainable strategy into properly aligned programs and projects. Adopting a sustainable strategy requires a clear mission, effective communications, and impactful change management. As sustainability programs expand within an organization, program and project managers have impactful roles helping to drive change to promote sustainable strategy.

Notes

[1] Project Management Institute, "PMI's Pulse of the Profession: The High Cost of Low Performance," February 2014, http://www.pmi.org/-/media/PDF/Business-Solutions/PMI_Pulse_2014.ashx.
[2] Ibid.
[3] Ibid.
[4] EPA, "Sustainability Analytics," accessed July 23, 2015, http://www.epa.gov/sustainability/analytics.
[5] Ibid.
[6] R. Rehan, M. Nehdi, and S. P. Simonovic, "Policy Making for Greening the Concrete Industry in Canada: A Systems Thinking Approach," *Canadian Journal of Civil Engineering* 32, no. 1 (February 2005): 99–113, DOI:10.1139/L04-086.
[7] Ibid.
[8] Ibid.
[9] Ibid.
[10] David Gardiner & Associates, LLC, et al., "Power Forward: Why the World's Largest Companies Are Investing in Renewable Energy," accessed February 20, 2015, http://www.ceres.org/resources/reports/power-forward-why-the-world2019s-largest-companies-are-investing-in-renewable-energy.
[11] Mieko A. Ozeki, Project Management of the First STARS Report for the University of Vermont, September 10, 2014.
[12] SASB, "Vision and Mission," *Sustainability Accounting Standards Board*, accessed June 2, 2014, http://www.sasb.org/sasb/vision-mission.
[13] SASB, "Determining Materiality," *Sustainability Accounting Standards Board*, accessed October 27, 2014, http://www.sasb.org/materiality/determining-materiality.
[14] Ibid.
[15] William Boston, Mike Spector, and Amy Harder, "VW Scandal Threatens to Upend CEO," *Wall Street Journal*, September 23, 2015.
[16] William Boston and Sarah Sloat, "Volkswagen CEO Martin Winterkorn Resigns over Emissions Scandal," *Wall Street Journal*, September 23, 2015, sec. Business, http://www.wsj.com/articles/volkswagen-ceo-winterkorn-resigns-1443007423.
[17] SASB, "About SASB," *Sustainability Accounting Standards Board*, accessed June 2, 2014, http://www.sasb.org/sasb.
[18] SASB, "SASB Materiality Map™," *Sustainability Accounting Standards Board*, accessed September 30, 2014, http://www.sasb.org/materiality/sasb-materiality-map.
[19] Ibid.
[20] SASB, "Determining Materiality."
[21] Dow, "Sustainable Chemistry," February 11, 2015, http://www.dow.com/sustainability/goals/chemistry.htm.
[22] Dow, "2015 Sustainability Goals, Sustainable Chemistry, The Sustainable Chemistry Index" (U.S.A., January 2015), http://storage.dow.com.edgesuite.net/dow.com/sustainability/goals/50409-SustainableChemistry-WPaper-Digital.pdf.
[23] Dow, "Sustainable Chemistry."
[24] Dow, "2015 Sustainability Goals, Sustainable Chemistry, The Sustainable Chemistry Index."
[25] Dow, "Partners for Change," accessed February 11, 2015, http://www.dow.com/sustainability/change.
[26] Project Management Institute, "PMI's Pulse of the Profession: The High Cost of Low Performance."
[27] "Sustainability Nears a Tipping Point," *MIT Sloan Management Review*, copyright © Massachusetts Institute of Technology and 1977-2015 All rights reserved, accessed February 13, 2015, http://sloanreview.mit.edu/reports/sustainability-strategy.
[28] Ibid.
[29] IBM, "The Customer-Activated Enterprise: Insights from the IBM Global C-Suite Study," September 30, 2013, http://www-01.ibm.com/common/ssi/cgi-bin/ssialias?subtype=XB&infotype=PM&appname=GBSE_GB_TI_USEN&htmlfid=GBE03572USEN&attachment=GBE03572USEN.PDF#loaded.
[30] Ibid.
[31] "Sustainability Nears a Tipping Point."

Chapter 5

Project Management Techniques Inform Sustainable Strategy Development

In most organizations, new strategies begin with programs and projects to support and implement desired changes. Project management is instrumental in effecting change to policies, processes, and procedures within an organization. Project managers are on the forefront of creating and managing change. Adoption of sustainable strategy focuses on identifying a goal and then adopting a series of actions to achieve that sustainability goal. Using the methodologies of program and project management to implement sustainable strategy improves program and project outcomes. The structure and rigor of project management provide a framework through which to engage stakeholders, identify priorities, manage risks, and effectively allocate and manage resources for success. Too often, sustainability projects are either developed or implemented as "feel good" initiatives, without the structure and rigor provided through project management methodology. Sustainability champions who partner with project management professionals within an organization and who use project management methodology to develop standards and protocols are more successful in moving their organization forward on its sustainability journey.

5.1 Project Manager Barriers to Effectively Managing Sustainability Projects

With 72% of S&P 500 companies reporting on sustainability issues, sustainability requirements are increasingly becoming challenges for projects and program managers.[1] In a survey of the global project management community, project managers reinforced that 90% of respondents work for organizations that have some level of sustainability, ranging from energy-saving programs to a full triple bottom line (TBL) approach.[2] A majority of respondents indicated that they face challenges in integrating sustainability into projects and that they are looking for education, tools, and case studies to better address sustainability issues in projects.

In terms of training needs, the most-identified areas are developing a culture of sustainability, embedding sustainability into programs and projects, and engaging stakeholders and management.[3] These survey results clearly identify sustainability as an area of interest for those in the project management community. While the existing project management discipline provides guidance in a number of areas such as stakeholder engagement, change management, communication, risk management and response, and reporting standards, other areas need further refinement to make project managers more effective in their management of sustainability projects. One of the unique aspects of sustainability project management is that initially the project manager is also the sustainability champion, and she must take on the role of sponsor as well as project manager. As a result, a great deal of time is spent engaging the C-suite and obtaining support. Sustainability also requires some subject-matter expertise about product life-cycle and supply chain management. While this knowledge can be obtained through engagement of a subject-matter expert, program and project managers need to be familiar with the process in order to develop a realistic time management and resource management plan. While project management methodologies include stakeholder management, many sustainability projects require complex collaborations and engagement of external stakeholders that necessitate a multifaceted approach.

Project managers are being asked to take on sustainability projects, but they are finding the process challenging. Sustainability champions can help by providing subject-matter expertise, engaging the C-suite on sustainability issues, and creating a organizational culture that promotes sustainability.

5.2 Project Management Maturity and the Sustainability Journey

As an organization moves through its sustainability journey, sustainability programs and projects evolve and grow increasingly complex (refer to Figure 3.2). In the beginning, exposure stage, of the sustainability journey, management usually focuses on compliance projects such as a regulatory or customer requirements. A customer compliance project might be responding to a customer's environmental and labor practices questionnaire. Energy efficiency projects that provide a quick payback are another type of project undertaken in this early phase. As the sustainability champion, take full advantage of these opportunities to demonstrate the impact of sustainability-based projects on the organization's strategic business mission. For example, gathering data and creating systems and processes to address a customer's sustainability questionnaire ensures business continuity with that client. The energy efficiency project saves money and may even qualify for points on a customer compliance sustainability scorecard. Focusing on organizational benefit and incorporating project management methodology into sustainability projects adds credibility and rigor to the process. Leverage the knowledge base of project managers within the organization to provide structure and metrics to clearly identify a successful outcome. Incorporating project management techniques into sustainability projects provides better engagement, implementation, and outcomes. Using this approach, the sustainability champion creates a credible foundation to move the organization forward on its sustainability journey. Project management plays an important role in the process of integrating sustainability into an organization at all stages of an organization's sustainability journey (see Figure 5.1).

Moving into the integration phase, sustainability becomes part of the business process. The vision of sustainability is incorporated into an organization's values and how work is performed. In order to effect this change, program and project managers are tasked with integrating the organizational sustainability vision into operations by changing processes, products, and people. By undertaking a life-cycle assessment, a sustainability project team identifies that the majority of a product's greenhouse gas (GHG) emissions composition comes from their supply chain processes. In order to have a meaningful impact on the product's GHG emissions, their business policies and processes need to be adapted to provide suppliers with a sustainability code of conduct reinforced by a survey requesting information on the suppliers' sustainability policies, protocols, and procedures. Based on the feedback, a vendor

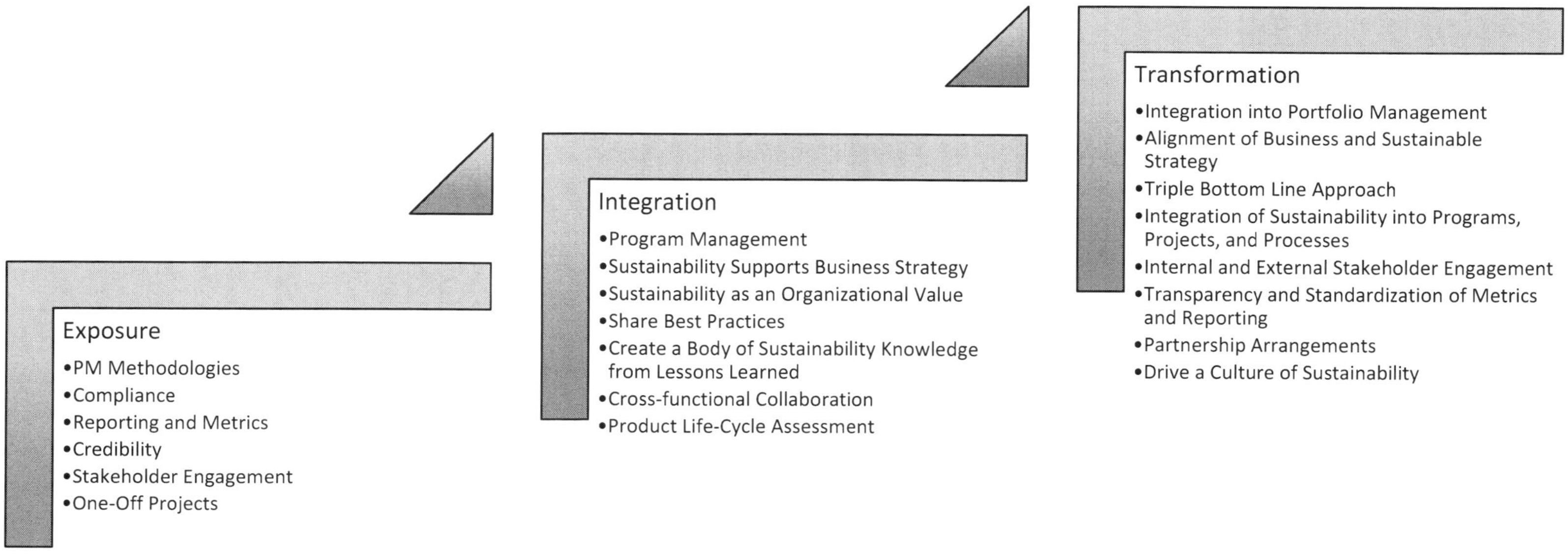

Exposure
- PM Methodologies
- Compliance
- Reporting and Metrics
- Credibility
- Stakeholder Engagement
- One-Off Projects

Integration
- Program Management
- Sustainability Supports Business Strategy
- Sustainability as an Organizational Value
- Share Best Practices
- Create a Body of Sustainability Knowledge from Lessons Learned
- Cross-functional Collaboration
- Product Life-Cycle Assessment

Transformation
- Integration into Portfolio Management
- Alignment of Business and Sustainable Strategy
- Triple Bottom Line Approach
- Integration of Sustainability into Programs, Projects, and Processes
- Internal and External Stakeholder Engagement
- Transparency and Standardization of Metrics and Reporting
- Partnership Arrangements
- Drive a Culture of Sustainability

Figure 5.1 Project Management's Impact on the Sustainability Journey

training project may be launched to help educate suppliers on the benefits of a sustainable approach and to help them improve their own environmental and social performance. To further reinforce the significance of the project, sustainability compliance thresholds are established for suppliers, which must be met in order to continue to conduct business with the firm. The areas of focus might be on improved labor standards, greater energy efficiency, or alternative ingredient sourcing. From a business perspective, this approach reduces reputational and brand risk by addressing labor standards and reducing the likelihood of having an organization's brand tied to a supplier labor violation. Other business benefits include creating a product that has more appeal to a green consumer and even some reduced costs from resource savings. As sustainability projects are incorporated into operations, organization leaders are better able to manage reputational and brand risk, identify opportunities for costs and resource savings, and provide credible information to customers about the environmental, social, and governance (ESG) impact of their products.

In order to effect the magnitude of change that occurs in the integration phase, a foundational body of knowledge on sustainability needs to be created and disseminated throughout the organization. The Sustainability Office (SO) or a Program Management Office (PMO) assembles a body of knowledge based on lessons learned and best practices to share with project management professionals in order to provide foundational resources for sustainability programs and projects. This foundational body of knowledge allows an organization to move incrementally forward by building on shared experience and flattening the learning curve for project management professionals. Although sustainability projects have unique aspects, best practice is to utilize standard proposal formats and templates. Adopt the proposal templates and standards of the organization whenever possible, so that senior management knows where to look for key information. Presenting senior management with a familiar project proposal format allows for easier identification of key data and improves the probability of approval. The project review process is constrained by time; as a result, proposals presented in nonstandardized formats may not get full consideration. As the maturity level of sustainability grows within the organization, the SO develops policies and protocols for projects that facilitate alignment between sustainable strategy and business process.

Moving into the transformation stage, sustainable and business strategies converge into a unified strategy. Sustainability is an integral part of the strategic planning process. New opportunities and business models are viewed through a sustainability lens. New solution platforms are developed to address both core business and sustainability goals. Table 3.1, Reconceiving Products and Markets, outlines new solutions platforms for organizations such as Wells Fargo, Intel, and GE to meet the needs of their clients for sustainable solutions. These are organizations that have entered the transformative stage. Portfolio strategy and component selection align to support transformative strategy. Establishing portfolio criteria to align sustainable and business strategy, portfolio management establishes a framework for alignment standards and a methodology for choosing the most impactful programs and projects for funding approval. Sustainability programs are not distinct; rather, sustainability becomes part of all program and project charters. One is not conceived without the other. While the sustainability office may act as a coordinator, projects proposals originate from the business functions. Solutions to challenges are approached from a cross-functional perspective, with compensation tied to performance scorecards that incorporate sustainability goals and targets. In the transformation stage, sustainability is integrated into programs, projects, and processes across the organization. Organizations that have mature project management structures and integrated sustainability agendas are able to create impactful programs and projects that drive the organization forward towards its sustainable vision.

5.3 Sustainability Project Charter

In the previous chapter, we focused on engaging the C-suite and making the business case for sustainability. Despite the clear business benefits, engaging the C-suite and garnering their support and allocation of resources remains a challenge for sustainability champions. One of the reasons is that sustainability initiatives have suffered from being presented as "feel good" rather than business-imperative projects. The link between sustainability and business strategy needs to be identified and explained. If the perception of senior leadership is that sustainability programs are optional or something that is done when there is extra time or money, then sustainability will never become part of the organization's culture. In order to present sustainability projects with the same credibility as other projects, use project management methodologies to make the business case and to clearly lay out the plan of action. Begin the development of a sustainability project with a project charter, and work through the various components to ensure that all aspects of the project have been considered and that it aligns with the corporate vision and mission. A template for a sustainability project charter is provided in Table 5.1.

The project charter should clearly state the goals of the project, including the benefits to the organization in terms of meeting business and sustainability goals. Stakeholder engagement is a key aspect of implementing a sustainability project successfully, especially identification and requirements gathering from both internal and external stakeholders. Community groups and local citizens can have a major impact on decisions about facility location, expansion, and closure. Ensure that you are including their feedback in your requirements and the project deliverables. Establish a realistic time frame and budget.

Table 5.1 Sustainability Project Charter

Title	Use the title to convey meaning.
Executive Summary	High-level description of the project; include scope & strategic benefits.
Sponsor(s)	Identify person(s) & functional area(s).
Project Goal & Desired Outcome	Consider: Long-term impacts Internal & external impacts Environmental & social impacts Behavior change Policy & process change
Project Manager	Identify person & level of authority.
Benefits	Outline the goals of the project and alignment with business & sustainable strategy.
Stakeholders	Identify internal & external stakeholders. Outline material issues. Outline expectations & engagement strategy.
Milestones & Key Deliverables	Create a timeline. Identify key deliverables & milestone schedule.
Budget	Identify source of budget & high-level detail.
Resources	Identify people, technology, & other resources required.
Project Risks	Describe threats and opportunities, including environmental & social impacts.
Success Criteria	State clearly how success will be defined. Identify tools/sensors for monitoring. Select clear metrics for success. Define reporting protocols.
Assumptions & Constraints	Clarify assumptions & identify constraints.
Location	Country of operation, work rules, culture, regulatory & legal requirements.
Organizational Templates & Standards	Use PMO resources. Follow proposal templates.

Too often, sustainability projects are thought to be free of cost. Absolutely make use of all available free resources, but include reasonable costs for planning and implementing the project. Conduct a complete risk assessment. Use the rigors of project management to carefully consider and lay out both the positive and negative risks. Understand that beginning a process of engagement with external stakeholders requires commitment, and that leadership may receive feedback they weren't expecting. Once the box is open, the issues raised must be addressed in a manner that is satisfactory to the stakeholders. Otherwise, the efforts might be perceived as not authentic or transparent, resulting in a negative rather than positive stakeholder impact. Utilizing project management methodologies to clearly define business purpose, project methodologies, resource requirements, and the definition of success improves sustainability project outcomes.

Going through the process of completing a detailed project charter provides a sound foundation for programs and projects designed to help an organization meet its sustainability goals. It provides clear and concise information on project goals, business value creation, and time and resource commitments. It allows for a meaningful dialogue among the sustainability champion, the project manager, and the decision-making individual or body. Unlike many other projects that are carried out within a function, sustainability initiatives tend to be long-term and cross-functional in nature. Following are some questions to consider when integrating sustainability into projects:

1. How are sustainability issues integrated into strategic planning?
2. What areas of sustainability are being addressed through project selection?
3. How can sustainability be incorporated into project concept formulation?
4. Do we have sufficient stakeholder commitment to launch a sustainability project?[4]

Sustainability issues are futuristic in focus and span organizational lines, requiring collaboration and complex resource integration that may not be an area familiar to project managers. Organizations that have experience with sustainability programs and project often create a Sustainability Office, with the Chief Sustainability Officer (CSO), or sustainability champion, functioning as a sponsor championing the programs and projects with senior management. While this structure facilitates coordination, business-line leader need to be co-sponsors in order to garner support for the resources and budget necessary to complete a project successfully. Best practice suggests establishment of a steering committee comprised of cross-functional leadership in order to inform and engage business functions across the organization.

Sustainability goals tend to be long-term and require significant operational and cultural change. Ensure that project success is defined in terms of both outcome and progress along the way. For instance, with a project to increase the percentage of women on the board of directors, it may take time to change the composition of the board. In order to monitor progress activity project metrics should measure incremental progress such as the number of women included in the slate of candidates considered for an open board seat or the number of board members receiving diversity & inclusion training. Project outcome measures are also important, such as raising the ratio of women on the board of directors to 25%. Ensure that the project has both progress and outcome metrics to follow project directional progress and ultimately to measure success.

A project physical location and sphere of impact are important, as countries around the world vary in their requirements for sustainability reporting and compliance. In April 2014, for instance, the European Parliament passed legislation requiring large publically traded companies to report on sustainability. In 2014, 2500 companies voluntarily issued sustainability reports. In 2017, when the law goes into effect, the number of companies expected to be required to report will rise to 7000. In order to comply with this requirement, companies will need to rethink how they conduct, track, and report on their business operation and sustainability practices. Organizations will need to create a structure to gather information on environmental, social, human rights, anticorruption, bribery, and employee-

related issues. Business decisions will need to be evaluated using criteria that address these i
all functions. Business processes will need to be adapted and restructured to address th
In addition, new reporting protocols and technologies will need to be established. The EU legislation encourages organizations to report using global reporting protocols such as the Global Reporting Initiative (GRI) Sustainability Reporting Guidelines and the UN Guiding Principles on Business and Human Rights.[5] It is important to note that this EU requirement is not limited to organizations that have existing sustainability programs. It will impact any organization with operations within the EU; so, it is a big impetus for companies to adopt recognized sustainability reporting protocols and standards. This legislation will have a broad-reaching impact, requiring affected organizations to launch a number of programs and projects in order to comply with the requirements.

5.4 Sustainability Drivers

Identifying the drivers of sustainability and the areas of significant impact within the operation provides a sound foundation for an organization's sustainability program. An organization's drivers will be a function of a variety of factors including industry, function, product or service, sourcing structure and strategy, distribution, customers, and geographic location.

These sustainability drivers will have different levels of relevance depending on the industry. Figure 5.2 shows an array of sustainability drivers in order to stimulate discussion and brainstorming on

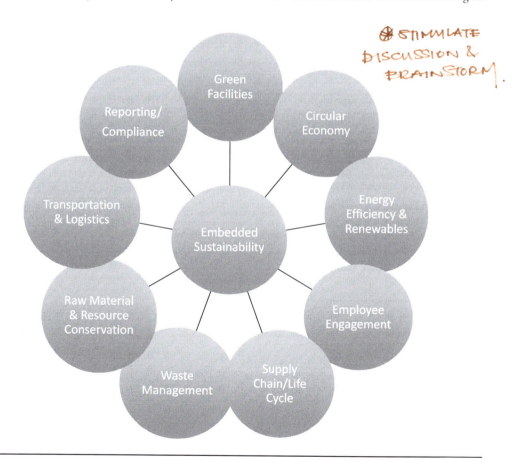

Figure 5.2 Sustainability Drivers

the best approaches for addressing sustainability within the organization. Program and project selections are based on which drivers are relevant given the industry, structure, and operation of the organization. From the previous EU example, regulation and compliance are drivers for those impacted organizations. For a service organization, the driver might be attracting top talent and engaging employees.

Based on the sustainability drivers for an organization, an effective sustainability program can be developed. A manufacturer may focus on achieving a smaller environmental footprint through energy and resource conservation, less waste through recycling, reusing, and reformulating, and a greener product offering through reduced packaging, more environment- and human-friendly ingredients, and product innovation. In order to address these issues, management may undertake a project to evaluate the impact a circular economy assessment has on its product life cycle. The concept of the circular economy, which was developed by the Ellen MacArthur Foundation, is a product development approach that is restorative by design and strives to maintain products, components, and materials at their highest value throughout the process.[6] As shown in Figure 5.3, the circular process begins with designing

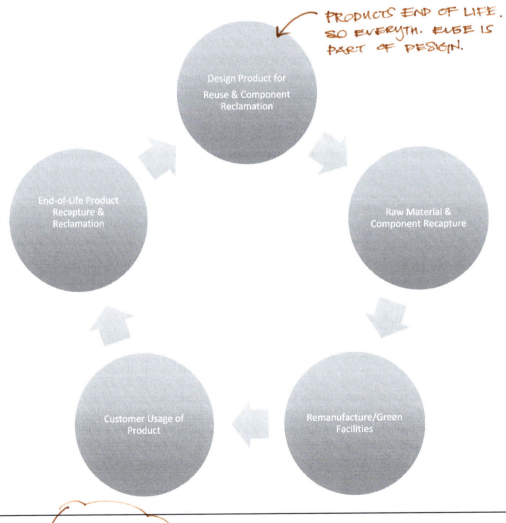

Figure 5.3 The Circular Economy

a product to protect natural capital. The product's end of life is considered at the design phase so that material selections, construction, and component reuse are all part of the process. Thought is given to the manufacturing process in terms of reducing virgin raw material usage as well as the component reclamation and consumer usage processes. Product recapture and resource reclamation allow for materials and components to be reused in the manufacturing process. Organizations that adopt a circular economy approach are driving sustainability values deep into organization processes.

Caterpillar Inc. has become a world leader in remanufacturing and has found ways to reduce, reuse, recycle, and reclaim components, parts, and materials that previously headed to landfills. Engineers design products so that parts can be refurbished to the same quality as new, at a significantly reduced cost. The environmental benefits include less waste, lower manufacturing-generated GHG emissions, and less virgin raw material requirements. In addition, customers are happy to receive parts and components with the same performance as new at a significantly lower cost. This long-term program developed by Caterpillar to unleash the benefits of a circular economy approach has resulted in improved customer service, cost savings, and environmental benefits. At each stage in the process, this approach seeks to reduce raw material usage, energy, water, and labor requirements through planning and designing for long-term reusable parts. The process creates value throughout the product life cycle through lower resource and production costs that can be shared by producers and customers.

Another sustainability driver for facility-intensive organizations is management's desire to build greener facilities, requiring new construction to meet Leadership in Energy & Environmental Design (LEED) certification requirements. In the United States, buildings account for 38% of CO_2 emissions and are significant users of potable water and electricity.[7] A LEED-certified building utilizes 25% less energy and generates 34% lower GHG emissions.[8] These savings are very attractive to real estate investors and green-oriented tenants.

Yet another sustainability driver is upgrading current facilities to improve energy efficiency and operating effectiveness. Often, one of the first sustainability projects undertaken as an organization begins its sustainability journey is improvements to building facilities, such as more efficient heating and cooling systems. While a facility driver may be an impetus for sustainability for a building-intensive operation, it may not be a driver for an organization that leases its space, such as a consulting firm. A driver for a consulting firm may be curtailing unnecessary business travel to reduce GHG emissions. Their focus may be on educating their clients and their staff about the benefits of technology such as teleconferencing to reduce on-site visits and the corresponding air and vehicle travel that generate GHGs.

Other organizations may be faced with compliance standards for reporting on sustainability efforts from either regulators or customers. An example is a supplier scorecard requirement from large retailers such as Walmart. The scope of these requirements will direct the project efforts in developing a reporting standard. Once the drivers of sustainability are identified for your organization, programs and projects can be selected to address these challenges. While drivers may vary, best practice is to establish an organizational process to prioritize projects based on relevant sustainability criteria to allocate resources to the most effective programs and projects.

5.5 Sustainability Assessment

Drilling down into organizational functions and assessing their environmental, social, governance, and human capital materiality impacts the transfer of sustainable strategy into sustainability programs and projects. Our discussion of materiality, to this point, has been at a strategic level rather than an operational level. Initiating a project to assess the impact of operations on materiality criteria provides a framework for identification and action. Functions can be broken into internal functions such as

100 Becoming a Sustainable Organization

= EVALUATE SUST. RISKS, IMPACTS & BENEFITS

Table 5.2 Matrix for Assessment of Material Issues by Function

Functions	Specific Product, Activity, or Service	Environmental	Social	Governance	Human Capital	Ranking Scale 1–5 (1=high, 5=low)
Manufacturing Activity, Product, or Service						
Sales & Marketing						
Distribution/Warehousing						
Facilities						
Administration						
Employees						
Human Resources						
Information Technology						
New Product Development						
Finance						
Supply Chain						
Customers						

⟨ SUSTAINAB. GOALS ⟩

Environmental Capital, Social Capital, Governance, and Human Capital Considerations			
Environmental	Social	Governance	Human Capital
Electrical usage	Community impact	Code of conduct	Diversity & inclusion
GHG emissions	Communication	Ethics policy	Labor standards
Natural gas usage	Volunteerism	Ethics & code training	Training & development
Fuel usage	Humanitarian standards	Supplier standards	Recruitment & retention
Water usage	License to operate	Regulatory & legal compliance	Alignment of compensation
Wastewater treatment	Community engagement	Compensation tied to sustainability goals	Incentives
Raw material usage	Stakeholder engagement	Reporting standards	Health & safety
Recycled raw material usage	Access to product/service	Transparency	Wellness
Waste hazardous & non-hazardous	Customer satisfaction	Environmental policy	
Product design	Ethical marketing	Board structure	
Package design	Product quality & safety	Shareholder engagement	
Environmental training	Entry into new markets		
Environmental goals			
Product life cycle impact			

manufacturing, sales, marketing, distribution, facilities, IT, finance, and human resourc nal functions such as the supply chain and customers. Identifying material issues, ran based on organizational relevance, and developing a rating scale facilitates the sustainabili creating a list of sustainability priorities.

Table 5.2 is a template to facilitate identification of sustainability issues across the various functions of an organization. The headings can be altered to reflect the relevant environmental, social, governance, and human considerations for an organization. The table is designed to facilitate drilling down into the function, product, and process levels to evaluate their sustainability risks, impacts, and benefits. Through this process, organizational sustainability goals can be aligned with operations. Environmental issues include energy, water, raw materials, waste, product and package design, GHG emission targets, and recycling. Social issues focus on communities, voluntarism, stakeholder engagement and communications, access to services, customer satisfaction, product quality and safety, ethical marketing, and development of new markets. Governance considers policies and standards concerning ethical behavior, codes of conduct, supplier engagement, reporting standards, and transparency in the entire process. Human capital criteria consider the organization's human capital policies and procedures, addressing issues such as diversity and inclusion, compensation tied to sustainability goals, labor standards, the safety, health, and wellness of the workforce, and training and development in support of ethics, environmental standards, and stakeholder engagement.

While most management teams can identify "low-hanging" fruit, creating systemic change within an organization requires a holistic organizational evaluation that considers each business function's role. Table 5.3 provides an example of how two business functions, manufacturing and human resources, might utilize the matrix.

Table 5.3 Sample Matrix for Sustainability Assessment of Materiality

Function	GHG Emissions	Water Usage	Raw Materials	Community Impact	Diversity & Inclusion
Product Manufacturing	Life-Cycle Assessment Energy & Fuel Usage Product Transportation Waste	Supply Chain Usage Life-Cycle Assessment Wastewater Treatment	Supply Chain—Virgin Raw Materials Usage Opportunity to Recapture Raw Materials from End User	Plant Emissions Transportation—Traffic & Emissions Waste Disposal Hazardous Waste Employment	Potential Employee Referral Source Skills & Knowledge Base
Human Resources	Sustainability Competencies Employee Commuting Business Travel Training & Development Alignment of Compensation with GHG Targets	Job Descriptions Employee Empowerment Training & Development Alignment of Compensation with Water Targets	Green Product Ingredients Tied to Compensation	Voluntarism Employee as Ambassador Stakeholder Engagement	Recruitment & Retention Innovation Improved Staffing Improved Customer Service Employer of Choice

The key is to include all functional areas in the scope of the assessment as all functions, products, services, and processes are impacted by sustainability. Once the material issues have been identified, priority ranking focuses resource allocation toward projects designed to address key targets. These projects will often be interrelated or subprojects, as sustainability issues and solutions require an integrated cross-functional approach. As depicted in Table 5.4, priority project–based solutions such as a product life-cycle assessment and identification of sustainability competencies are interrelated and impact multiple business functions, creating material impacts on environmental, social, and human capital. While not all of the identified sustainability issues can be addressed simultaneously, creating this roadmap facilitates ongoing engagement with functional business leaders on the subject of sustainability impacts within their areas of responsibility. It starts the conversation and provides a framework for a plan to move the sustainability agenda forward.

Table 5.4 **Materiality Project Priority Matrix (Sample Projects)**

Function	Project	Material Issue	Impact	Ranking Score	Priority
Manufacturing Warehouse & Distribution Procurement Sales & Marketing Human Resources	Product Life-Cycle Assessment	ESG Goals & Targets GHG Emissions Energy Usage Water Usage Raw Material Usage Material Recapture Product Reclamation Labor Standards Human Rights Recruitment & Retention Community Impact	Embedding Sustainability into Job Functions Cost Savings Reduced Waste Stream Reduced Supply Chain Risk Identify Sources of GHG Emissions Identify Required Sustainability Competencies Enhance Reputation & Brand Improved Supply Chain Risk Management	1	1
Human Resources All Other Business Functions	Development of Sustainability Competencies for the Workforce	Sustainability Mission Priority ESG Goals Areas of Impact Identified in Product Life-Cycle Assessment Human Rights Ethics Employee Safety, Health, & Wellness Diversity & Inclusion	Developing Sustainability Knowledge Base Embedding Sustainability into Job Functions Recruitment & Retention Alignment of Compensation & Incentives Improved Safety Ratings	2	2

As part of this process, ensure that both internal and external stakeholders are surveyed for their responses on how material issues impact operations. Within the organization, engage multiple levels to gain input from a variety of managerial and employee perspectives. While it might be most expeditious to have this matrix created solely by internal stakeholders, it will lack perspective from the organization's key external stakeholders. Their input is extremely valuable in developing a sustainability roadmap, especially as it reflects the corporate image projected to your customers and communities. In order to develop the most effective sustainability policies, procedures, and products, it is important to have key stakeholder input. (In Chapters 8 and 9 we discuss specific strategies for engaging internal and external stakeholders.)

The priority rankings included in the assessment inform management on the most pressing and impactful sustainability issues. Using the assessment matrix, create a Materiality Priority Matrix to identify priorities for programs and projects proposed to address sustainability challenges in selected functional areas. A sample Materiality Project Priority Matrix is shown in Table 5.4. identifying a product life-cycle assessment and development of sustainability competencies as priority projects.

Assessing and ranking potential projects provides a framework for developing project priorities over the near- and longer-term horizon designed to address sustainability challenges within business functions. This process takes sustainability from the strategy phase into the adoption phase through planning and implementing at the business level. The selection should be based on a project's relevance to the organization's sustainability mission, goals, and targets.

5.6 Sustainability Continuum

Once an organization begins its sustainability journey, it is on the sustainability continuum. Organizations that are in the early phase of their sustainability journey will often select low-cost and high-impact projects, such as a lighting retrofit project to reduce energy usage. These are known as "low-hanging fruit." While these types of projects are not the most impactful in terms of creating long-lasting organizational culture change, they generate a quick win in order to get senior management focused on the value proposition of sustainable strategy. These types of projects represent the first steps in the sustainability journey and are impactful in terms of introducing the concepts of sustainability and getting the organization onto the sustainability continuum.

The sustainability continuum reflects an organization's relative progress on its sustainability journey. It reflects an organization's maturity level with the concepts of sustainability. As shown in Figure 5.4, indicators help organizations identify their position on the continuum. The exposure stage includes energy savings, building efficiency, and compliance projects. The integration stage begins a more holistic approach to sustainability and may include a life-cycle assessment project. Organizationally, management develops a greater understanding of the positive relationship between sustainability and business strategy. Other projects might focus on expanding revenues through new product solutions designed for green consumers. The transformation stage represents substantial organizational change, where sustainability is part of the core strategic vision. Sustainability is part of all decision, processes, and metrics for success.

A project such as a flexible work project might be undertaken in the late integration or early transformation stage. Because of its cross-functional nature and broad scope, this type of project transforms an organization's culture and moves it toward greater sustainability. As a result, it is a more complex project requiring broad-based organizational collaboration, senior management commitment, and significant resources. As an organization moves forward on the sustainability continuum, its maturity with sustainability grows. This progress is reflected in the types of programs and project proposals that are undertaken as well as the greater awareness and understanding that managers and employees reflect in their comprehension of the organization's sustainability mission.

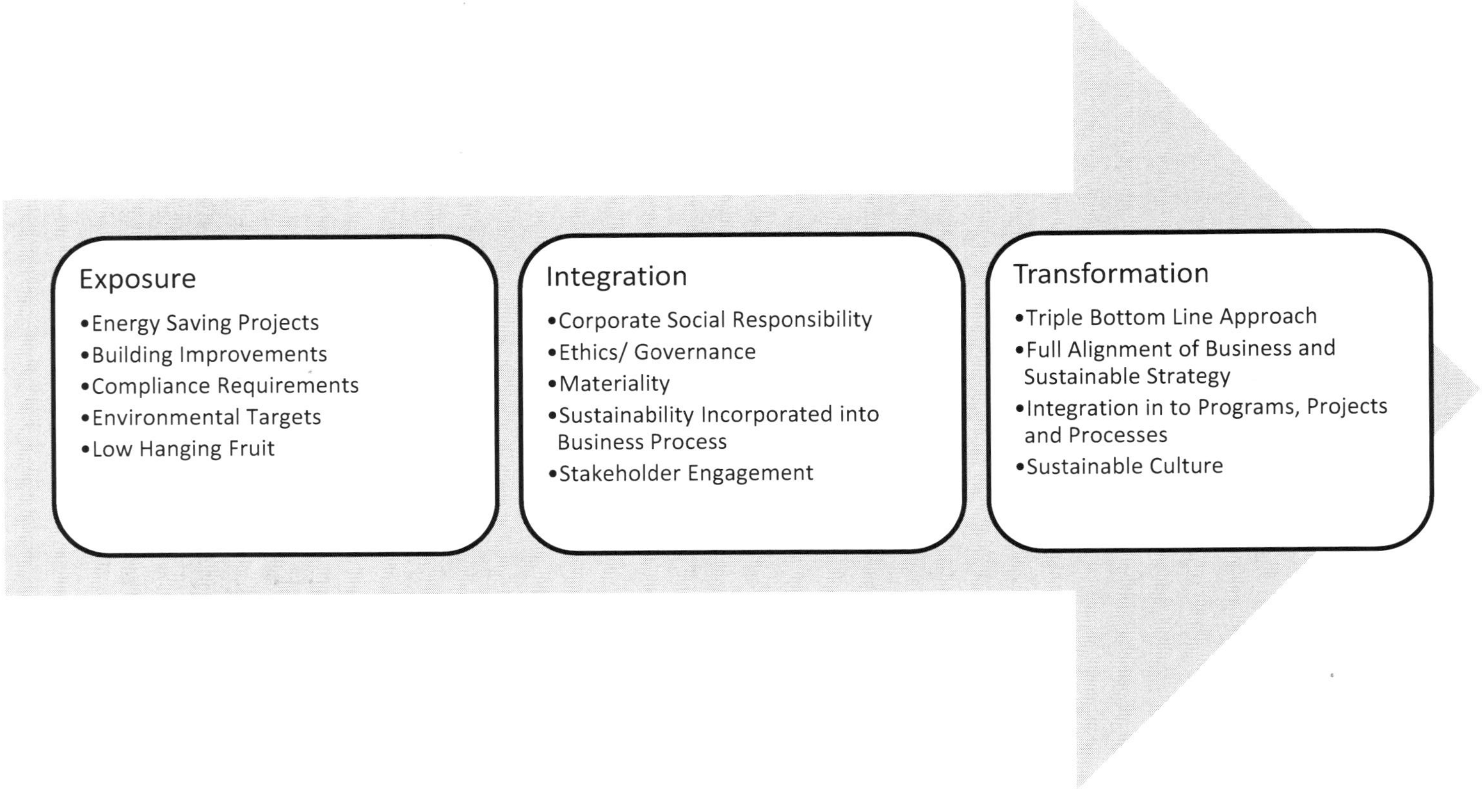

Exposure
- Energy Saving Projects
- Building Improvements
- Compliance Requirements
- Environmental Targets
- Low Hanging Fruit

Integration
- Corporate Social Responsibility
- Ethics/ Governance
- Materiality
- Sustainability Incorporated into Business Process
- Stakeholder Engagement

Transformation
- Triple Bottom Line Approach
- Full Alignment of Business and Sustainable Strategy
- Integration in to Programs, Projects and Processes
- Sustainable Culture

Figure 5.4 The Sustainability Continuum

The Materiality Project Priority Matrix also provides guidance to sustainability project sponsors to better gauge the probability of their project receiving funding relative to other projects vying for resources. As an organization moves through its sustainability journey, project impacts and priorities will realign to reflect organizational growth and maturity with the sustainability process. Lessons learned, stakeholder expectations, and a competitive marketplace lead to changes in priorities and ranking of projects.

The following case highlights the importance of aligning sustainability projects with core business value creation. Management at a mid-sized, privately held company noticed that other buildings in the area were installing solar panels, and they thought that it might be a good idea for themselves. Given the favorable tax incentives, they decided to consider installing a solar array on their headquarters building to reduce energy costs. They performed due diligence, modeling, and prepared a financial projection to measure the internal rate of return (IRR) and payback period. The project pricing and performance model was based on the assumption that electricity rates will rise significantly over the next 15 years. While the project was under the final tollgate review, shale natural gas exploration and development boomed, dropping the price for electricity generation. As a result, the cost assumptions based on a rising rate for traditionally generated electricity came into question. In addition, the project required a significant capital investment that was not clearly aligned with core business value creation. The project status went from green to red when management assessed the reduced projected energy savings and significant capital allocation. While this project reduced energy usage and GHG emissions, the project outcome did not create new business opportunities, improve core processes, or address a client concern. When the payback period moved outside the acceptable zone, the project was cancelled and the funding was reallocated to projects that more directly impacted business value creation. The lesson learned is that sustainability projects are evaluated against other projects for resource allocation. In order to receive funding, projects must align with the strategic mission.

Successful sustainability programs and projects must have solid foundations in terms of creating core business value. Use the rigors of project management to facilitate developing a meaningful business case with measurements of success. From a sustainability project perspective, it is important that your project address issues that are material to your organization. If your project is not material, it will not garner the necessary support from the sponsor and senior management that it needs to succeed. As a sustainability champion, it is imperative to champion and encourage the project sponsor to build a solid business case for supporting sustainability projects.

5.7 Harness the Power of Project Management

In order to have sustainability projects benefit from the strength of project management, seek out project management professionals within your organization. A great place to begin is your Project Management Office, PMO, to understand the protocols and requirements for projects within your organization. The PMO is a repository for templates, technology-based tools, best practices, and lessons learned to start sustainability projects on the best path forward. Resources might include templates, forms, work breakdown structures, benchmarking data, research, contracts, and quality standards. If you are new to an organization, the PMO is a good resource for getting a view of the organizational framework, including culture, protocols, appetite for risk, and regulatory and compliance issues. For sustainability projects to be given the appropriate resources, they must use the same language and business standards as other projects within your organization. Understanding both your organizational culture and structure and its operating protocols gives project management professional significant advantage in moving a sustainability project forward. In addition, it is important to include the PMO early on in the sustainability roadmap process so that you can create a process for documenting best practices and lessons learned to develop a body of knowledge for future sustainability projects.

> **The Project Management Institute**
>
> The Project Management Institute (PMI) is a professional association for project management practitioners that provides standards, resources, and certification for project management professionals. PMI has recently acquired ProjectManagement.com and moved their Global Sustainability Community of Practice (GSCoP) to that platform. The ProjectMangement.com platform offers resources including webinars, tools, blogs, case studies, templates, and a network of project, program, and portfolio management practitioners. For sustainability practitioners, it is a great resource for information and tools to promote sustainability in project management within your own organization. The new website is www.projectmanagement.com.

Tapping into the pool of project management professionals within your organization improves the expected outcome of sustainability projects. Project managers are trained in project management methodologies and techniques such as developing detailed project plans and managing complex projects. They understand the rigors of the process, including establishing metrics, benchmarking performance, delivering projects on time and within budget, and reporting on progress and deliverables. The structure and methodologies of the project management process add credibility to sustainability projects. A core tenet of project management is identifying the business benefits of the project and managing the benefits processes. Focusing on benefit management and tying actions to benefits delivery focuses stakeholders and project team members on the desired outcome. Utilizing the practices of project management more clearly defines a project's benefits, facilitating reporting on status and deliverables to keep stakeholders engaged. Project management methodologies advocate defining clear metrics and selecting measurement sensors, which addresses a common problem with sustainability projects. Sustainability project outcomes often are hard to measure or "fuzzy" in demonstrating their goals achievement because success was not clearly defined at project inception. Benchmarking current performance and agreeing on metrics to define project success has a significant benefit in terms of demonstrating the beneficial impact of a sustainability program.

5.8 Scope, Time, and Cost

Focusing on the three areas of scope, time, and cost is a good start to improving the outcome of sustainability projects.

5.8.1 Scope

Incorporating sustainability into the scope of projects encourages inclusion of diverse internal and external stakeholders, resource allocation, human capital, community, ethics, organizational sustainability targets, and short-term and long-term deliverables. Project managers must include educating the project team and all stakeholders on the organization's sustainability mission and the material issues being addressed by the project.

While stakeholder engagement is important to define the project scope in all projects, it is crucial in sustainability projects. Cast your net wide in the stakeholder identification process. For sustainability projects, stakeholders will include both internal and external stakeholders. Initially, it is better to be too inclusive rather than miss a group that needs to be part of the requirements-gathering process. Take time to gather stakeholder information. Make sure that enough time is spent with your stakeholders to gather all relevant issues to ensure that an issue of concern doesn't come up later. It is best to get all of the requirements on the table at the beginning of the process. Embrace skeptics and those known to be difficult. It is better to include them in your stakeholder management plan rather than back-track once the project is underway.

As stakeholder requirements are gathered, it is helpful to agree on terms and definitions so that you can communicate effectively and agree on the project scope and deliverables. Many of the traditional project management techniques, such as brainstorming, Delphi techniques for subject-matter expertise, and mind mapping techniques, are useful in the requirements-gather process. Using these approaches and developing a requirements traceability matrix ensures that deliverable are well defined and come from a specific source such as a stakeholder, legal, compliance, ethical, or contractual requirement. Often with sustainability projects, deliverables are not well defined, and demonstrating success becomes difficult. Utilizing project management techniques such as matching requirements against the initial project charter, including subject-matter experts, and facilitating workshops refines the project scope and clarifies the scope definition.

Once the scope is clearly defined, then the project manager is able to develop a work breakdown structure (WBS) decomposing the project deliverables into smaller components to ensure that the work packages are small enough to be assigned to a group or individual and that time and cost estimates for completion can be made accurately. While a project manager should be as inclusive as possible in the requirements-gathering phase, once the project scope has been determined, controlling the scope and avoiding "scope creep" is just as important in sustainability projects as in other projects.

5.8.2 Time and Cost

Effective project managers are always talking about bringing their projects in on time and within budget. They are skilled at monitoring resource inputs and productivity outputs. Their expertise in developing a plan, defining required activities, laying out the sequencing, and estimating the timing and resources provides a structured approach. Focusing on the critical path, they understand that changes and resource reallocations can impact bringing a project in on time. Without this type of structure, the completion of a sustainability project may lag behind original estimates because of scope creep and resource delays. A project plan clearly identifies responsible parties, deliverables, and due dates, keeping all parties accountable for their part of the project. Adoption of sustainable strategy is a long-term vision that breaks down into programs and projects that are undertaken based on lessons learned and feedback from previous projects as project managers seek to continuously improve the process. The long-term nature of this change must be considered when establishing milestones and benchmarking changes.

Organizational sustainability visions are often long-term, such as reaching net-zero GHG emissions within the next 10 years. Projects are planned and annual performance targets are set for business units in order to move the organization toward this significant target. Project sponsors and managers must establish realistic incremental goals for projects and programs to continue to move the organization forward to reach a significant sustainability goal.

The process of allocating scarce resources is an ongoing challenge for senior management. Sustainability projects must be structured to deliver tangible business value with measurable outcomes. Using the guidelines of project management to establish verifiable goals with metrics to measure and control the process facilitates sustainability projects success. Sustainability initiatives require the same level of resources as other projects in terms of financial, human, natural, and technical resources. One of the core tenets of sustainability is to consider organizational impact on natural and social capital. Sustainability projects often focus on improving natural capital impacts through design change, product reformulation, recycling, reuse, and reduction of resources. Social capital can be improved through projects that focus on voluntarism, employee benefits, human rights, and community engagement. While management may agree with the purpose of a sustainability project charter, if the project plan does not follow project management methodologies, the approval process is significantly more challenging. When resources become tight, programs that lack clear definition are the first to be cut, because they haven't been structured to demonstrate strategic alignment and value creation.

In order to improve program outcomes, use the sustainability cycle for success, shown in Figure 5.5. It is based on the Deming cycle for continuous project improvement. As we have discussed, sustainability is a journey, with organizations learning from their initial forays and adapting programs for improved outcomes. As a team embarks on its sustainability project, lessons are learned, feedback gathered, and refinements made for future projects.

The process begins with developing a sustainability program that demonstrates the organization's sustainability commitment to specific environmental, social, and governance goals. Clearly outline goals, targets, timelines, and plans for success. Create a plan to implement this strategy, including identifying roles, responsibilities, and timelines.

Identify metrics to measure progress and success. Create a team and allocate resources. Ensure that the goals are tangible, so that action steps can be clearly delineated and tracked. For example, a goal may be to reduce energy consumption by 10% across the organization within the next year. Gather energy data to create a baseline of energy usage by facility, vehicle, process, product, and service. Engage with internal and external stakeholders to better understand challenges and possible solutions. Then create a plan and implement actions designed to reduce energy usage, such as building monitoring systems, behavior change, production changes, energy-efficient equipment, or fleet changes. While it is not recommended that all of these conservation actions be undertaken simultaneously, these are areas that might be considered for a project. Communicate the project results to both internal and external stakeholders. Then gather feedback and evaluate how you have done. Is the 10% goal reasonable, or can the project achieve a higher standard? Is 10% too much of a hurdle over the allotted timeframe because significant process changes or capital improvements are required? Evaluate options for further energy efficiency, which might include installing energy-efficient lighting, new building systems, or purchasing electric fleet vehicles. The approach will depend on the industry and the areas of energy usage that are material to the particular organization. As a team goes through this process, they learn what works and doesn't work within their organization. While senior leadership sets the overall strategy for a sustainable approach, managers and employees closer to the business function often have the best ideas on how to adopt energy-saving ideas into their daily processes. Include them in the process and feedback loop. Motivate participation through incentives. After verifying and reporting on results, use the lessons learned to modify the program in order to better address organizational sustainability challenges. If the project was a pilot, address the opportunities and challenges of scaling the project throughout the organization.

In order to maximize the success of sustainability programs and projects, incorporate the following best practices into the project management process.

1. Educate the project team and other stakeholders on the relevant sustainability issues to be addressed by the project.
2. Ensure that all project team members and stakeholders are using common terms and definitions to discuss and to set targets for the project.
3. Align your projects with your organization's sustainability goals and demonstrate alignment through expected outcomes.
4. Address environmental, social, and governance issues early in the project life cycle and establish sensors to monitor results relative to ESG criteria throughout the project.
5. Schedule tollgate reviews of ESG criteria impact as part of the continuous improvement process.
6. Ensure that the project sponsor and business function have ownership and accountability for ESG as well as business requirements.[9]

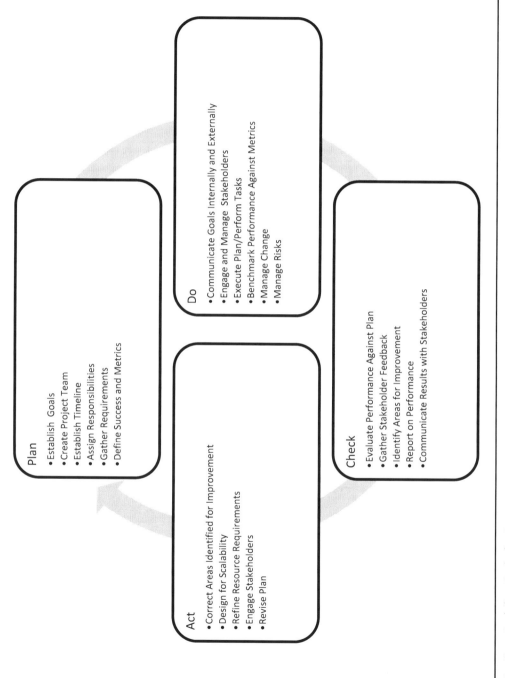

Figure 5.5 Sustainability Cycle for Success

Table 5.5 Sustainability Framework to Promote Workforce Safety

Goals	Metrics	Key Actions
Improve Workforce Safety. Achieve Zero Incident Rate in the Next 12 Months.	Total Number of Incidents. Lost-Time Rate. Workers Compensation. Employee Survey Results. Safety Audit Results.	Create a Culture of Safety. Senior Leadership Engagement. Board of Director Agenda Item. Identify Roles, Responsibilities, & Accountability. Dashboard Reporting. Develop Training Programs. Schedule Training Days. Conduct Regular Inspections. Develop Mentoring Program. Identify Corrective Action. Recognize Safety Performance.

The following example incorporates some of these best practices. Management's goal is to address a human capital sustainability issue around workplace safety. They have concerns over the level of work-related accidents. The level of incidents is not only impacting the workers directly involved, it is also impacting worker morale, productivity, and the organization's corporate image as an employer of choice. In order to address these concerns, management has adopted a sustainability goal of achieving zero work-related incidents over the next 12 months. The following project has been designed to specifically improve worker safety and move toward a zero incident rate. Rather than relying solely on historical indicators as they have in the past, the project manager institutes real-time measurements including engaging with the workforce to gather information on their view of unsafe conditions. The framework, as shown in Table 5.5, focuses attention on metrics designed to create accountability at all levels of the organization and to measure risk areas in real time. The action list includes incorporating the goals of zero work-related injuries as a board of directors initiative with dashboard reporting to give this concern senior-level attention. The goal is to identify risk areas and apply corrective action prior to the occurrence of a safety incident in order to move toward eliminating work-related injuries. This project approach requires the business units to accept accountability and responsibility for the process by requiring tangible actions to avoid injuries. Specific goals are identified with metrics defined to measure progress and success, and key actions are identified for specific groups in order to generate the desired outcome.

Establishing tollgate reviews to assess the impact of the key actions allows for adjustment and refinement throughout the project. Most crucial is that incidents will appear regularly on the board's agenda, keeping it a top priority for all levels of management. When safety becomes part of senior managements' responsibilities, not just the responsibility of the human resource department or the plant manager, it gains the resources necessary to eliminate work-related accidents. If the leadership team is committed to zero incidents, the organizational policies, training, and protocols will reflect this goal. Creating alignment between sustainability and business issues and developing a clear set of actions measured by definable metrics results in desired project outcomes. Moving forward on the sustainability continuum is an ongoing process, and utilizing project management methodologies provides structure and builds rigor in building the case for the business benefits of sustainable strategy.

5.9 Building Sustainability into All Projects

While the discussion to this point has been on sustainability projects and programs, to embrace sustainability fully as an organizational strategy means incorporating sustainability criteria, metrics, and targets into all programs and projects. According to an analysis of global projects from Goldman Sachs, over 70% of capital project delays are caused by problems that arise because of unanticipated sustainability issues.[10] Interestingly, sustainability concerns delay capital projects more frequently than commercial or technical factors. While the boardroom may have bought into sustainability, the process of adopting rigorous environmental and social sustainability requirements as part of organization-wide projects such as capital projects has lagged behind. This statistic reminds us that the benefits of a sustainable strategy aren't confined to sustainability programs and projects. Rather, sustainability threshold requirements need to be considered in all programs and projects. The portfolio component selection process needs to include sustainability requirements, as ESG issues have a significant impact on business functions and the successful outcomes of programs and projects. If these considerations are not included in portfolio component proposals, then the true risks and opportunities of a project are not being accurately evaluated. Project management professionals need to consider sustainability impacts as they create and implement project plans. Table 5.6 provides examples of the impact that a sustainable approach has on the project management process.

While the list in Table 5.6 is not meant to be exhaustive, it gives perspective to the broad reach of sustainability across an organization's functions and the corresponding impact on the project management process. If an organization has a policy to ensure that approved suppliers complete supplier scorecards and that threshold performance levels must be achieved to conduct business with the organization, then project managers must select project vendors from a list of approved suppliers. This sustainability compliance requirement may alter long-term vendor relationships if a past supplier is unable or unwilling to comply with sustainability protocols.

Table 5.6 Examples of Sustainability Project Impacts

Project Plan Component	Project Impact
Stakeholder Engagement	Broader Perspective of Project Impact & Deliverables
Communication	Share Sustainability Mission Internally & Externally Gather & Share Information with Internal & External Stakeholders Sustainability Marketing Message Alignment Provide Customers Information on Sustainable Products/Services Sustainability Policies & Practices Reporting Standards
Risk	Climate-Change Risk Physical Risk Operational Risk Reputational Risk Compliance & Regulatory Risk Site Selection
Human Resources	Team Composition Diversity & Inclusion Cultural Adaptation Sustainability Training Employee Engagement
Vendor Management	Supplier Code of Conduct Supplier Sustainability Reporting Supplier Sustainability Surveys Supply Chain Management Approved Vendor List
Quality/Design	Product Life Cycle Circular Economy Ingredient List Product Designed for Reuse Packaging Design Restricted Raw Materials

In forming project teams, are the sponsor and project manager using a network of associates that reflects the organization's broad-based diversity and inclusion goals? From a product quality standards perspective, does a new product formulation meet both the functional requirements and the sustainably sourced ingredient requirements? Project managers need to be aware of organizational sustainability requirements as they undertake projects. The sustainability office or PMO often maintain the sustainability body of knowledge so that project management professionals have a resource to be aware of organizational requirements, enterprise environment factors, as well as the available organizational assets. Project management professionals must be aware of organizational commitments to industry, governmental, and global standards in order to manage organizational projects effectively.

> **Ford Motor Company: Journey to a More Efficient Truck**
>
> In the United States, the federal government has mandated improved new fuel efficiency standards for passenger vehicles. Under the revised standards, passenger and light-duty trucks are expected to reach fuel efficiency standards of 54.5 miles per gallon (mpg) by 2025.[11] In order to address this challenge, Ford Motor Company's team created a project to develop a more gas-efficient pickup truck with all the functionality of its current offering. With their aluminum-bodied F-150, they have achieved the goal and delivered the best fuel efficiency of any gasoline-powered full-size pickup truck. The truck gets 22 mpg and 26 mpg on the highway. Overall, this fuel economy represents a 5–29% improvement over the previous model.[12] How were they able to achieve this goal? Management focused on redesigning the truck body using aluminum rather than the traditional steel, as they felt that reducing the weight of the vehicle would be the most effective means to meet rising regulatory standards and consumer concerns. The switch required a significant investment, including construction of two aluminum sheet mills to meet demand. The new truck weighs 700 pounds less than the previous model, which significantly improves fuel efficiency. Because of its lighter weight, a smaller motor powers the vehicle, and performance such as accelerating and braking is improved. An added bonus of the lighter weight is more towing capability and increased weight load.[13]
>
> While incorporating aluminum into the design of the F-150 created a lighter vehicle, aluminum is also a better raw material than steel to recycle. As part of the project design, Ford has set up a recycling program that converts scrap aluminum into $300 in value per vehicle.[14] They have created a closed-loop process by collecting scrap, cleaning it, and returning it to the aluminum plant on the same truck that delivers the aluminum for stamping of the body panels. Ford has been willing to invest in the technology to separate and clean the scrap because of the high recycling value of aluminum. As a result of this closed-loop recycling process, Ford is able to recoup about 20% of the incremental cost of using aluminum in the F-150 trucks.[15]
>
> While Ford originally selected aluminum as a material to reduce the weight of its vehicle and improve overall fuel efficiency, the new product development project to design and manufacture a truck with reduced GHG emissions has had a much broader impact. The production process has improved its resource utilization through recycling. In addition to better fuel efficiency, the consumer is getting a better vehicle with improved handling and enhanced functionality for towing and hauling. One of the benefits of incorporating sustainability into projects is that the often they begin with one goal such as GHG emission reduction but end up with multiple benefits such as lower raw material costs and a better product for the consumer. This example highlight the benefits of incorporating sustainability goals across the organization and embedding them into core projects such as new product design and manufacture.

Adoption of sustainable strategy creates fundamental change in people, process, and policies across the organization. As a result, project management professionals need to be given resources and training on the organization's sustainability standards and commitments in order to incorporate them effectively into projects. The further along on the sustainability continuum an organization resides, the broader the reach of sustainability into programs and projects.

5.10 Program Management

As an organization grows in its sustainability maturity, program management professionals are increasingly important in facilitating sustainability strategy alignment across the organization. Leveraging existing organizational assets and systems, program management professionals ensure that benefits are identified and realized horizontally across business function and vertically cascaded within functions.

As C-suite support for sustainability grows, managements' desire increases to achieve sustainability benefits in programs and projects across the organization. Embedding sustainability into an organization requires engagement of business function leaders, their influencers, stakeholders, and project management professionals.

We have already discussed the sustainability office or PMO acting as the repository for resources and data relative to sustainability programs and projects. In addition, the sustainability officer can recommend reporting protocols, technology-based tools, and other standards to promote a unified approach and baseline standards for execution of sustainability projects. These recommended processes include preproject planning and development, project planning, implementation, verification, and closure. The SO maintains sustainability compliance requirements, targets, standards, and protocols such as vendor compliance standards, emissions targets, diversity targets, ingredient standards, raw material sourcing requirements, quality standards, water protocols, labor standards, and community impact protocols. In addition, the SO maintains the information on past projects so that effective stakeholder engagement techniques, communication plans, and other successful outcome assets are available to be leveraged by all project management professionals.

The board of directors and senior leadership develop the organization's sustainability vision and then create goals and targets to reach that long-term vision. Goals such as carbon emission reduction, human rights protection, water conservation, biodiversity impacts, zero waste, or health and safety standards provide management and employees with tangible targets to work toward. Moving from sustainability goals to behavior and process change requires that these goals are cascaded throughout the organization. Program and project management methodologies and actions are an important link to effectively integrating sustainability goals, principles, policies, and actions into an organization. Through creation of a sustainability office, policies, guidelines, tools, and techniques are developed to guide project management professionals through the process of integrating sustainability throughout the organization.

While the sustainability officer champions an organization's sustainability vision, she can't manage every sustainability program and project. Business functional group leaders must embrace sustainability and launch their own programs and projects for sustainability to become embedded into the organization. The sustainability officer acts as a resource, facilitator, and subject-matter expert, but to create the broad-based change that is required to fully adopt sustainable strategy, sponsorship of programs and projects must come from across the organization. The sustainability officer collaborates with the various functional leaders to promote cross-functional solutions to complex sustainability challenges and opportunities.

The following product stewardship case illustrates how SC Johnson developed a program to change its product formulation ingredients and practices in order to improve the environmental footprint of its products. At SC Johnson, management sets environmental goals every 5 years and reports regularly on their progress. One of the cornerstones of their program is the SC Johnson Greenlist™, which

Table 5.7 SC Johnson Greenlist™ Process Scores[16]

Ranking R =	Materials Impact on Environment or Humans
Best R = 3	Little or no impact
Better R = 2	Minimal
Good R = 1	Acceptable
0-Rated R = 0	Unacceptable

was developed to improve the environmental footprint of its product development, formulation, and packaging. Over the past decade, they have evaluated raw materials and scored ingredients based on environmental and functional impacts. The result is the Greenlist™, which is a list of approved materials ranked by their environmental and human impact. Based on this standard, they have created the Greenlist™ Process to continuously improve ingredient choices for product formulations based on their impact on the environment and humans. The ranking hierachy as shown in Table 5.7 clearly delineates product ingredient classes and assigns numeric values to facilitate end-product scoring.

The program's success is based on cross-functional collaboration, organizational acceptance, multi-level adoption, and systematic implementation. Performance standards and scorecards have been adapted to make this initiative a priority at both the product development and management levels. Through this program, SC Johnson has reformulated their products to achieve more sustainable results, moving from 18% of their products in the "Best" and "Better" categories in 2000/01 (baseline year) to 47% in 2013/14.[17] By establishing the Greenlist™ program with clear standards for continuously improving their product formulations, SC Johnson has developed products that are better for the environment and safer for customers. The program also promotes clear communication with both internal and external stakeholders. SC Johnson promotes transparency by listing product ingredients using industry-standard nomenclature that is meaningful to consumers. The entire program was supported by meaningful changes within their hierarchy of compensation. At SC Johnson, sustainability is embedded into their core operation and is a driver of brand quality.

Sustainable program management provides a framework to address environmental, social, and governance issues through projects designed to impact organizational policies, processes, and people. It creates a structure and repository for an organization's body of knowledge on sustainability. Developing recognized sustainability protocols, processes, and policies through a sustainability office provides legitimacy for sustainability. Sustainability is a recognized strategic initiative of the organization that is receiving ongoing management support. Through developing this body of knowledge and framework, the sustainability office, acting as the program management office for sustainability projects, takes up the mantle for educating and engaging senior management. As a result, one of the most significant barriers to sustainable strategy adoption and related project commitment is addressed through this ongoing engagement process.

Developing a program management protocol ensures that the C-suite's sustainability vision is consistently incorporated into projects. Desired outcomes in terms of organizational goals and specific targets relative to environmental, social, governance, human capital, and profitability are clearly identified, ensuring that these types of considerations are included in preproject planning. Material issues are identified and included appropriately into projects. Projects are structured using lessons learned from previous projects. All risk categories are identified, and reporting requirements and standards are considered at the project planning stage so that these requirements can be addressed from the project's inception. Program management impact is summarized in Figure 5.6.

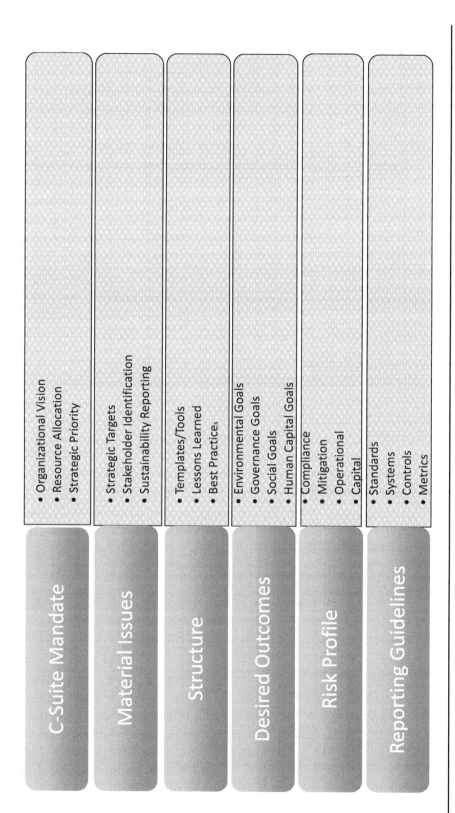

Figure 5.6 Program Management Impact

Table 5.8 Program Management Requirements

Environmental & Social Commitments	Decision-Making Protocols	Subject-Matter Experts
Sustainability Risk & Reporting Standards	Stakeholder Management Strategies	Business Value Assessments
Customer & Regulatory Requirements	Governance Standards Organizational Principles & Values	Data Management Systems

Utilizing the questions presented earlier, in Figure 4.10, Project/Program Alignment with Sustainable Strategy, promotes strategic alignment in the project chartering process. This process includes identifying business benefits, goals, and targets impacted, key players, project structure, and alignment of resources. For a more detailed list of the ESG criteria considerations, refer to Table 5.2. As the organizational body of knowledge is built for sustainability, program management requirements are developed to better manage and proactively address the environmental, social, and governance impact on organizations.

Project management facilitates building project management professionals' skills set to better address the requirements for delivering programs and projects in support of an organization's sustainability mission. Sustainability project success is defined differently. Traditional measures such as product profitability, cost savings, or process improvement are expanded to incorporate environmental, social, and governance metrics. Project goals reflect management's commitment to economic and social agendas. Project managers' technical competencies to promote sustainable strategy need to be developed and enhanced in areas such as product stewardship, green design, product life-cycle assessment, energy efficiency, alternative energy, and stakeholder engagement. Stakeholder management is more collaborative, developing long-term relationships with both internal and external stakeholders to addresses sustainability issues. Project managers need to focus on cross-functional problem resolution rather than siloed functional solutions. Personnel management, reporting standards, and requirements protocols are adapted to promote the sustainability agenda. These types of requirements are summarized in Table 5.8.

Integrating these requirements into a program management framework facilitates embedding sustainability into an organization. It ensures that management's vision of sustainability is incorporated into policies, guidelines, templates, and standards for all program and project managers. Sustainability practices are both internally and externally focused and require redesigning program and project protocols to incorporate this broad range of requirements. Considerations include sustainability practices of suppliers, vendors, and customers, as well as regulatory and other compliance requirements. The organization's principles and values may include participation in voluntary industry sustainability protocols or other global sustainability principles or reporting standards. As a result, new technical competencies, behavior changes, process and protocol changes are required for management of projects' sustainability requirements.

5.11 Conclusion

Project and program managers have a significant impact on the adoption of sustainable practices within an organization. An organization's maturity on the sustainability continuum informs the impact that project management professionals have in terms of embedding sustainability into programs and projects and impacting the culture of the organization. As sustainability becomes more embedded into the program and project management process, the organizational impact increases.

Sustainability offices facilitate adoption of sustainability through creating a body of sustainability knowledge and selecting standards and protocols to raise sustainability awareness, compliance, and adoption across the organizations. While project management methodologies add credibility to the sustainability process, the adoption of sustainable strategy requires that project management professionals learn new technical, managerial, behavioral, and procedural competencies. As a result, the project management community is seeking guidance on tools, techniques, and best practices on how to incorporate sustainability into the project management model. Challenges include insufficient C-suite engagement, complex stakeholder engagements, redesigning systems and processes and changing employee and manager attitudes and behavior. In the coming chapters, we address these challenges and provide an array of solutions and case studies.

Notes

[1] Governance & Accountability Institute, Inc., "Seventy-Two Percent of S&P Index Publish Sustainability Reports in 2013," June 2, 2014, http://www.ga-institute.com/nc/issue-master-system/news-details/article/seventy-two-percent-72-of-the-sp-index-published-corporate-sustainability-reports-in-2013-dram.html?tx_ttnews[backPid]=1&cHash=8e53ff176eb49dc3b7442844c65833ac.

[2] Tom Baker, Pedro Echeverria, and Kristina Kohl, "The Voice of Project Managers on Sustainability Projects and Requirements," June 14, 2015, http://www.projectmanagement.com/white-papers/303957/The-Voice-of-Project-Managers-on-Sustainability-Projects-and-Requirements.

[3] Ibid.

[4] Marcus Ingle, "Project Sustainability Manual: How to Incorporate Sustainability into the Project Cycle. . . ," Portland State University, July 2005.

[5] Sustainable Business News, "It's the Law: Big EU Companies Must Report on Sustainability," *GreenBiz*, April 17, 2014, http://www.greenbiz.com/blog/2014/04/17/eu-law-big-companies-report-sustainability.

[6] Ellen MacArthur Foundation, "The Circular Model—An Overview," *Ellen MacArthur Foundation*, accessed November 24, 2014, http://www.ellenmacarthurfoundation.org/francais/leconomie-circulaire/the-circular-model-an-overview.

[7] LEED, "LEED | Leadership in Energy & Environmental Design," accessed September 24, 2015, http://leed.usgbc.org/leed.html.

[8] Ibid.

[9] Bruce George Global Solutions Architect DuPont Sustainable Solutions, "Improving Environmental Performance and Value Capture in Capital Project Execution," *Environmental Management & Sustainable Development News*, accessed November 10, 2014, http://www.environmentalleader.com/2014/08/11/improving-environmental-performance-and-value-capture-in-capital-project-execution.

[10] Ibid.

[11] "54.5 MPG and Beyond: Fueling Energy-Efficient Vehicles," *Energy.gov*, accessed March 5, 2015, http://energy.gov/articles/545-mpg-and-beyond-fueling-energy-efficient-vehicles.

[12] Mike Ramsey, "Ford's New Aluminum-Bodied F-150 Truck More Gas Efficient," *Wall Street Journal*, November 21, 2014, sec. Business, http://www.wsj.com/articles/fords-new-aluminum-bodied-f-150-truck-more-gas-efficient-1416591524.

[13] Ibid.

[14] Mike Ramsey and John W. Miller, "Recycling Twist Cuts Ford Truck Costs," *Wall Street Journal*, December 16, 2014, sec. Business, http://www.wsj.com/articles/how-recycling-shaves-the-cost-of-fords-new-pickup-1418753315?KEYWORDS=Ford+150+aluminum+recycle.

[15] Ibid.

[16] SC Johnson, "Sustainability: The SC Johnson Greenlist Process," accessed July 21, 2015, http://www.scjohnson.com/en/commitment/focus-on/greener-products/greenlist.aspx.

[17] SC Johnson, "SC Johnson Sustainability Infographic," 2014, http://scjohnson.com/en/commitment/report/infographics.aspx.

Chapter 6

Creating a Culture of Sustainability

Leadership is charged with creating organizational vision and developing strategy to deliver results in line with this vision. How that strategy is designed and implemented is a function of the organizational culture. Culture is a hot topic in the C-suite, as CEOs recognize the impact that their culture has on the organization's long-term success. It is what makes organizations different from one another. CEOs are embracing the idea of culture as a competitive advantage. Strategy can be duplicated, but culture is unique. Creating a culture of sustainability takes time and significant change management. Once an organization has embraced sustainability, the creation of a sustainable culture serves to propel the organization even further toward its sustainability vision. Leadership establishes the strategic vision, but for sustainability to be impactful and become part of an organization's culture, all levels of employees must embrace the vision and make it part of their roles and responsibilities. Creating a culture that embraces sustainability is necessary to move an organization forward in its adoption of more impactful sustainability-related policies, programs, and projects. While management sets the sustainability agenda, project management professionals play a key role in creating the organizational change required for sustainability adoption and integration. Moving an organization toward a culture of sustainability requires an understanding not only of organizational vision but also of current capabilities and capacity for change. An organization's culture determines the approach, resource commitment, process, and time frame involved in transforming into a sustainable organization.

6.1 Organizational Culture

The quotation attributed to Peter Drucker, "Culture eats strategy over breakfast,"[1] demonstrates the impact of culture on successful performance. Companies with great cultures outperform their peers. Research suggests that culture accounts for 20–30% of the differential in corporate performance between companies with strong cultures and their peers with less impactful cultures.[2] CEOs frequently discuss the importance of their organization's culture because it impacts how their people respond to customers, opportunities, and challenges on a daily basis. Culture impacts how organizations share information, make decisions, deal with change, and ultimately function in the marketplace. In order to understand how a strategy is going to be adopted and implemented, one must know whether managers and employees are equipped to implement this strategy. Do they have the education, expertise, resources, capabilities, and capacity to introduce sustainable changes within the organization? Introducing a similar sustainable strategy into two different companies generates different results because of their unique cultures.

Organizational culture is defined as: "The values and behaviors that contribute to the unique social and psychological environment of an organization."[3] It embodies the collective experiences, values, philosophy, and future expectations of the people within the organization. It is comprised of leadership's vision, shared core values, and a reflection of the organization's attitudes and beliefs (see Figure 6.1). Organizational culture is deeply rooted and reflects the shared experiences of the organization's owners, managers, and employees.

An organization's culture is expressed through the actions of its leaders and employees. These behaviors and actions impact relationships with both internal and external stakeholders. It is demonstrated by how management treats its employees, how employees treat one another, and how management and employees interact with customers and the community. Culture is the everyday embodiment of organizational core values and their incorporation into functional operations. Leadership sets the

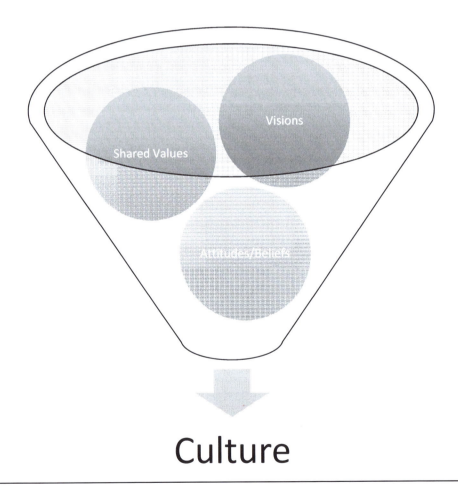

Figure 6.1 Organizational Culture

Figure 6.2   Reflections of Culture

tone and models behavior for the employees. Is the organization more of a command-and-control structure with very regimented processes? Or is it more collaborative, with policies that promote independent thinking and creativity? Culture dictates how power and influence flow within the organization. The decision-making process, reporting requirements, systems and controls, guiding principles, routines, processes, and internal stories all impact and reflect an organization's culture. Figure 6.2 considers some organizational attributes that reflect culture.

Culture is reflected in goals but also in how activities are conducted as it relates to customer relations, environmental stewardship, community service, marketing messages, business practices, new product innovation, supplier selection, and work habits. Culture is reflected in desired outcomes, the type of programs and projects that are selected, and how work is performed. It is apparent in work group structures, flexibility of schedules, and even team formations. Are project teams inclusive, and do they represent the diversity of the organization? Or are the same core people handling all the important projects? Communication plans and methodologies reflect culture. Some are open and direct, with a goal of keeping all employees in the loop. Others are circuitous and secretive, with information being used as a way to maintain power. Some organizations speak in a language of acronyms so that anyone new has no idea of what is being said. How an organization functions, its values, structure, and policies and procedures, all are part of the culture. Are the principles of sustainability part of the core values, shared experiences, and stories of the organization?

In assessing an organization's culture, consider both the verbal messaging and the actions taken by leadership. Verbally supporting sustainability sends the right message, but acting as a role model in terms of supporting decisions, programs, and projects to effect organizational change is significantly more impactful. While adopting a culture of sustainability requires significant change, it is a necessary step in the journey to becoming a sustainable organization.

6.2 Culture of Sustainability

Understanding how organizational values and principles incorporate the values of sustainability is crucial to adopting a sustainable business strategy. The gap between the present organizational environment and the desired sustainability vision forms the roadmap for adopting a culture of sustainability. The process begins with identifying existing organizational principles and values and how these impact business decisions that relate to environmental stewardship, corporate social responsibility, governance, and ongoing profitability. In order to infuse sustainability values into an organization, programs and projects must change core policies, processes, and behaviors. Mark Brownstein, Associate Vice President & Chief Counsel of the U.S. Energy and Climate Program, during remarks at the Environmental Defense Fund Closing Key Note for the 2014 Wharton Initiative for Global Environment conference, made a valuable comment about creating meaningful and long-term environmental change within organizations: "Environmental performance is very much about people and process."[4] In order to have a real impact on operations, environmental, social, and governance principles, policies and practices must drill down within the organization and be reflected in how work is performed on a daily basis. In other words, sustainability goals, standards, and protocols need to be adopted as part of the corporate culture. Senior management's vision and support for adoption of sustainability as an organizational pillar is an important first step in creating a sustainable culture. In order to be impactful, sustainability needs to become part of organizational culture, translating vision into employees' decisions, actions, and behaviors. Leadership's and employees' actions and behaviors directly impact the perception of the organization as a sustainable entity by both internal and external stakeholders.

To better understand organizational culture, take a look at the actions of your management and workforce. Identify how they are engaging with each other as well as with customers, suppliers, and community members. How is senior leadership demonstrating commitment to sustainability as a core value to external stakeholders? Look at your programs for engaging with your local communities and identify who from the organization is involved and how they participate. Does the organization have a sustainability story that is shared with stakeholders? Consider leadership's internal and external communications: Is the organization's sustainability story part of the messaging? Think about the alignment between business and sustainability strategies. Have these initiatives been aligned through policies and procedures, and are they supported by training and incentives?

Creating a sustainable culture is about engaging stakeholders on relevant issues to develop a sustainable strategy and then translating organizational sustainability vision into the governance structure,

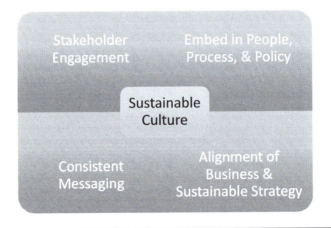

Creating a Sustainable Culture

processes, and values of the organization. Engage your people on the issues and demonstrate the benefits of adopting a strategy that incorporates environmental, social, and economic goals. This approach creates long-term value. Sharing the organization's sustainability vision and progress toward these goals with both internal and external stakeholders further embeds sustainability into the organizational culture. Management's sharing the organization's sustainability stories reinforces the importance of sustainability to employees, industry peers, suppliers, customers, and community members. Figure 6.3 is a graphical representation of this process.

The following assessment provides a baseline to measure an organization's sustainable strategy and how aligned the organizational culture is with the vision of sustainability. Answering these questions helps to identify your current position as well as gaps between the actual and desired level of sustainability awareness and action within the organization.

Assessment of an Organization's Sustainable Cultural Foundation

1. What is our organizational business mission?
    a. Is sustainable strategy part of the mission, or is it a separate initiative?
    b. Is sustainability everyone's job, or is it the job of one person such as the Chief Sustainability Officer (CSO)?
2. What is our sustainability strategy?
    a. Can it be clearly defined?
    b. How is this strategy reflected in our decision process and actions?
    c. Is there alignment between business and sustainability strategies?
    d. Can employees identify the organization's sustainable strategy?
    e. Is the organization involved in partnerships to promote sustainability with global agencies, industry members, academia, or NGOs?
3. How is our organization supporting its sustainable strategy?
    a. Where does sustainability fall in terms of corporate priorities? Is it on the board or C-suite agendas?
    b. Does the CEO speak frequently about aspects of business sustainability at internal events? At industry, customer, or supplier events?
    c. Is middle management informed and engaged on the business risks and opportunities of sustainability?
    d. What type of organizational structure exists to promote cross-functional collaboration?
    e. Are sustainability goals included as part of senior management performance scorecards? Workforce performance ratings?
    f. Do employees understand the organization's sustainability mission and how it impacts their function?
    g. Are all employees given training on ethical, corporate responsibility, and environmental issues? Are controls in place to enforce governance standards?
    h. What type of budget has been set aside to support sustainability?
    i. Is sustainability programming part of the annual budget process?
4. Who are the leaders and most active supporters of sustainability?
    a. What are their roles in the organization?
    b. What are their skill sets and backgrounds?
    c. How are they involved?
5. What activities or programs are our employees involved with to support environmental stewardship or community investment?
    a. Are these activities being performed in the name of the corporation?
    b. What types of resources does the corporation provide for volunteer activities and programs?
    c. How are initiatives promoted internally and externally?

6. How are internal and external stakeholders engaged and informed about sustainability activities?
   a. Who is involved in the engagement process?
   b. Who is responsible for external and internal communications?
   c. Are internal and external messages consistent?
   d. How is the feedback incorporated into the decision-making process?

The profile created from this assessment provides a picture of the organization's sustainability values and the level to which these are embedded within the organizational culture. Answers to these questions help identify successful initiatives and areas for improvement. The further along an organization is on its sustainability journey, the more complete and informed will be the answers to the assessment questions. This assessment provides an information roadmap for specific areas for program and project development to support becoming a more sustainable organization. Each organization's culture will have its own strengths and weaknesses and be on its own path toward embedding sustainability.

6.3 Defining Sustainability as a Core Value

Understanding the core values of an organization is a process of peeling back the layers and identifying the values that are promoted and rewarded by the management team. When senior management adopts sustainability as a core value, it is embarking on a complex change management project. Determining an organization's environmental, social, and business values provides a foundation for building a change management plan which incorporates these considerations into management's and employees' decisions and actions. Figure 6.4 defines environmental, societal, and business values to better depict the impact on operations.

Environmental values include an organization's areas of impact on natural capital and a region's biodiversity. Through a life-cycle assessment, employees, suppliers, and customers are informed about each of their roles in impacting these areas. Specific goals and targets focus attention on material areas, and agreed-upon reporting standards and metrics facilitate communication with all stakeholders. Social values focus on issues such as an organization's partnership with the community, local investment, and job creation. Business values focus on creating economic impact, developing new sustainable solutions, serving underserved markets, mitigating risk, and developing new partnerships. Incorporating sustainability as an organizational pillar means changing policies, systems, processes, and priorities to incorporate sustainability into the value creation process across all levels of an organization.

Identifying the role of sustainability in an organization's value structure sets a clear determination of the organization's level of sustainable culture. Consider the organization's environmental impacts and its ability to protect natural capital such as water, air, and raw materials. How active is leadership in promoting protection for natural resources? Do policies concerning protection of natural capital impact community relations favorably or negatively? The long-term implications of organizational strategy, policies, and practices should be matched against sustainable values. Gaps between desired and actual organizational decisions, actions, and behaviors provide opportunities for project management professionals to effect change.

To better understand how programs and projects create sustainable change, let's look at Skanska AB, a global construction and development company based in Sweden. Skanska is a large organization with 57,000 employees across three continents. Including their subcontractors further expands the workforce, bringing their total impacted population to about 250,000. Creating a culture of sustainability within this diverse and decentralized workforce is a real challenge. Ensuring that both employees and subcontractors understand and act in accordance with Skanska's corporate sustainability values is a complex program management challenge. Learning from past challenges and missteps, Skanska's management adopted sustainability as a core value, making it part of their mission to move their decentralized organization along the sustainability continuum in a cohesive manner.

Environmental Values
- Targets, Goals, Standards
- Policies & Procedures
- Life-Cycle Approach
- Biodiversity/Ecosystems
- Natural Capital

Societal Values
- Community Engagement
- Employee Investment & Empowerment
- Safety, Health, & Wellness
- Human Rights
- Labor Standards

Business Values
- Sustainability Integrated into Brand & Customer Value Proposition
- Product & Service Innovation
- Risk Mitigation
- New Partnership Opportunity
- Access to Capital

Figure 6.4 Defining Sustainable Values

As leadership developed their strategic plan to embed sustainability, they considered their organizational drivers for sustainability as well as their organizational culture. Skanska management was not always convinced of the value of a sustainable approach. In 1996, they had a rail tunnel project near Helsingborg, Sweden. The project used an industrial sealant that resulted in contamination of nearby water, which harmed fish and livestock.[5] Acknowledging the negative environmental and community impacts that this project caused, leadership became determined to ensure that this type of environmental and social degradation was not repeated. They experienced a senior leadership culture shift toward valuing sustainability as part of core operational values.

As they began the process of evaluating their operation, leadership became aware that its construction operations contributed significantly to materials going to landfill. In addition, their end products, buildings, generate approximately 50% of global carbon emissions. Leadership at Skanska believed that creation of a culture of sustainability would provide positive impact not only on the environment and society through less emissions and lower resource utilization but also on business value through cost reduction and higher contract awards. Increasingly, tenants around the world are looking for properties that meet their own sustainability standards, providing a competitive advantage to firms meeting this need. In addition, pension fund investors are seeking real estate assets that attract high-quality tenants and generate high yields. Buildings that are constructed in compliance with LEED standards attract these types of tenants and provide the returns required by investors. While their drivers of sustainability included market-driven forces, regulatory and compliance requirements for buildings and energy efficiency disclosures remained important as these issues continue to drive tenants, governments, and planners to request more sustainable facilities.

In order to address Skanska's regulatory, financial, and client-driven sustainability requirements, management focused on programs and projects to embed sustainability into the culture. In developing a more sustainability-focused culture, Skanska's leadership focused on clarifying its core values. The result is the "Five Zeros," which facilitate communication of their core mission both internally and externally. They serve as goals for their own operations as well as goals for their clients' projects. As a result, Skanska is able to help their clients move their own sustainability programs forward, creating added business value for both themselves and their clients. The "Five Zeros" are

1. Zero Loss-Making Projects
2. Zero Environmental Incidents
3. Zero Accidents
4. Zero Ethical Breaches
5. Zero Defects[6]

These goals provide a framework and reference point to guide both their own employees and their subcontractors in their daily activities. This framework provides clear guidelines for employees to make decisions, undertake actions, and demonstrate behaviors that align with the "Five Zeros." To further elaborate the vision of sustainability, management created a vision statement centered on a triple bottom line approach—people, profit, and planet—to share with both internal and external stakeholders. The resulting vision statement is detailed in Table 6.1.

In order to implement this vision, management's mission focused on the following areas of materiality: Human Capital, Health and Safety, Community Involvement, Ethics, Energy, Carbon, Materials, Water, Local Impacts, Project Selection, Supply Chain, and Shared Value. Translated into everyday language, this means that they care about the environment, their employees, and both their local and global communities. They respect the local environment and use natural capital with care. Organizationally, they align with business partners that share these values and select projects that align with these goals. With this strategy, they are looking to create value for themselves as well as their clients, partners, and local communities.

Table 6.1 Skanska's Sustainability Vision[7]

People	"We have to be a responsible employer and aim to achieve the highest standards of fairness, inclusion, ethics, and safety."
Prosperity	"We're doing some of our best work when we recruit small local businesses to work on our projects."
Planet	"Our commitment to the environment is an important piece of our part of our sustainability strategy, and one that Skanska has been focused on for decades."

To support this vision, they have developed the Skanska Leadership Profile, which promotes building leadership skills in their 22,000 professional workers. They have a Corporate Ethics Committee to promote best practices and review ethical issues in order to achieve their goal of zero ethical breaches. Skanska's Journey to Deep Green™ is designed to move not only their own organization but also their customers and suppliers toward improved environmental performance. This program includes promoting products and construction protocols that meet current laws and regulations as well as embracing newer techniques and materials that result in near-zero environmental impacts.[8] Through the use of simulation technology known as building information modeling, they are better able to predict quantities of materials required for projects. As a result, they order the right amount of materials and have been able not only to reduce waste going to landfill but also minimize project costs.

Part of the process has been to embed protecting the environment and engaging communities into their culture. In order to demonstrate their commitment to sustainability principles, Skanska became the first Scandinavian company to establish a whistleblower hotline for global operations that is operated by an independent third party to ensure confidentiality. Skanska has been included in the FTSE4Good index, a sustainability investment index, for over a decade.[9] Drawing from lessons learned, Skanska leadership has aligned their business and sustainable strategies and created a culture around their core values that promotes decision, actions, and behaviors that protect the environment, value society, and generate economic value.

6.4 Benefit Corporations

While existing organizations must go through a change management process to transform into a sustainable organization, newly formed organizations have the opportunity to create a sustainable culture from inception by incorporating it into their organizational charter. A benefit corporation is a new type of corporate structure that has been legally adopted in 26 states in the United States. Benefit corporation directors are mandated to consider the impact of their decisions not only on shareholders but also on other stakeholders, society, and the environment. The reasons to become a benefit corporation are many and include attracting socially responsible investors, benchmarking your sustainability story, differentiating your organization from the competition, protecting your brand image, cost savings, attracting customers, and attracting and engaging top talent.[10] Beginning the organizational journey with sustainability as a core pillar creates a culture that supports a business model designed to provide long-term perspective and promote innovative solutions to environmental, social, and governance challenges and opportunities.

As sustainable strategy is adopted by more organizations, forward-thinking business leaders are founding new companies with the concepts of sustainability embedded into their operation from the time of inception. Becoming a Certified B Corporation is a means to announcing to investors, customers, suppliers, and other stakeholders that your organization is founded on triple bottom line principles of people, planet, and profit.

Rubicon Global, a technology-based waste reclamation company, is a Certified B Corporation, and its management has made these principles part of their core operating principles.

Dealing with the garbage created by society has always been a challenge. Disposal of garbage is expensive for businesses, municipalities, and not-for-profits, and the environmental impact of landfill on natural capital is significant. As we entered the twenty-first century, the amount of municipal waste grew while the number of landfills fell because of more stringent EPA requirements. In 1988 there were 6500 landfills; by 2002 the number had dropped to 2500.[11] A new solution needed to be found. The traditional waste management model is based on an asset-intensive model, with investments in equipment such as trucks, landfills, and recycling centers. Revenues come primarily from customer fees and material reclamation. While traditional firms have expanded and streamlined their material reclamation service to provide a more sustainable solution, the basic business model remains the same.

Rubicon Global has taken a totally different approach to solving the garbage challenge through embedding sustainability into their business model. Their primary goal is to move waste material out of the landfill stream into a more sustainable option; however, their customer pitch is based on saving the client money. Their model monetizes the waste stream to reduce disposal costs while reducing the environmental impact. Rubicon doesn't own significant assets, as traditional waste management firms do. Their approach is to use technology to create a database of customers' waste streams and suppliers who recycle those materials. Through this networked solution, local small business suppliers can bid on material reclamation in their area even if the facility is part of a much larger company. Their approach is to reduce, re-use, and recycle waste in order to create a sustainable solution for clients. According to a client, 7-Eleven, they have been able to reduce their waste management spending by 25% through this process.[12] With average savings of 20% and a goal of zero waste for clients, Rubicon offers a business model that is founded on the principles of sustainability.[13] Rubicon's innovative business model aligns its business and sustainability goals through offering cost-effective solutions for clients to address their own environmental and social challenges. Sustainability has been baked into the corporate culture from inception.

B Lab, a nonprofit organization that certifies and supports B Corporations, offers an assessment to become a Certified B Corporation. In order to obtain the certification, an organization must achieve a threshold score on the B Impact Assessment, which measures environmental, social, and governance performance. B Corporations voluntarily meet higher standards of transparency, accountability, and environmental, social, and governance performance. It is a growing movement, with more that 1000 Certified B Corporations in 60 industries.[14]

The B Impact Assessment (bimpactassessment.net) provides a useful framework around which to engage your team as you strive to create a sustainable culture. There are five parts to the assessment: Governance, Workers, Community, Environment, and Impact Business Models. (Project managers should allocate 2 hours to conduct the initial assessment.) Once the assessment is complete, the B Impact Report provides a baseline on an organization's social, environmental, and governance performance. Most first-time assessors receive a score between 40 and 60.[15] If the organization has already adopted sustainability as a strategy, it may score higher. The entry score for becoming a Certified B Corp is 80 out of a maximum of 200 points.[16]

Even if an organization is qualified as a Certified B Corp, engaging in the assessment process provides guidance and structure to help prioritize issues and programs to move forward on the sustainability journey. In addition, the B Impact Report provides useful benchmark comparisons against other companies and industries. Using the B Impact Assessment provides a tool for sustainability champions to identify areas for project priorities in order to move toward a more sustainable culture within their organization. For example, engaging your workforce in a sustainable manner in terms of a living wage, fair labor standards, and a safe working environment has a meaningful impact in terms of creating a sustainable culture. The focus and impact on a culture of sustainability is far-reaching in terms of the successful outcomes it can generate within an organization.

CASE STUDY: ICESTONE

By Jennifer A. Haugh

Jennifer A. Haugh is CEO of Iconic Energy Consulting and a member of the Emerging Leaders in Energy and Environmental Policy (ELEEP) network. ELEEP is a joint project of the Ecologic Institute and the Atlantic Council of the United States. Launched in fall 2011, ELEEP is a dynamic, membership-only forum for the exchange of ideas, policy solutions, best-practices, and professional development for early and mid-career American and European leaders working on environmental and energy issues.

When Superstorm Sandy ripped through New York City, in October 2012, few were expecting such an epic assault. After all, hurricanes are a Caribbean phenomenon. At least they used to be.

But this storm seemed determined to teach the northeast a harsh lesson in climatology: that atmospheric conditions are unpredictable, and they can be devastating. The "new economy" is defined less by market confidence, international trade, or fluctuations in the business cycle. It is defined by who can survive the weather.

Sandy was far and away the largest natural disaster to strike New York, costing the state $41.9 billion in damage, with $19 billion in losses to New York City alone, according to CNN.[*] Although accurate reporting is challenging, the storm potentially affected roughly 1.5 million businesses that employ 9.3 million individuals, all of whom were vulnerable to the effects of Hurricane Sandy.[†] Nearly three-quarters of New York businesses have fewer than 50 employees, making survival difficult without the added blow of a hurricane.

B Labs, a national nonprofit organization, would argue that it's businesses that focus on sustainability that can survive these setbacks. Their certification process guides for-profit corporations through a sustainability checklist to achieve a bottom line that is *stakeholder*-driven, not just shareholder-driven.

What this means is management commits to a triple bottom line: people, profits, and planet. The new economy depends on an equal focus on all three to weather storms, literal and figurative.

B Labs' philosophy was put to the test when the factory floor of one of its member corporations, IceStone, was found submerged in water after Sandy. A small company operating out of Building 12 in the Brooklyn Navy Yard, IceStone's recycled-glass countertops were finding their way inside the spaces of such clients as Starbucks and the Gates Foundation after only seven years on the market.

Now a Category 3 hurricane had threatened to sink the business and put IceStone's 39 employees out of work.

How did IceStone survive?

One of the first clues could be IceStone's commitment to developing its employees. For instance, Jana Milcikova responded to a Craigslist ad in 2004 for an internship. When the Slovakian started at IceStone seven months later as a research engineer, she discovered

---

[*] CNN Library. "Hurricane Sandy Fast Facts." November 5, 2014. From http://www.cnn.com/2013/07/13/world/americas/hurricane-sandy-fast-facts/, accessed December 1, 2014.

[†] D&B. "After Hurricane Sandy: Hurricane Impact Report on Business." From http://www.dnb.com/lc/credit-education/after-hurricane-sandy-hurricane-impact-report-on-business.html, accessed December 1, 2014.

> that the company treats its employees like family—even hanging original portraits of employees on the office walls. Milcikova worked in the research and development, quality, and customer service departments before the company restructured and she found herself managing the whole company as president.
>
> With the economic crisis starting in 2008 threatening the company's survival, Dal LaMagna—the inventor of Tweezerman—took over as IceStone's CEO in 2011. Under his leadership, employees were paid a minimum of $15 an hour (twice New York's minimum wage) and 70% of their healthcare costs, among other benefits. IceStone's employees would be embedded in all levels of decision-making and taught how the company worked, particularly in finance. "Everybody wore many hats and contributed with the best of our abilities," said Milcikova.
>
> During the course of 2012, employees ran operations and brought the company from losing $3 million a year to the door of breaking even.
>
> That is, until Sandy. "The factory was under five feet of water," LaMagna said. "More than 3,000 electrical components—motors, transformers, circuit breakers, controllers—were doused in salt water. As far as I could tell, it was game over."
>
> All told, IceStone had an estimated $2-3 million in equipment damage—with no flood insurance—and $1.5 million in revenue losses during cleanup.
>
> But empowerment within the workforce paid off. Doubts began to fade when LaMagna saw a group of employees pushing water off the factory floor with squeegees. They could rebuild the factory, they said. Rather than scrapping and replacing all their equipment, IceStone's employees went through each part and assessed the damage, only replacing what was necessary. Small business disaster recovery loans helped to cover the rest. One employee initiated a Kickstarter campaign, raising $12,000 to supplement cash flow.
>
> Six months, 70 restored equipment motors, and 2,000 hand-washed machine parts later, IceStone was back in business. Prior to Sandy, the company was operating at $4 million in revenue; two years after Sandy, revenues have climbed back to $2.5 million and continue to grow.
>
> "All of us knew that it took all of us to recover from Sandy," said LaMagna. "Employees feel like they are part of a team—that they are valued for what they contributed, that they do better when the company does better."

The IceStone case highlights the impact of a sustainable culture on the ongoing successful operations of an organization. The inclusive culture that was created by IceStone's management allowed the organization to survive and thrive in the case of adversity. Investing in employees and demonstrating their value to the organization creates a culture that can overcome significant obstacles, creating value for all stakeholders.

6.5 Creating a Culture of Sustainability

Whether an organization is created with sustainability as a foundational pillar or transforms itself to incorporate sustainability, the key to becoming a sustainable organization is alignment between organizational values and sustainable strategy.

In Chapter 4, a materiality priority checklist was introduced as a tool to focus strategy on the areas of most significant organizational impact. Once material issues have been identified, organizational goals with specific targets are set. Clearly identifying goals and quantitative targets facilitates clear communication on organizational intention both internally and externally. The Culture of Sustainability Matrix (see Table 6.2) expands these goals and targets to identify drivers that impact embedding sustainability into an organization. Understanding specific drivers facilitates selecting programs and projects to deliver the desired outcome. Identifying the drivers for sustainable change facilitates defining metrics to measure progress toward the selected targets. For example, a social capital goal for a telecommunications firm may be to improve access to service in an underserved market. The target may be a 10% increase from the present level. The driver is creating new solutions that deliver service at a lower price point or in remote areas to meet this population's requirements. In order to measure progress, a metric—number of underserved customers/total customers—is selected, tracked, and reported.

Once specific programs and projects are identified, project management professionals are able to take up the mantle, managing programs and projects that impact drivers of sustainability. This process becomes part of a continuous cycle through which sustainability becomes part of the project management process. As projects and programs are chartered, planned, and implemented, sustainability becomes part of the process not just for sustainability projects but for all projects. Through this process, the concepts of sustainability become incorporated into the DNA of the organization. Once sustainability becomes a core value, then programs and projects change operational processes, impacting how work is performed, products are designed and manufactured, and employees are rewarded. The goal is to create alignment between organizational outcomes and sustainable business strategy.

Table 6.3 is a checklist to help program and project managers embed sustainability into the project management process. This checklist provides a framework for project management professionals to incorporate sustainability considerations into programs and projects. The checklist promotes strategic alignment with the sustainability mission. It focuses project impact on stakeholders, employees, risk management, and operations. Then it considers the effectiveness of the communication strategy, monitoring and control, and vendor management. Utilizing this assessment helps project management professionals plan their programs and projects in a manner that promotes organizational sustainability vision. Sustainability-based programs and projects are designed to create organizational change and to promote sustainability-driven decisions, actions, and behaviors. In order for an organization to adopt a culture of sustainability, it must infuse sustainability into all functions, reaching internally as well as externally. As sustainability becomes part of the project management process, it begins to permeate across the organization and becomes part of the collective thought process.

6.6 Misalignment of Values and Actions

Let's take a look at what happens when words and actions don't align around sustainability mission and goals. The headlines are full of news of companies that have had their business value impacted by environmental, social, and governance missteps. We have already mentioned the Volkswagen AG scandal and the related loss of reputation, significant fines, and legal implications. In discussing the failure, management cited both individual misconduct and a culture that allowed rules to be breached in pursuit of a goal to expand diesel sales in the United States. VW's downfall was a result of a culture that promoted misalignment of values and actions.

Culture is reflected in the decisions that employees make, including their adherence to codes of conduct and ethical standards—not just the creation of policies, but training, information sharing, and rewarding behavior that complies with these standards. Having an ethics policy has little impact if the culture of the organization is to reward behavior that produces short-term results but violates ethical standards. Both the VW and the following BNP and JP Morgan cases highlight this issue.

BNP Paribus, a major French financial institution, announced a settlement in June 2014 with U.S. authorities concerning U.S. dollar transfers made to restricted countries such as Iran and Sudan, in

Table 6.2 Culture of Sustainability Matrix

Issues	Goals	Target	Driver	Metric (annual)
ENVIRONMENTAL CAPITAL				
Climate change risk	Zero GHG Emissions by 2020	10% Reduction	Energy Usage	GHG from Fuel Utilized
Environmental spills/accidents	Zero Environmental Accidents by 2017	Reduction: 25% Year 1, 50% Year 2, 75% Year 3, 100% Year 4	Accidents/Errors	Occurrence/Unit of Product
Water usage	Water Eduction	50% Employee Training Year 1	Facility Water Usage	Water/Unit of Product
Energy usage				
Fuel usage				
GHG emissions				
Waste management				
Biodiversity				
SOCIAL CAPITAL				
Stakeholder engagement	Improved Internal Stakeholder Satisfaction	25% Rating Improvement	Stakeholder Engagement Plan	Annual Employee Survey Results
Community relations	Increase Community Investment	20% Increase in Volunteer Hours	Community Programs/ Voluntarism	Community Volunteer Hours/ Employee
Facilities impact				
Customer impact				
Customer safety				
Product safety				
Marketing message				
Market access	Increase Service to Underserved Market	10% 2016, 20% 2017	New Business Opportunity	# Under Served Customers/Total Customers
Client privacy	Zero Data Breaches	Reduction: 50% 2017, 75% 2018, 100% 2019	Personal Information Protection	# Data Security Breaches/Annum
Underserved markets				
HUMAN CAPITAL				
Diversity & inclusion	Increase Board Diversity	Women Comprise 25% of Board 2020	More Diverse Slate of Board Candidates	% of Women on Board
Training & development				

Topic	Initiative	Target	Practice	Metric
Recruitment & retention				
Compensation & incentives				
Labor practices				
Health, safety, & wellness				
Child labor				
BUSINESS PROCESS				
Economic viability				
Incorporating externalities				
Research & development				
Product's social impact				
Circular economy	End of Life Product Recapture	100% by 2020	Customer Health and Safety	% Product Taken Back for Reuse/Disposal
Packaging				
Pricing				
Product quality				
GOVERNANCE				
Regulatory & legal requirements				
Policies & codes				
Ethics	Compliance with Regulation	100% of Employees Trained by 2018	Employee Education and Compliance	Annual Amount of Regulatory Fines / # of Settlements
Investor relations				
Board composition				
Executive pay				
Political lobbying				
Sourcing policies				
Value chain standards	Tier I Suppler Code Compliance	100% Tier 1 Supplier Compliance by 2020	Tier 1 Suppliers' Policy and Compliance Protocols	% Suppliers Participating in 3rd-Party Audits
Supply chain engagement				

Table 6.3 Project Checklist to Promote a Sustainable Culture

	Yes	No

Project Charter
Is the project in alignment with both business and sustainability goals?
Is the project cross-functional in scope?
Are sustainability goals part of the project?
Are goals SMART-Specific, Measurable, Achievable, Relevant, Timebound?
Has the project sponsor included performance metrics for sustainability goals?
Is the project owned by a business function(s)?
Is the project initiation global?
Is the project execution local?
Was the project selected based on sustainability portfolio standards?

Stakeholder Engagement
Have both internal and external stakeholders been identified?
Have external stakeholders been included in the requirements gathering process?
How will stakeholder concerns be addressed in the project?
Are there opportunities for stakeholder partnerships?
Are volunteer and community outreach programs in place?

Human Capital
Has diversity been addressed in the composition of the project team?
Has cultural diversity been considered in scheduling and communication?
Is sustainability part of all team members responsibilities?
Are members of the project team familiar with organization's sustainability goals?
Do they understand their role relative to project goals and metrics?
Have incentives been developed to support project sustainability metrics?
Is the compensation strucutre in line with living wage standards?
Is the working environment safe?
Are employees empowered to identify opportunities to enhance sustainability performance?
Does the project address issues identified by sustainability value mapping?

Communication
Is a reporting structure in place for sponsor and or steering committee?
Has a communication plan been created for external stakeholders?
Has a communication plan been developed for internal stakeholders?
Has the organizational communication style been considered in the plan?
Has the use of acronyms and company terms been considered?
Is internal and external messaging consistent?

Monitor and Control
Are ESG metrics incorporated into projects?
Are ESG metrics part of tollgate reviews?
Is the project team educated on the code of conduct?
Is a code of ethics in place? Is the project plan in compliance?
Is technology or an information management system being used to track data?
Is a recognized reporting standard being used?
Do PMO standards include sustainability criteria, templates, lessons learned, best practices?

Vendor Management
Is there a supplier code of conduct?
Are vendors chosen that meet this code?
Is there an evaluation and remediation process for vendors?

Risk Management
Has the risk tolerance of the organization been considered?
Has climate change risk been incorporated into the project?
Has brand and reputational risk been considered?
Has country compliance risk been considered?
Has existing and pending regulatory risk been considered?

Table 6.3 Project Checklist to Promote a Sustainable Culture (*Continued*)

	Yes	No
Customers		
Does this project open new markets?		
Does this project address a customer sustainability opportunity?		
Does the project address a customer compliance requirement?		
Does the project improve our competitive ranking?		
Operations		
Does the project help to meet environmental, social, or governance goals?		
Does the project reduce resource requirements?		
Does the project impact product/service life cycle?		
Does the project promote employee education on sustainability?		

violation of U.S. law. Jean-Laurent Bonnafe, CEO, in an open letter appearing in the *The Wall Street Journal*, apologized for the behavior of BNP employees involved in violating the bank's business and ethical principles. He committed the organization to strengthening their processes and controls to ensure that the code of conduct is understood and that employees behave in compliance with those standards. The total cost to BNP in terms of brand and reputation is not yet known; but the immediate costs have already been expensive. In addition to pleading guilty, BNP was fined $9 billion and banned from clearing U.S. dollar transactions for one year.[17] The full impact to their financial performance, reputation, ratings, and customer retention is yet to be determined, but the cost arising from misalignment between stated and actual culture is clearly pricey.

For an indication of financial cost from an ethics breach, we can look at the case of JPMorgan and the London Whale. In 2012, JPMorgan was impacted by a governance and ethics misstep that originated with a significant trading loss (over $6 billion) in its London portfolio. As sizeable as the loss sounds, it had a relatively small impact on the company's performance. Management's handling of the event and the perceived weaknesses in governance and ethical procedures cost the company much more. The trade took place in a unit under the direction of the Chief Investment Office (CIO), whose role is to manage risk for the bank's portfolio. Instead, monies assigned to the CIO group were used to trade for the bank's own returns. As a result of this breach of compliance, the bank admitted violating securities law and agreed to a $1 billion fine. While the fine is not insignificant, the major loss for the bank was the damage to its stellar reputation and its leadership position in the industry. Under the guidance of Chairman and CEO Jamie Dimon, JPMorgan had successfully navigated the 2008 financial crises. Mr. Dimon was well regarded and admired by his peers, regulators, and governmental leaders. Because of the scandal, a great deal of institutional credibility was lost, resulting in a formal criticism of the institution by the U.S. Senate. In addition, Mr. Dimon found himself in a battle for his Chairman title as a shareholder proposal sought to split the CEO and Chairman roles. Further, the fines and legal costs resulted in a quarterly loss being reported in October 2013. The trading loss, while sizable, was not the most significant contributor to the firm's loss of market capitalization. The perceived lack of management's enforcement of governance and ethical policies had a much larger impact on valuation.[18]

Moving forward to create a culture of sustainability, "walking the walk" is just as important as "talking the talk." Alignment of core values and organizational actions and behaviors is crucial to developing a culture of credibility.

6.7 Establishing Goals

As we saw in the Skanska example, it is important to establish goals in order to communicate organizational vision to your diverse stakeholders. Defining and measuring sustainability goals helps promote a sustainable culture, because both internal and external stakeholders understand your mission and are given information on targets and progress. Goals that are "SMART" facilitate the process of embedding sustainability through clearly and quantifiably stating what your organization hopes to achieve.

SMART Goals

Specific—clear and unambiguous
Measurable—clearly defined so you understand when they have been achieved
Achievable—reasonable and attainable within the scope of the organization
Relevant—relate to material issues
Time-bound—set a timeframe for completion[19]

SMART goals identify clear and measurable metrics, such as 10% reduction in GHG emissions by 2016. Goals can be absolute or relative. Absolute means achieving the goals despite growth in the organization in terms of sales, new people, new facilities. Relative goals are usually expressed as a percentage of something else, such as GHG/revenues or waste/production unit. Relative goals are useful for measuring progress, but absolute goals have a greater overall impact in terms of creating long-term impacts and moving toward zero impact goals.

Some companies establish aspirational absolute goals, sometimes known as "big hairy audacious goals" (BHAGs). These types of goals push management to develop new ways of conducting business that require a major change in the operation and culture of the business. BHAGs such as zero emissions, carbon neutrality, or water neutrality often take more than a decade to achieve. BHAGs are valuable for creating a culture around sustainability and giving a framework to drive long-term change. However, they are challenging in terms of assigning responsibility and tying results to management compensation. In order to reach these stretch targets, the journey needs to broken into more manageable targets and objectives that can be reached in a shorter time frame. Management and employees need to be directed by SMART goals in order to take the daily steps needed to reach the ultimate BHAGs.

Examples of SMART Goals

- Achieve an Energy Star Portfolio Manager Rating of 75.
- Reduce energy consumption by 20% in the next 24 months.
- Reduce paper usage by 30% in the next 12 months.
- Identify suppliers with and without sustainability programs by 2017.
- Increase employee participation in community volunteer program to 75% by 2017.
- Receive an invitation from the DJSI to participate in their 2018 survey.

Depending on your organization, its industry, supply chain, and customer base, the goals will vary. If your organization is facility-intensive, focusing on moving your organization into the highest category of Energy Star is a SMART goal, and it may qualify your building for an Energy Star certification. On the other hand, if you lease space in an office building, the Energy Star Portfolio Manager isn't a relevant tool, as it requires whole-building data for meaningful analysis. The key is to select goals that are relevant to your operation and that support the business mission and goals of the organization.

6.8 Create a Team

While your organization might have an Environmental, Health & Safety (EH&S) department or a Sustainability Office, one group alone cannot change the culture of an organization; it requires a cross-functional team approach. For a successful transformation it is important to have the correct mix of people, including both those who are interested in environmental/social issues and those who are key to the decision-making process within your organization. Team functionality is further delineated between strategy development and action-oriented groups, as both strategy and operational aspects are important to the process. Working with internal human capital experts such as HR professionals facilitates the process of building an effective team, engaging stakeholders, and developing incentives to build a sustainable culture. (Chapter 10 discusses this topic in greater detail.)

Most sustainability initiatives begin with one-off projects known as "low-hanging fruit." Often, these projects focus on issues of energy use, waste disposal, or resource utilization. "Low hanging fruit" projects have a quick but measurable impact on the organization and are relatively easily accomplished by a single function or group. Generating early successes builds team momentum and provides tangible evidence to ensure continued senior management support. As an organization moves forward on its sustainability journey, programs and projects become more complex and interconnected. In order to address these issues, project managers need to develop cross-functional teams to facilitate the process of moving sustainability from a standalone function to becoming part of the culture and embedded into operations. As the sustainability journey continues, teams need to become even broader in scope and include external stakeholders to address significant sustainability challenges.

6.9 Internal Stakeholder Adoption

Stakeholders, especially employees, need to understand not only the corporate vision of sustainability but also how they are going to be impacted by it from both a professional and personal perspective. Understanding and adopting the mission is much more effective if there is an overlap between corporate and personal values. Are your employees interested in supporting local community groups such as schools or parks? Create a program that highlights your organization's skill set and promotes engagement with the local community. Employees will be proud to participate. Through participating and projecting a positive image of themselves and the organization, employees take on the role of ambassadors for your organization. Sustainability value mapping (see Figure 6.5) focuses on engaging

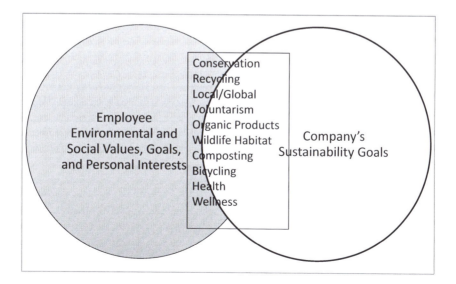

Figure 6.5 Sustainability Value Mapping

employees through identifying common ground between an organization's sustainability goals and an individual's interests and values concerning environmental and social issues. Providing opportunities for employees to participate in environmental and community activities that match their areas of personal interest improves engagement. As a result, employees gain a better understanding of the positive impact that a culture of sustainability can have in their workplace and community.

Employee participation and engagement improves as employees identify personally with sustainability goals and programs. Some will be engaged through alternative commuting options such as biking, others will be engaged through community investment such as resource contributions to local schools. As employees begin to understand an organization's sustainable mission and how sustainability programs will impact their roles and lives, employees begin to embrace sustainability change. Through the change process, employees gain a better understanding of sustainable strategy and their role in the process. Once employees buy into the organization's sustainable strategy, they can recommend local action to further sustainability within their own functions and roles. Through this process, employees become change agents further embedding sustainability into organizational culture.

Shannon Schuyler, CR Leader at PWC, shared at a Green Biz Forum that developing volunteer opportunities that align with staff interests, talents, and skills increases participation, promotes employee engagement, and improves customer satisfaction. She reported that employees who participate in Corporate Responsibility programs have a 5% higher retention rate and that the profit impact of the CSR programs is $165million.[20] Employees need to feel personally empowered and identify with the programs in order to really make it part of the culture.

6.10 Metrics to Drive Change

In order to drive sustainable change, it is important to identify, measure, and reward desired outcomes. Management and employees focus on metrics included in their performance scorecards. If managers are not being evaluated on sustainability metrics, they will not have the right incentive to spend time and resources on sustainability programs and projects. Performance measures need both to focus on past performance and to encourage decision making that aligns with sustainability goals. (Chapter 14 takes a more in-depth look at metrics.) Metrics such as GHG emissions/employee track the impact of sustainability programs over time and record progress against a baseline. However, to encourage long-term sustainability decision making that drives significant change, management needs metrics that are grounded in the organization's core mission and values. In order to facilitate this process, organizations are using shadow pricing for carbon emissions to help decision makers understand the long-term impact of selecting projects that depend on fossil fuel usage. While technology continues to evolve and the price of using renewable energy continues to drop, projects built with renewable energy often have higher up-front capital costs and may, at least initially, have a higher cost of energy. As a project sponsor making a decision about building a facility, it can be a difficult conversation with a steering committee to discuss the up-front costs to build a green facility with renewable energy sources versus a traditional build. Leading organizations are using the concept of shadow pricing per tonne of carbon emissions to reflect more accurately the true long-term cost of making a decision to use fossil fuels. They are providing a tool for project managers to make a more effective analysis of the costs of their options and recommendations. As a result of this more accurate comparison, this tool is driving more decisions toward selecting options that reduce carbon emissions.

In order to drive change with an organization's culture, a combination of resources needs to be utilized, from clear and concise goals from the leadership team to metrics and incentives to reward desired behaviors. Creating a culture of sustainability takes time and resources, but the beneficial impact to the organization, community, and natural environment is worth the effort.

> **Carbon Shadow Pricing at IKEA**
>
> As IKEA pursues its mission of creating a better everyday life for people around the globe, sustainability is a driver of innovation in terms of both products and processes. IKEA is on plan to sustainably source their wood by 2020. Almost 88% of its cotton in 2014 came from sources with standards for water consumption, pesticide, and fertilizer usage. They anticipate investing an additional $1.5 billion in renewable energy sources over the five-year period 2016–2020. IKEA tracks its product life-cycle GHG emissions from raw material purchase through product disposal.
>
> After the UN Climate Change Summit in New York, which took place in September 2014, IKEA committed to evaluating internal carbon shadow pricing. According to the Carbon Disclosure Project (CDP), about 150 companies use internal pricing of carbon emissions, mainly for long-term planning. Some of the companies currently using the model are Google, Exxon Mobil, Walt Disney, and Wal-Mart. The CDP reports that shadow carbon prices vary from $6 to $60 per tonne.[21] The purpose of the internal pricing is to accurately price the cost of using fossil fuels through incorporating the cost of emissions in the cost equation. In effect, including the cost of carbon considers the full costing impact and makes renewables more cost-effective. From a planning perspective, the carbon shadow charge can impact a variety of decisions including material sourcing, supplier partnerships, and facilities investments. The goal in using shadow pricing for carbon is to encourage leadership to focus long-term on the impact of their decisions. It also serves as a tool to identify revenue opportunities and risks. The ultimate goal is to protect the environment while driving business efficiency.[22]

6.11 Best Practices

In order to facilitate embedding sustainability into your organization's culture, here are examples of best practices from "best in class" companies as selected by Ceres, a nonprofit organization that advocates for sustainability leadership. At Alcoa, executive compensation is tied to safety, diversity, and environmental stewardship. At PepsiCo, sustainability strategy and goals are discussed at the annual shareholders meeting. In addition, the company discloses key sustainability challenges in its financial reporting. GE has teamed with its HR department to embed sustainability in its company's culture. Ford has created requirements for first-tier suppliers to promote GHG emissions and energy efficiency in its supply chain.[23] Each of these examples focuses on a different technique for embedding sustainability, because each organization is unique and will have different needs and experiences in terms of creating a culture of sustainability. The following are some universal best practices for embedding sustainability:

1. Make sustainable strategy an integral part of the business mission.
2. Give senior management responsibility, and tie compensation to sustainable performance.
3. View sustainability holistically, both internally and across the value chain.
4. Make sustainability part of the corporate brand and the customer value proposition.
5. Ensure consistency between internal and external stakeholder messages.
6. Offer outreach and volunteer programs that support the core business mission and build upon the strengths and values of employees.
7. Adopt recognized reporting standards such as GRI to promote transparency and facilitate understanding performance by all stakeholders.

6.12 Portfolio, Program, and Project Manager Perspective

As an organization moves through its sustainability journey, the way in which projects are selected and implemented evolve. In the beginning, projects are one-off sustainability projects focusing on regulatory compliance or energy efficiency, often called "low-hanging fruit." Continuing on the sustainability journey, projects move from one-off projects to cross-functional sustainability programs and finally to sustainability standards becoming part of all projects and programs. Through a continuous improvement process, best practices and lessons learned are incorporated into the project management process and become standards for future projects. Both successes and failures contribute to the lessons learned and become part of the organization's story. The Assessment of an Organization's Sustainable Cultural Foundation (discussed in Section 6.2) identifies the type of culture the organization currently has and the type of culture desired. This process allows sustainability champions and sponsors to assess their organization's readiness to adopt sustainability and to develop protocols for project development and implementation that address shortfalls and maximize strengths. This approach allows for identification and selection of programs and projects to embed sustainability into the organization.

Organizational cultural awareness helps program and project managers better understand planning and implementation requirements for sustainability projects, leading to better alignment between goals and outcomes. This process identifies organizational readiness and capacity for change. Using the Project Checklist to Promote a Sustainable Culture (see Table 6.3) facilitates alignment of programs and projects with an organization's sustainability mission driving organizational change. Through this process, project management professionals drive adoption of a sustainable culture, impacting people and process within the organization and creating long-term business value.

6.13 Conclusion

In order to reap the benefits of a sustainable strategy, a culture of sustainability needs to be created and promoted, facilitating organizational understanding of core environmental, social, and governance values. CEOs value organizational culture because they know that a strong culture of sustainability provides a competitive advantage that is not easily duplicated. Creating a sustainable culture means adopting sustainability into operations and creating policies and procedures that align employee actions with the sustainability vision. Misalignment between the sustainability vision and actions results in costly missteps for leadership teams. Project management professionals provide a link between organizational sustainability vision and the selection and implementation of programs and projects to drive sustainability into organizational culture.

Notes

[1] Eduardo P. Braun, "It's the Culture, Stupid!," *Huffington Post*, June 25, 2013, http://www.huffingtonpost.com/eduardo-p-braun/its-the-culture-stupid_2_b_3487503.html.
[2] Deidre H. Campbell, "What Great Companies Know About Culture," *Harvard Business Review*, accessed March 12, 2015, https://hbr.org/2011/12/what-great-companies-know-abou.
[3] "What Is Organizational Culture? Definition and Meaning," *BusinessDictionary.com*, accessed December 9, 2014, http://www.businessdictionary.com/definition/organizational-culture.html.
[4] Mark Brownstein, "Closing Keynote," The Energy/Water Nexus in Unconventional Oil & Gas, Wharton Initiative for Global Environmental Leadership Conference, Philadelphia, PA, October 29, 2014.
[5] Andy Sharman, "How Skanska Aims to Become the World's Greenest Construction Company," *Financial Times*, March 23, 2014, http://www.ft.com/cms/s/0/73a1bea4-a61a-11e3-8a2a-00144feab7de.html#axzz3G2sx7KHF.

[6] "The Five Zeros—Our Core Value," *Skanska*, accessed December 15, 2014, http://www.usa.skanska.com/pages/standardwide.aspx?id=3371&epslanguage=en.

[7] "Sustainability—About Skanska," *Skanska*, accessed October 13, 2014, http://www.usa.skanska.com/pages/standardwide.aspx?id=3678&epslanguage=en.

[8] Skanska, "Introduction to Skanska Sustainability Agenda," accessed December 15, 2014, http://www.usa.skanska.com/cdn-1cebae0c45c50cf/Global/sustainability/Intro%20to%20Skanska%20Sustainability%20agenda%20Sept12.pdf.

[9] Sharman, "How Skanska Aims to Become the World's Greenest Construction Company."

[10] "What Are B Corps?," accessed October 15, 2014, http://www.bcorporation.net/what-are-b-corps.

[11] Initiative for Global Environmental Leadership, "Disrupting the World's Oldest Industry," March 2014, http://d1c25a6gwz7q5e.cloudfront.net/reports/2014-03-06-Disrupting%20the%20World's%20Oldest%20Industry.pdf.

[12] Ibid.

[13] Rubicon, "Rubicon, Company, Overview," accessed January 23, 2015, http://rubiconglobal.com/company.

[14] Benefit Corp Information Center, "Benefit Corp. vs. Certified B Corp," accessed March 16, 2015, http://benefitcorp.net/what-makes-benefit-corp-different/benefit-corp-vs-certified-b-corp.

[15] "How to Become a B Corp in Six Weeks or Less," *Triple Pundit: People, Planet, Profit*, accessed October 15, 2014, http://www.triplepundit.com/2014/10/become-b-corp-six-weeks-less.

[16] Ibid.

[17] Jean-Laurent Bonnafe, "Taking Responsibility for the Past and Turning to the Future," *The Wall Street Journal*, July 2, 2014.

[18] Patricia Hurtado, "The London Whale," *QuickTake*, October 17, 2013, http://www.bloomberg.com/quicktake/the-london-whale.

[19] "SMART Goal Setting Examples," *OnStrategy*, accessed December 9, 2014, http://onstrategyhq.com/resources/smart-goal-setting-examples.

[20] Ellen Weinreb, "PwC: 10 Tips to Bolster Employee Engagement," *GreenBiz.com*, accessed July 18, 2014, http://www.greenbiz.com/blog/2014/04/09/pwc-10-tips-bolster-employee-engagement.

[21] Alister Doyle, "IKEA May Tighten Carbon Rules to Protect Environment," Reuters, October 13, 2014, http://www.reuters.com/article/2014/10/13/us-climatechange-summit-ikea-idUSKCN0I211020141013.

[22] Ibid.

[23] Jo Confino, "Best Practices in Sustainability: Ford, Starbucks and More," *The Guardian*, sec. Guardian Sustainable Business, accessed October 17, 2014, http://www.theguardian.com/sustainable-business/blog/best-practices-sustainability-us-corporations-ceres.

Chapter 7
Portfolio Management Supports Strategic Sustainability Alignment

Portfolio management provides a framework to facilitate embedding sustainability into an organization. Using portfolio analysis allows senior leadership to better select components (i.e., projects, programs, and subportfolios) that have strategic fit with the organization's vision of a sustainable future. This process focuses on project benefits, assessment, and realignment to ensure that sustainability goals are achieved. Through a project matrix approach, portfolio managers identify project benefits and areas for new component opportunity in support of sustainable strategy. Developing a process of component criteria and assessment to promote sustainable strategy guides the processes toward alignment with strategic vision. Supporting a portfolio process through effective organizational structure and communication helps to drive sustainability into an organization. Creating a body of sustainability knowledge and a portfolio process provides a means to incorporate lessons learned from early adoption components, assess component impacts, and realign resources to support priority components. Identifying a matrix of portfolio criteria against which to assess project proposals facilitates selection of projects that deliver benefits in line with strategic sustainability vision. This process facilitates selection of components that are material to the organization and align with long-term sustainability targets. Establishing project standards also facilitates communication of sustainability and business goals throughout the organization by providing a framework that focuses project sponsors on challenges and opportunities that align with strategic interests. It also provides a framework to address governance issues about policies, supplier standards, a code of ethics, and compliance standards such as client and regulatory requirement thresholds. For example, the directive on disclosure of nonfinancial and diversity information reporting required in 2018 by the EU member states will have a significant impact on large corporations operating within EU borders. For companies with operations that will be impacted by these requirements, portfolio standards need to be established to ensure that existing and future projects comply with these requirements. By incorporating a portfolio management approach, management aligns resource allocation with component selection in support of sustainable strategy and vision.

144 Becoming a Sustainable Organization

7.1 The Portfolio Analysis Process

The portfolio analysis process begins with understanding the organization's strategy and goals. From a sustainability perspective, goals are long-term and impactful and may take a decade to reach. To achieve sustainability goals often requires significant change in operations, capital investments, and stakeholder engagement. In order to reach the sustainability vision, strategy needs to be divided into components that build on previous programs or projects in order to deliver the desired result.

A gap analysis of the organization's current performance relative to the long-term sustainability goal identifies strengths and weaknesses in current operations, component inventory, resources, and capabilities. The gap analysis provides a roadmap for moving the organization from its current state toward its strategic sustainability vision and goals. Map the organization's strategic sustainability vision against an inventory of current projects, programs, and portfolios in order to determine benefit alignment. Identify shortfalls between these components and the challenges and opportunities facing the organization. Consider internal competencies such as product knowledge, talent pool, customer relationships, culture, production, distribution, design, processes, and systems. Identify areas of strength and weakness to better inform program and project proposal and selection in order to drive desired change.

In order to address both internal and external stakeholders requirements, create a process to gather input and feedback. Internal stakeholders impact the organization directly and include owners, investors, and employees; external stakeholders are affected by actions and decisions of the organization and include customers, suppliers, creditors, local, state, and national governments, labor unions, NGOs, and community groups. Stakeholder input reframes portfolio criteria to support new solutions through partnership collaborations and to minimize potential threats to operations. In addition, it promotes ongoing stakeholder engagement.

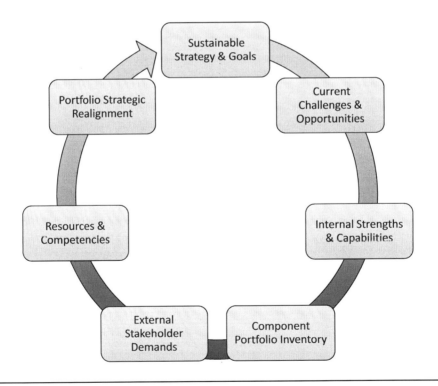

Figure 7.1   Sustainable Strategy Portfolio Alignment

As indicated in Figure 7.1, strategic portfolio alignment begins with identification of goals and strategies. Overlaying organizational challenges and opportunities with strengths and capabilities against the current inventory of portfolio components provides insight into existing portfolio alignment. Add in stakeholder and resource requirements and required competencies to identify programs and projects to improve strategic portfolio alignment.

Given an organization's current component portfolio inventory and its set of capabilities, consider the necessary resource requirements in terms of money, technology, and skills that need to be brought to the table in order for projects and programs to deliver benefits in line with desired sustainability outcomes. Using an inclusive approach in the portfolio alignment and resource allocation helps to promote acceptance across the organization as well as among external stakeholders. An effective communication plan and transparency of process supports internal and external stakeholder buy-in to the portfolio decision process.

7.2 Identifying the Portfolio Management Process

Research suggests that aligning people, programs, and resources with portfolio components that support sustainable strategy delivers on business value creation. However, project management professionals are looking for frameworks, tools, and resources to facilitate this process. In Figure 4.6, Project Portfolio Alignment Mapping, we introduced the concept of selecting projects that align with both business and sustainable strategy. This tool forms a foundation for alignment between sustainability and business benefit. However, as an organization progresses on its sustainability journey, further refinement is required in order to move toward transformational sustainable strategy.

Most organizations begin their sustainability journeys focusing projects inward on regulatory or customer compliance programs that seek to minimize their environmental footprint. Next, they tend to look at product stewardship projects such as a life-cycle assessment. While these types of sustainability projects provide benefits in terms of environmental impact, cost savings, and risk mitigation, transformational programs require a future-oriented focus and external stakeholder engagement to create new sustainability-based business opportunities. In order to create a platform for transformational sustainable solutions, management needs to focus on portfolio components that promote new technologies and drive innovation.

Stuart Hart, a pioneer in the field of creating sustainable business strategies, has developed the "Sustainable Value Framework." This framework facilitates evaluating global sustainability challenges and opportunities through a business lens in order to improve environmental and social performance as well as to contribute to shareholder value.[1] The Sustainable Value Framework considers the impact of programs and projects in four key strategic areas:

Pollution Prevention: Minimizing waste and emissions from current operations
Product Stewardship: Engaging stakeholders and managing the product life cycle
Clean Technologies: Identifying, creating, and implementing new sustainable technologies that provide new solutions
Base of the Pyramid: Identifying and creating new products and services to meet the needs of the poor and underserved.[2]

Evaluating current portfolio components against these strategies reveals opportunities to launch new programs and projects aimed at managing risk, generating cost savings, protecting reputation and brand image, improving process and productivity through innovation, and seeking new markets and partnership opportunities. Hart suggests, through implementing this framework, management finds that most of its existing portfolio components for sustainability fall under the strategy of pollution prevention. Given that organizational sustainability programs are often driven by regulatory compliance, the focus on pollution prevention is not surprising. However, to move beyond this limited scope, management must look outside its internal operations to its value chain suppliers and customers and to other stakeholders to leverage the value creation benefit of a sustainable approach.

Adopting this approach is especially useful for organizations that are on the sustainability continuum and are seeking to refine their strategy to improve results. Evaluating sustainability-driven portfolios annually provides the following strategic benefits:

1. It provides a forum to bring leadership together to discuss sustainable strategy accomplishments and challenges.
2. It facilitates the development of strategic initiatives in order to address future challenges and opportunities for both internal and external stakeholders.
3. It identifies governance changes that are needed to support programs and projects.
4. It forms the foundation for a common language of sustainability in order to facilitate understanding across the organization.
5. It establishes clear strategic objectives and demonstrates alignment between component selection and business strategy and value creation.
6. It provides a roadmap of programs and projects and their anticipated impacts.

Figure 7.2 lays out a sustainability portfolio alignment matrix that facilitates analysis of the gaps between existing sustainability portfolio components and requirements to meet strategic sustainability visioning. The process calls for the existing inventory of sustainability projects and programs to be assigned to the quadrants best aligned with the components' benefit. The matrix facilitates developing a roadmap of component strategic benefits and identifies strategic areas with excess resource allocation and areas of deficiency.

In addition to giving characteristics of each sector's strategic focus, the quadrants include an example of a project type. Those organizations that are relatively new to the sustainability journey will find that many of their projects focus on current environmental and social challenges. These are projects such as energy efficiency, emissions reduction, water conservation, community engagement, and volunteer programs. Once management has successfully implemented projects focused on environmental and social impact, they tend to look externally at their value chain for other opportunities. Considering the impact of the entire product life cycle from raw material sourcing through the supply chain, manufacturing, distribution, and consumer usage and disposal gives rise to the next level of projects. These types of projects are characterized by collaboration with other stakeholders such as customers, suppliers, academics, and industry groups. While programs that focus on internal operations and product governance are impactful for environmental and social impact, they focus primarily on an organization's current sustainability challenges. In order to transform your operation and create new markets and new solutions, and to leverage new technologies, management needs to focus on sustainability visioning and the types of programs and projects that will move the organization toward that vision.

In order to drive toward a vision of sustainability, management needs to think expansively and consider disruptive technologies and solutions that provide new platforms from which to offer sustainable solutions for customers. In addition, management needs to consider new markets and partners for these types of solutions.

John Viera, Global Director of Sustainability for Ford Motor Company, shared how Ford is addressing megatrends such as growing urbanization, an increasing global middle class, air quality deterioration, and changing consumer attitudes about car ownership. In order to address these challenges, Ford has undertaken 25 mobility experiments, or projects, around the globe in order to determine which ones enhance lives by reducing congestion, improving parking options, and bettering air quality. One such project is "Go Drive," a car-sharing service based in London. Using a smart phone app, customers reserve, access, and unlock shared vehicles. The platform is a pay-as-you-go service and allows one-way

Portfolio Management Supports Strategic Sustainability Alignment   147

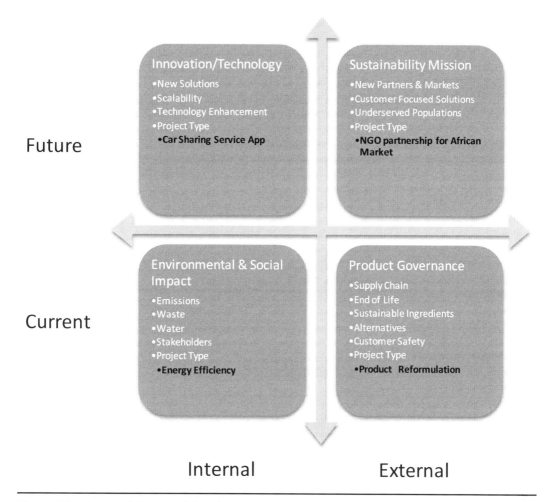

Figure 7.2   Sustainability Portfolio Alignment Matrix[3]

journeys. The vehicle fleet is made up of electric and energy-efficient vehicles. The project has been very successful with Millennials in the London market and now has 20 locations. Another project is "Spotlight on Parking," which seeks to address congestion in megacities, of which 30% is caused by people looking for parking. Using GPS mapping and a parking space database, vehicle sensors detect and share open spot information in real time. The benefits include fuel savings, GHG emission reductions, and reduced congestion. Cross-functional project teams are focusing on the nexus of car, technology, data, and analytics to provide scalable and impactful solutions to move Ford forward in meeting its sustainable strategy.[4] Management has created a portfolio of components to determine which projects are the most impactful in addressing these stakeholder-driven challenges. This approach allows Ford to move forward with its transformational goals of creating new products and services to address changing markets and global megatrends.

7.3 Sustainability Portfolio Assessment

The following Sustainability Portfolio Assessment focuses attention on the existing portfolio inventory and its effectiveness in delivering results in line with sustainable strategy. The assessment considers organizational capabilities, sustainability definitions, measures of success, progress relative to plan, and additional resources required to deliver results effectively. Answers to these questions identify organizational strengths and weaknesses as well as threats and opportunities.

Sustainability Portfolio Assessment

1. How effective have we been in achieving our sustainability and business goals?
2. How well does our current portfolio inventory align with our sustainability goals?
3. What are our organizational project strengths?
4. Do we have the skills and expertise to reach our sustainability vision?
5. How are we engaging with external stakeholders such as customers, suppliers, government, community, NGOs, and academia?
6. Are we considering all avenues of cost and risk reduction?
7. Are there needs in our client base that are not being served from an environmental and social impact perspective?
8. Are there actions that we can take to improve or protect our organization's image and brand reputation?
9. Do we have the opportunity to operate in new markets or with new partners as a result of our sustainability agenda?
10. What are our most significant environmental and social project impacts?
11. Do our governance policies and processes support our sustainability vision?
12. How are we tracking and measuring our sustainability impacts?
13. What would it take for us to be carbon-neutral?
14. What does net zero waste mean for us? Net zero water?
15. Where are most GHG emissions created in our value chain?
16. Where can we most impact biodiversity in our product life cycle?
17. How are we engaging our supply chain?
18. What does it mean for us to consider and assume responsibility for our full product life cycle?
19. What must change within our organization to meet our sustainability goals?
20. Are there new technologies or innovations that could help us offer more sustainable solutions?
21. How are we progressing with our sustainability agenda relative to our peers?
22. How aggressively are we pursuing our sustainability vision?

This assessment provides a roadmap for an organization's portfolio assessment that highlights areas of strength and opportunities for improvement. Working from this assessment allows for a determination of resources, technical and specialty skills, and technologies that are needed to augment core competencies in order to achieve the organization's sustainability vision. This assessment facilitates the development of innovative solutions and leverages new technologies, and sustainability-based platforms to address customer sustainability challenges and to serve new markets. It facilities a clearer picture of strategic vision and the portfolio priorities to achieve the vision,

7.4 Communication of Goals and Drivers to Stakeholders

Once the strategic vision has been formulated, specific goals and action need to be defined. Clarifying goals and targets allows leaders, managers, and employees to understand the top organizational priorities and to adapt their function, processes, and policies to create alignment between corporate actions and desired goals. In addition, creating a structure helps to clearly and concisely communicate an organization's sustainability vision to internal and external stakeholders. While organizations may have similar goals such as reducing GHG emissions, the approach that each takes will differ to reflect the organization's own unique set of drivers. Programs and projects are developed to impact these drivers in order to reach specific sustainability goals.

Table 7.1 outlines strategic priorities and specific programs targeted to help achieve these goals for a telecommunications firm. Notice that the strategic goals reflect all aspects of creating sustainable business value, focusing both internally and externally as well as on both current and future challenges.

Establishing measurable sustainability goals and identifying key drivers for change provides a clear set of portfolio component priorities, allowing for clear messaging of organizational sustainability priorities to employees, suppliers, customers, and communities. The goals and drivers will vary depending on organizational core competencies as well as the stage of the organization's sustainability journey. Creating such a list of priorities provides leadership with a focal point from which to direct strategy and to make tough decisions on program and project funding and on resource allocation. Each business function should be made aware of the guidelines in order for them to make the best recommendations for portfolio component selection. Projects that support sustainable strategic vision and provide tangible business benefits in line with these priorities should receive funding.

Table 7.1 Composite of Sustainability Goals and Drivers for a Telecommunication Company

Goals	Drivers
Continue development of our sustainable strategy to promote our recognition as a sustainable organization and an employer of choice.	• Monitor employee engagement and alignment of work with sustainable strategy. • Promote diversity and gender equality. • Report on sustainable performance using recognized standards. • Engage with key stakeholders.
Promote safe and responsible use and disposal of our products and services.	• Provide education on best practices for use and disposal. • Provide recycling opportunities for end consumers. • Create a product reclamation program.
Reduce carbon footprint by 25% by 2020.	• Develop and implement energy conservation plans for facilities, logistics, and employees. • Evaluate fleet conversion to alternative fuel sources. • Establish reporting protocols. • Share best practices.
Reduce impact on biodiversity and human rights.	• Develop supply chain protocols and measure via scorecards. • Redesign process to promote recycling and component reuse. • Monitor and integrate criteria to eliminate human rights violations from sourcing of rare earths and minerals.
Bridge the digital divide and promote inclusion in developing countries.	• Design a digital education program. • Deploy high-speed networks to underserved areas. • Identify and address the needs of underserved populations.
Develop new market opportunities and lead in customer service and support.	• Partner to develop innovative sustainable solutions such as smart cities, collaborative economy, and circular economy. • Mine big data for trends and potential innovative solutions.
Promote sustainable principles within our industry sector.	• Participate in global groups such as UNGC and CDP. • Report using industry standards such as GRI. • Participate in industry sustainability initiatives.

Orange, a French telecommunications firm with operations that span the globe, has 236 million customers in 30 countries with 165,000 employees. It is a large organization but one that is determined to promote a culture of sustainability. CEO Stephane Richard describes a key role for the firm as putting digital technology at the service of society.[5] This commitment begins with conducting activities in a responsible and exemplary manner. To this end, the Governance and Corporate Social Responsibility Committee reports directly to the Board of Directors to ensure that ethical and corporate social responsibility challenges are considered at the most senior decision level and in turn inform corporate policies. In order to address their goals for continuous improvement across the organization, each country, entity, and function creates its own programs to address ethical goals raised at the group level. Table 7.2 provides a list of some Orange ethics goals and specified programs to help achieve these targets.

Table 7.2 Orange Ethics Program

Ethics Goal	Programs
Increase awareness of protocols and best practices for addressing conflicts of interest from clients, fellow employees, suppliers, government representatives, and others. Develop an ethical customer service standard for client data protection, conduct toward competitors, social network standards, marketing, and sales.[6]	• Modification to the ethics guide and the e-learning tool to include conflicts of interest and updated case studies in a variety of languages. • Development of a self-assessment tool on ethics of service relations. Organization-wide rollout slated for 2014. • Publication of the Orange data protection charter in November 2013.[7]

As part of Orange's assessment of its internal strengths and capabilities, management identified further strengthening its ethics guidelines and training as a top priority. Senior management supported rolling out a new ethics e-learning module across the organization in order to further bolster their commitment to ethics education as it relates to customers, employees, officials, and suppliers. Orange leadership is clearly signaling the importance of governance within their organization and has made it a priority for resource allocation and competency building. Through the process of selecting sustainability goals that serve to fill gaps between their current capabilities and their vision, they have identified specific programs and projects to impact key drivers for achieving these goals.

7.5 Portfolio Management Alignment

As portfolio components are developed, mapping the link between sustainability programs and business vision and goals is imperative. As sustainability programs and projects are recommended, consider the core business strategy components listed in Table 7.3 and how the recommended programs impact commitments to stakeholders, sustainability targets, future plans, and resources. Use the qualifying questions to better develop alignment between programs/projects and core sustainable business strategies. The goal is to align the portfolio management process with both business value creation and sustainability vision, thereby ensuring that portfolio component selection reinforces both agendas.

7.6 Creating a Supportive Organizational Structure

The portfolio component management process benefits from an organizational structure that promotes senior-level support and cross-functional collaboration. As a first step, consider your organization's structure and its effectiveness in supporting sustainable strategy. In order to have policy and decision

Table 7.3 Portfolio Strategic Alignment Qualifier

Business Strategy	Qualifying Questions
Alignment	• Are we focusing on the right goals? • How well do programs and projects align with our sustainability vision? • Do management and employees understand the organization's sustainability vision? • Do key stakeholders understand our sustainability vision?
Productivity	• Are employees encouraged to behave and act in a manner that is consistent with our sustainable strategy? • Are resources being allocated to promote desired behaviors and outcomes?
Investments	• Are we investing in both capital improvements and workforce training and development to maximize ESG impact? • How are investment decisions being determined?
Scalability	• Are sustainability programs and projects being planned and implemented under a structure that allows best practices and processes to be adopted throughout the organization?
Governance	• Do our sustainability programs address risk, security, and compliance issues? • Are our policies and codes of conduct in compliance with sustainable strategies?
Long-Term Impact	• Are our sustainability programs viable and relevant for reaching our long-term sustainability goals? • Are we focusing on forward thinking and creative solutions?

making at the most senior level in alignment with sustainable strategy, the Board of Directors in conjunction with the C-suite need to set the tone. They create the vision and then approve strategic sustainability initiatives and policies to support achievement of the vision. Once the strategy is defined, the organizational structure needs to be aligned to facilitate cascading C-suite vision and guidelines into the organization for action and implementation. Figure 7.3 shows an example of such a structure.

Developing an organizational structure to promote a portfolio management approach involves creating cross-functional committees that move across traditional business functions. This networked approach facilitates collaboration and sharing across the organization. If possible, create a Sustainability Office, headed by a Chief Sustainability Officer (CSO), charged with deploying the strategic sustainability vision, policies, and procedures throughout the organization. In addition, the Sustainability Office functions as a repository to gather and maintain the body of sustainability knowledge, providing technical and methodology support to guide the lines of business in the process of identifying, selecting, and recommending projects.

The Office of Sustainability is an ideal group to establish and maintain the sustainability goals, targets, and standards. Acting as a central clearinghouse for sustainability projects, it facilitates the idea generation process by communicating corporate sustainability priorities throughout the organization. In turn, the CSO gathers ideas from employees and prioritize projects against corporate sustainability objectives. By serving as a central repository, the Sustainability Office facilitates a cross-functional approach by encouraging collaboration among various groups with similar or related ideas. Once these ideas have become fully formed projects with sponsors, the CSO prioritizes proposals by performing a portfolio assessment ranking. Through this process, proposed portfolio components are evaluated

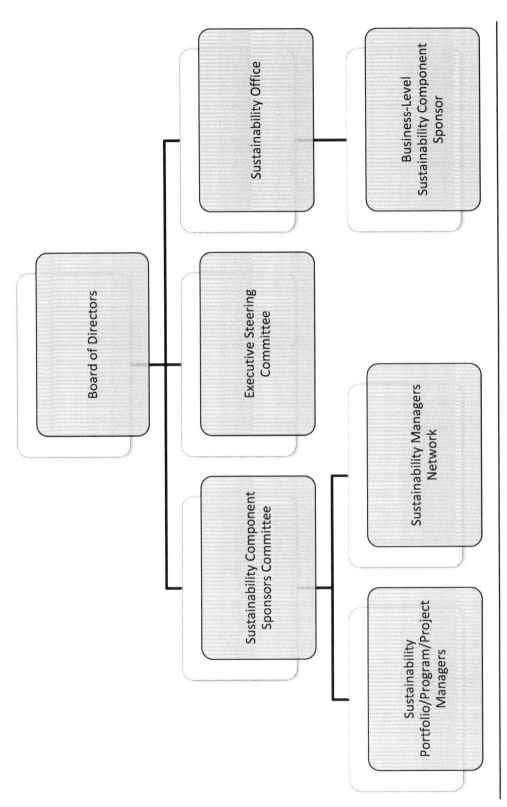

Figure 7.3 An Organizational Structure to Align Sustainable Strategy

against sustainability goals, drivers, and other criteria. The CSO provides an analysis identifying those components that are most likely to deliver strategic benefits.

Further support comes from establishing an Executive Steering Committee, a cross-functional group comprised of senior business leaders, tasked with reviewing portfolio component strategic alignment. Their role is a high-level review of portfolio component status, strategic alignment, and resource allocation. In addition, they review and recommend governance requirements for policies and guidelines to support program and project initiatives.

Create a link between the Sustainability Office and the Executive Steering Committee by having the CSO report on the status of sustainability projects across the organization. This process ensures that projects are monitored regularly and are performing in accordance with projected benefits. In addition, it keeps key business leaders informed of program and project status as well as the business benefits of a sustainable approach.

As the sustainability journey accelerates, the number of sustainability portfolio components submitted for consideration grows. While the increased interest and participation in sustainable strategy means that sustainability is becoming more embedded in to the culture, it also means that more and more projects are vying for limited resources. Creating a portfolio component selection structure and communicating the process and criteria to all participants will result in improved engagement and better alignment with sustainable strategy. Those components that are selected go through modeling and analysis to determine organizational readiness to undertake the project. This process will identify competency requirements, resources, appetite for risk, and a quality assurance assessment. For projects that are ultimately accepted, project execution needs to be tracked and monitored to ensure that costs and benefits are in alignment with the plan.

This portfolio review provides leadership with a tool to monitor the current portfolio inventory against strategic sustainability goals. Evaluating components against sustainability goals and drivers facilitates identifying portfolio components that are misaligned with the current organizational vision. Components that are aligned with sustainable standards will continue, and those that are no longer delivering as planned are realigned or decommissioned. The sustainable portfolio management cycle is illustrated in Figure 7.4. Through this rebalancing process, funding and resources are made available for components that deliver on sustainable benefits.

To further embed the process (refer back to Figure 7.3), consider designating a Sustainability Project Sponsor within each business group who is tasked with implementing the strategic priorities and guidelines from the Executive Steering Committee. This role communicates with the Sustainability Office to ensure that sustainability priorities and standards are included in project plans. To provide a forum for information exchange and support for this function, create a Sustainability Project Sponsors Committee where these project sponsors can meet and share lessons learned and best practices so that portfolio standards are consistent across the organization.

Sustainability Project Sponsors work directly with Sustainability Program/Project Managers, who have operational responsibility for specific portfolios, programs, and projects to promote sustainability strategy within the business unit. The Sustainability Project Sponsor receives portfolio, program, and project updates from the Sustainability Manager and shares best practices and lessons learned from across the organization. Information is gathered and then fed back up the organization to inform and impact Executive Committee strategic and policy decisions.

Sustainability Program and Project Managers establish a network to meet regularly, sharing best practices and lessons learned as well as business functional expertise. This structure embeds strategic sustainability vision into programs and projects throughout the organization. As a result, people, policies, and processes are changed to adopt and promote the sustainability agenda.

This approach has the advantage of providing senior leadership direction, subject-matter expertise, and a network of resources for advice and guidance on implementation of sustainable strategy and specifically program and project selection. It provides a framework for portfolio component management

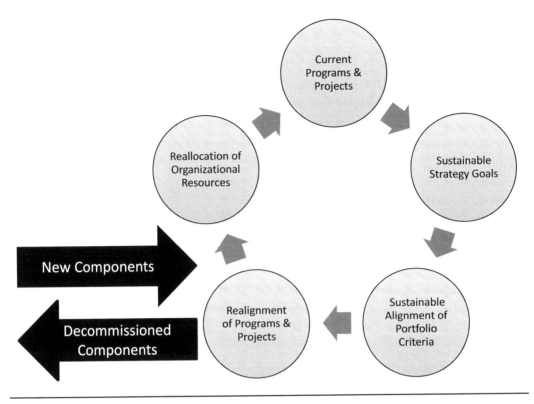

Figure 7.4 Sustainable Portfolio Management Cycle

and resources for project management professionals to tap into for guidance and support. While different organizations may structure themselves differently, sharing the responsibility for sustainable strategy across the organization is vital for long-term sustainability adoption and success. Isolating all sustainability functions within a Sustainability Office segregates those functions from core business processes. If the business functions are not involved, then it is extremely difficult to embed sustainability into the portfolio management process. One of the keys to successful adoption is for all parts of the organization to have responsibility for sustainability initiatives. Creating an organizational structure to promote the link between sustainable and business strategy is crucial for the success of the portfolio component management process.

7.7 Portfolio Component Selection

As sustainability becomes part of an organization's core values, the stream of component proposals will swell. The project portfolio management process focuses on gathering ideas and ranking components (i.e., projects, programs, and subportfolios) based on their alignment with sustainability and business strategies. The first step is to gain a clear understanding of the projects and the proposed benefits in terms of opportunities created or challenges addressed. Creating a component proposal questionnaire is a useful tool to help Sustainability Component Sponsors, the Chief Sustainability Officer, and the Executive Steering Committee better understand requirements and benefits, and the proposal's alignment with sustainability goals and drivers.

The component proposal questionnaire, as illustrated in Table 7.4, requests information on key points that the review committee uses to prioritize proposals. In addition to evaluating the proposal's fit with strategic alignment, it also seeks to identify organization readiness by asking about capabilities, competencies, and resources. There is also a place for the business unit's priority ranking, so that the committee understands how important this project is to the business unit. If the review committee feels that a proposal is not well aligned but a business unit ranks it as a top priority, this suggests that the proposal should be returned to the business unit for more work to better demonstrate strategic alignment. If, however, the committee feels that the proposal is not well aligned and the business unit ranks it as a low priority, it is usually safe to decline the project.

Gathering this information and sharing it with key stakeholders improves both the component selection outcomes and stakeholder engagement. The questionnaire provides a framework for sponsors to gather the requisite information and to evaluate their own proposed component's strategic fit before formal recommendation to the committee. With this structure, a common problem, which is the committee receiving proposals lacking sufficient information for evaluation or clear benefit align-

Table 7.4 Portfolio Component Proposal Questionnaire

Title (Project, Program)	Clarifications and Considerations
Project Type	*Maintenance, Compliance, Enhancement, Transformation*
Benefits Delivered	*Opportunities, Risks Mitigated, Productivity, Cost Savings, Engagement*
Challenges Addressed	
Sustainability Drivers	
Stakeholders	
Stakeholder Requirements	
Sponsor(s)	
Sustainability Goals/Targets	*Identify Areas of Impact*
Business Goals	*Identify Areas of Impact*
Project Impact on Sustainability Mission	*Low to High*
Project Impact on Business Mission	*Low to High*
Business Unit(s) Involved	
Priority Ranking by Business Unit	*Low to High*
Priority Ranking to Sustainability Office	*Low to High*
Budget	
Timeline	
Strengths/Capabilities	*Technical/Subject Matter Expert*
	External Resources
	Management Resources
	Internal Resources
Governance/Compliance	*Policies, Guidelines, Codes*
Employee Impact	*Training, Development, Incentives*
Financial Impact	*Return on Investment (ROI)*
	Net Present Value (NPV)
	Payback Period

ment, can be avoided. Without a clear process, the portfolio review committee often becomes frustrated because they receive a large volume of proposals that do not demonstrate alignment with strategic vision. Because of an insufficient prequalification process, the committee receives too many proposals and lacks the time or resources to provide feedback on every proposal declined. As a result, sponsors become frustrated because they do not understand the reasons that a program or project was declined. Creating a standardized component proposal questionnaire evens the playing field across the organization and allows for a more inclusive and transparent process. As a result, sponsors are encouraged to think of innovative solutions to sustainability challenges within a framework that provides value creation for the organization.

In addition, the questionnaire requests projected returns calculated as both return on investment (ROI) and net present value (NPV.) While ROI is the standard by which most organizations measure project performance, sustainability projects that often have significant upfront costs and long-term payback periods may not meet the ROI hurdle. The NPV approach, which compares the discounted value of the expected future cash flow to the present value of the initial investment, is more consistent with life-cycle costing and often demonstrates that projects will generate significant positive return when considered over the full useful life of the investment.

In calculating proposal returns, consider the social, environmental, and governance impacts of proposals. While ESG values do not appear on traditional income statements and balance sheets, they are important to value in order to calculate the financial return of sustainability programs. Noncash metrics such as GHG emissions, water usage, lives saved, living standard raised, accidents avoided, acres protected, or volunteer hours can be measured. Once a metric has been identified and measured, it becomes possible to identify a translation point to convert the metric into a financial value. As an example, a consulting firm wishing to value the hours spent by employees volunteering may use a translation point equal to a percentage of the firm's average billable rate. Selecting an average value allows the project manager to equalize the valuation for ease of calculation. Monetizing GHG emission reductions can be done through shadow pricing the cost of carbon. (An example is given in Chapter 6.) According to the CDP (formerly the Carbon Disclosure Project), corporations use a rate of $6–$60 per tonne of CO_{2e} to give a tangible value to projects and programs that reduce emissions.[8] In order to incorporate the social, environmental, and governance impacts, select a noncash metric and then monetize that metric to include it in the ROI or NPV calculation.

Another significant consideration is organizational competencies and resources for projects. While projects that support sustainable strategy incorporate standard business operation, some projects require specialty knowledge and subject-matter experts. Part of the portfolio selection process should consider the organization's strengths, weaknesses, and capabilities. If technical or specialty knowledge is required, are those resources available internally, or will external sourcing be required? How will the need for specialty resources impact the budget and timeframe for the project? All of these considerations are part of the component portfolio selection process.

7.8 Types of Components

Most programs and projects fall into one of the following categories: compliance, enhancement, or transformation. Compliance projects are required to comply with regulations, customer requirements, or to maintain operations. While most organizations are familiar with regulatory compliance for environmental or social issues, addressing customer compliance requirements for sustainability may be a new phenomenon for management. Increasingly, customers have supplier sustainability standards, codes, and scorecards to rank their vendors' sustainability programs and compliance protocols. These requirements can range from the source of raw materials to labor standards for plant operations. Complying

Table 7.5 Component Portfolio Selection Matrix

Component Title	Project Benefit	Sustainability Targets	Business Goals	ROI Target	NPV Target	Expertise & Competency	Total Alignment Score
Rating Scale	1=Low 5=High	1=Low 5=High	1=Low 5=High	1=No 3=Yes 5=Exceeds	1=Loss 3=Breakeven 5=Exceeds	1=Low 5=High	Maximum Score=33
Project A	3	4	2	3	3	3	18
Project B	5	3	5	3	3	5	24
Project C	2	3	2	1	5	2	15

with governmental and customer sustainability standards falls under the strategy category of addressing current, internal operations. Components proposed in this category are basic "keep the light on" actions that must be performed in order to keep the business going. They really fall into the same category as required maintenance and facility upgrades,

The enhancement category includes programs and projects designed to improve existing products, processes, or services. From a sustainable strategy perspective, this focus might be either external, looking at the value chain, or internal, looking at new processes or technologies. An example of an externally focused program in this category is improving product stewardship by ensuring that the supply chain is following fair labor standards such as a living wage, worker rights, and safety standards.

The final category of projects is transformative, which usually involve a significant change in how business is conducted. Examples include new platforms on which to develop solutions to address client social and environmental issues. Transformational portfolio components have the most profound impact on sustainability both within the organization and on external stakeholders. They are complex, requiring significant resources and senior management support. If an organization's management is not truly committed to the value proposition of sustainability, this category of portfolio components may appear optional to the Executive Steering Committee. Developing and requiring a robust portfolio management process that highlight the benefits and alignment of sustainable strategy with business value creation significantly improves the success rate of transformative portfolio components. Table 7.5 shows a matrix to facilitate portfolio component selection based on priority ratings of key criteria generating a total portfolio component alignment score.

Once all of the component information has been gathered, the portfolio manager translates the information into a portfolio selection matrix. The matrix has been developed to focus on component benefits, alignment with sustainability and business strategies, and anticipated project returns. Through a process of assigning values to each proposed component, the matrix provides a transparent process for selecting those projects that best align with the organization's strategic goals. Those projects with the best alignment are recommended for review by the selection committee. This approach gives the committee a framework for assessment and efficient asset allocation.

Once the portfolio component selection process is complete, the portfolio manager needs a methodology to track performance and to compare actual performance against projected outcomes. Through this process, programs and projects that are not performing as planned are realigned or decommissioned. Focusing resources on impactful components improves organizational agility in order to reach a sustainable vision. Using a component performance tracking matrix like the one shown in Table 7.6 allows for monitoring actual versus projected benefits, investment returns, rating by both the sponsoring business areas and the sustainability office, as well as gathering lessons learned for continued refinement of the process.

Table 7.6 Portfolio Component Performance Tracking Matrix

Portfolio Component	Projected Benefits	Realized Benefits	Projected ROI	Actual ROI	Lessons Learned	Business Project Rating	Sustainability Office Project Rating	Recommendations
Title 1								
Title 2								
Title 3								

The matrix tracks component performance against the initial proposal and portfolio selection process. Gathering this information provides a resource for portfolio managers as well as the sustainability officers and business leaders to learn from their decision and to adapt the portfolio management process iteratively. Gaining knowledge about contributors and barriers to success facilitates a better project portfolio development and planning process.

7.9 Incorporating Lessons Learned

Moving forward on the sustainability continuum requires a commitment to engage new partners and to innovate and experiment. The reason that project management professional refer to lessons learned is that despite diligent planning there will be unexpected outcomes that are both positive and negative. Part of the sustainability change management process is to empower people to try new solutions and to learn from successes and mistakes so as to build a stronger portfolio management process. Through an iterative process, these lessons learned are incorporated into the portfolio management process and shared throughout the organization in order to build and expand the foundational body of knowledge. As an organization moves forward on its sustainability journey, the level of internal and external expertise and experience will grow, facilitating improved alignment between sustainability and business strategies. In a fully transformed organization, management pursues a sustainable strategy which reflects full alignment of sustainability and business mission and goals.

Lessons learned from sustainability projects cover a broad spectrum, as sustainability impacts all areas of operation. One of the common lessons is taking a long-term view. Much of business is geared toward the short term, such as quarterly performance results that are tracked by investment analysts. Sustainability is a long-term vision, and the organizational changes that need to take place take time to develop and implement. While early initiatives focus on "low-hanging fruit" projects, significant transformations are complex and require significant resource commitments. In the Dow example, which appears in Chapter 3, the management team recommended adopting multiyear sustainability goals as business unit leaders struggled to deliver on annual goals. They found that impactful changes took time to implement and deliver, and that meaningful annual goals were difficult to reach.

Sustainability champions are important change agents. They must have vision and be creative and innovative in their approach. Initially this role is often held by the CSO, but as the sustainability program grows and evolves, success depends on C-suite and business unit champions to support the sustainability portfolio management process. Ideally, building a network of internal supporters at all levels throughout the organization provides the most impactful results. Often sustainability projects require senior leaders to move outside their comfort zone and to learn about new alternatives in order to support progressive projects. The first sizable sustainability project takes tremendous planning, education, vision, and commitment to deliver. Once the first one has been completed, take the lessons learned and adopt them as best practices, making them part of the program management and portfolio selection processes.

Lessons Learned from the Net Energy Neutrality Biogas Project[9]

This project reflects a transformative sustainability vision for the Philadelphia Water Department. It required thinking outside the box, leveraging current capabilities, as well as developing new stakeholder relationships and partnerships. While many parts had to come together for a successful outcome of this project, the ultimate key for success was the committed sustainability champion, who found a way to keep this complex multistakeholder project on time and on budget.

The Philadelphia Water Department (PWD), a water, wastewater, and storm water utility, formed a partnership with Bank of America, the financial partner, and Ameresco, the energy solutions partner, to design, build, and maintain a wastewater biogas-to-energy facility known as the Northeast Water Pollution Control Plant (NEWPCP) Biogas Project, with a goal of achieving net energy neutrality. While the PWD has been a leader in energy efficiency, this project presented a complex legal and financial structure that was outside the PWD area of expertise. Even though the project was forecasted to generate significant cost savings, it was not a top priority in the portfolio selection process because its focus was on energy efficiency rather than core business functionality related to water management. While the project made economic sense from a net present value assessment, it was vying for limited dollars against much-needed water-related capital improvement and infrastructure projects. In order to obtain funding to make this project viable, the sustainability champion had to be innovative.

At the same time that PWD engineers were thinking about the benefits of capturing the biogas created by the waste treatment process to generate energy, the political climate concerning sustainability was changing in Philadelphia. Mayor Michael Nutter became very interested in making Philadelphia more sustainable and launched his goal of making Philadelphia "the Greenest City in the U.S." (More information on Philadelphia's green initiatives is available at http://www.phila.gov/green/index.html.) In addition, deregulation of the energy utilities in Pennsylvania and removal of rate caps in 2011 was expected to have an impact on the PWD's future energy cost. As a result, the timing became right for the PWD to consider energy sourcing alternatives, and the NEWPCP Biogas Project gained leadership's interest and support. However, to move forward with this project, the sustainability champion had to find a way to fund the project without utilizing the traditional funding pool for infrastructure projects. While energy efficiency projects are eligible for an investment tax credit, a creative approach needed to be found to monetize the value of the tax credit, since the PWD is a governmental agency that doesn't pay taxes, Allowing that value to remain untapped would make the project significantly more expensive. The solution was a legal and financing structure never before used by the PWD and required establishing a public–private partnership (PPP) between the PWD, Ameresco, and the Bank of America. The structure was complex, with Bank of America, a taxable entity, owning the facility in order for the project to be eligible for the investment tax credit. Through a Power Purchase Agreement (PPA), the PWD was able to monetize the value of the recaptured biogas. Ameresco provided expertise to build, operate, and maintain the biogas recapture facility.

Undertaking the project demonstrated visionary leadership and commitment to the tenets of sustainability, mainly that projects that protect the environment and the community can also benefit an organization's financial performance. In addition, the project served as a model for other significant energy savings projects for the City of Philadelphia

(City) and its agencies. It offered an opportunity to garner lessons learned and best practices related to working with a public–private partnership. This type of deal structures was unique for the PWD and the City and required the sustainability champion to develop a network of internal sponsors across the organization in order to drive this project forward. He also forged partnerships with academic and governmental agencies to expand the resources available and to promote understanding and cooperation for the project.

In addition to the beneficial environmental impact, the project was expected to produce significant energy savings for the city. In terms of risk management, it would mitigate electricity price volatility and reduce peak grid capacity usage. It also would improve air quality in the surrounding neighborhood, making the PWD a better neighbor and significantly reducing the facilities GHG emissions, in line with Mayor Nutter's vision of a greener Philadelphia.

The NEWPCP Biogas Project was budgeted at $47.5 million for construction. Benefits from the project were expected to be a reduction in energy costs, a green source of energy, reduced GHG emissions, and improved community relations. The anticipated PDW energy cost savings was in excess of $12 million during the contract period (16 years), with additional savings accruing beyond this period. Biogas recapture provides a source of green energy for the PDW by capturing and utilizing as a source of energy the biogas that is currently being flared.[10] Eliminating the flaring process would significantly reduce GHG emissions and improve the odor level in the surrounding neighborhood.

In late 2013, the project went live. The facility was projected to generate 5.6 megawatts of energy for the NEWPCP while reducing the plant's carbon emissions by 22,000 tonnes per year. According to Philadelphia Mayor Nutter, "That's the equivalent of removing 4833 cars off the roads or planting 5390 acres of pine forests."[11] Paul Kohl, PE, Philadelphia Water Department, who served as the sustainability champion and project manager, reports that in the first year of operation, the facility has been able to deliver in terms of energy generation and cost savings. While the road to project success has had many detours and obstacles, the final project has delivered on the promised benefits of reduced GHG emissions, energy savings, and improved community relations. Keeping an eye on the vision and the related environmental, social, and economic benefits ultimately allowed this project to be completed successfully.

Lessons Learned

- **Take a long-term view.** In order to become a resource recovery facility, one must demonstrate a commitment to a set of long-term goals that call for sustainable development and growth that are similar to the energy net-zero goals. Environmental stewardship has an implied eye toward resource recovery.
- **Have a champion and build a network of internal advocates.** Successful energy programs and successful projects have at their heart a sense of ownership and entrepreneurial skill. "If there's nobody championing it, it's not going to get done."[12]
- **Develop a strategic energy plan with metrics.** Select metrics to measure indicators that track strategic alignment with the energy plan. Include tangible measures such as energy use key performance indicators (KPIs) and energy conservation metrics. Consider a portfolio management approach to energy management, which considers cost-effectiveness and function to reduce exposure to volatile markets and mitigate risk. The premium paid for alternative or renewable energy development can be seen as

a risk mitigation expense similar to paying insurance premiums. When energy prices become volatile, the price of protection is built into the energy portfolio through the use of alternative sources as an energy hedge
- **Identify responsibilities and expectations.** Energy use cuts across all departments and agency functions, from administration to operations. Responsibilities must be clearly delineated and delegated. Continued communication is a priority.
- **Innovate and lead.** Create governance processes and structure that promotes an environment for innovation, in which effort is rewarded and mistakes are accepted as part of the learning process. Reward can be a difficult concept for many utilities, but allowing someone to risk and fail is a great reward system—do not under estimate the positive impact. Be willing to acknowledge multiple drivers for sustainability initiatives. Good energy management can be simply that, good management; a green initiative does not have to be achieved by your traditional green person or visa versa. Merging the two can generate valuable economic, social, and environmental benefits.
- **Build stakeholder partnerships with academic institutions.** Developing a relationship with an academic institution facilitates stakeholder engagement and project outcome. Connect with researchers at an academic institution that specializes in your area of focus. Share information in order to develop the most efficient and value-added energy project. The combination of practical know-how coupled with a vision of what is possible, and creativity, can result in innovative solutions and even disruptive technological advancements. Jointly promote the project by sharing technology advancement, best practices, and lessons learned to promote other stakeholder engagement.
- **Leverage existing resources.** Use available experts and resources to understand energy efficiency and recovery opportunities. Take advantage of energy management resources at state agencies (NYSERDA, Wisconsin Focus on Energy), USEPA, DOE, and use WERF, WEF, and Water RF studies and tools. Sharing information between agencies and working together on policy matters is critical. Make connections. Tour similar facilities and engage in discussions with their operations staff in order to develop a foundational body of knowledge.
- **Engage with internal and external stakeholders.** From a public utility perspective, plan to educate politicians, board members, and the community. Your board and/or authority leadership need to be educated about the risks and the benefits of co-generation and the energy financing options available to achieve this goal. For most authorities, this is new territory, and the stakeholder management process needs to be carefully planned and implemented to achieve success. Engage with the community and the ratepayers on the utilities board. Plan local outreach to communicate the benefits and risks of the project and to addresses community concerns. Use consistent internal and external messaging about the green impacts of the project, and be visible and transparent in the process.
- **Manage the project to promote benefits.** Benefits include the positive impacts of the project on the utility, community, the city, and the environment, including:
  - Improved community air quality
  - Diversion of organics from landfill disposal to recovery
  - Reduction in miles driven by transporters of wastes, which reduces fossil fuel consumption and greenhouse gas emissions

- Convert waste into value by recapturing rather than flaring biogas (methane)
- Reduce the demand on the local power plant, which protects the grid and reduces the overall cost of electrical supply.
- **Net energy neutrality.** The advantages of an energy-neutral project go beyond energy efficiency and include operational risk mitigation. In the event of a grid power failure, the facility produces its own power and continues to treat wastewater "in island mode."
- **Training and education of staff.** Converting a waste treatment facility into a power generation facility requires different operator skill and attention to details to sell electricity on the market. Be prepared to spend time and money explaining the benefits, creating incentives, and developing the skill level to manage this increased operational complexity.
- **Explore alternative funding options.** In order to complete this project, both the legal and financial structure required outside-the-box thinking. Take advantage of outside sources of capital funds. Both the state and federal governments offer programs. A good source of currently available programs can be found through the DOE and NYSERA. Energy Service Companies (ESCOs) also help to identify and evaluate savings, develop engineering designs and specs, manage projects, arrange for financing, train staff, and offer guaranteed savings to cover costs. Consider alternative investors; many entities are ready to help launch projects that will create energy savings.
- **Life-cycle approach.** Consider the full project life cycle to determine benefit and strategic alignment. Include financial metrics, such as net present value (NPV), and compare the project to the status quo to make business case decisions on energy projects. While the NPV approach may not fully capture all of the social and environmental costs of capital, it is preferred to a ROI analysis.
- **Small wins matter.** Invest in actions that produce smaller-scale savings, such as lighting contracts, zone controls, and motion controls. These small wins have an impact and get people's attention and interest.
- **Designate a utility "rate geek."** Utility bills have a language all their own. Designate an individual to understand energy bills (electric, natural gas, fuel oil, steam, etc.) and rates in order to take advantage of cost saving opportunities and to develop a cooperative relationship with your power utility.
- **Value of waste.** Once value is identified in a waste stream, anticipate that others in the organization will change their view of this waste stream from a liability to an asset. Behavior and attitudes will change relative to this previously valueless waste stream.

Future Components and Benefits

As future energy efficiency projects are undertaken, the lessons learned from this project will serve as benchmarks and standards for inclusion and consideration in the project plan. While the use of an ESCO to fund the project was a new idea for Philadelphia, we are already seeing other projects adopt this structure because of the ability to make capital improvements without relying on the city's limited capital improvement budget.

Many of these lessons learned are making their way into the component selection process across the organization. From this experience, stakeholders are educated and informed and the PWD is prepared to address these types of opportunities in the future. Some of the most impactful lessons are to have a champion, think long-term, and be innovative in your thought process. Without these core drivers, the PWD project would not have been realized.

7.10 Conclusion

Each organization develops its sustainability vision and undertakes its own portfolio analysis, identifying internal strengths and weakness as well as additional requirements for capabilities and resources. Developing programs and projects to fill these gaps is part of the portfolio management process. Transforming an organization into a sustainable organization is an iterative process. Gathering information on programs and projects and documenting the lessons learned to create a body of knowledge allows for the development of best practices and project portfolio standards. Developing an organizational governance and management structure to support sustainability facilitates the portfolio management process. The sustainability officer, or designated portfolio manager, is able to embed lessons learned and portfolio criteria into project proposal, portfolio selection, and management processes. Creating a portfolio process aligns components and organizational resources, moving an organization forward along its journey toward sustainability.

The skill set required to create a portfolio strategy for sustainability projects aligns with a portfolio manager's capabilities and experience. Drawing on this body of knowledge to develop organizational criteria for component selection helps to drive an organization toward its sustainability vision. The process of identifying beneficial components, aligning components with organizational strategy and goals, resource requirements, and identifying gaps in components, benefits realization, and resources provides the framework for successful component planning and implementation. In addition, clear identification of organizational strategic priorities provides a platform for issues on which to engage internal and external stakeholders. Using portfolio management methodology to develop a body of knowledge for sustainability provides a resource for portfolio, program, and project managers so that lessons learned and best practices can be incorporated into future components. Through this iterative process, the organization's expertise and experience grows, allowing it to move from addressing internal present-day challenges to focusing on external and future opportunities to leverage the value proposition of sustainability.

Notes

[1] Stuart Hart, "Sustainable Value," accessed January 12, 2015, http://www.stuartlhart.com/sustainablevalue.html.
[2] Ibid.
[3] Stuart Hart, "Beyond Greening: Strategies for a Sustainable World," *Harvard Business Review*, January–February (1997): 66–76.
[4] John Viera and Ken Washington, "How to Scale Sustainability at a Global Level: Ford Motor Company Approach," Webinar, June 17, 2014.
[5] Orange, "Corporate Social Responsibility 2013 Complete Report," 2013, http://www.orange.com/en/content/download/23330/480379/version/3/file/Orange_2013_CSR_report.pdf.
[6] Ibid.
[7] Ibid.
[8] CDP, *Use of Internal Carbon Price by Companies as Incentive and Strategic Planning Tool*, 12/13, https://www.cdp.net/CDPResults/companies-carbon-pricing-2013.pdf.
[9] Paul Kohl, "Lessons Learned from a Combined Heat and Power Biogas Capture Project," Interview, 12/5/14.
[10] Ameresco, "Ameresco and Philadelphia Water Department Announce Northeast Water Pollution Control Plant Biogas Project," accessed March 23, 2015, http://www.ameresco.com/press/ameresco-and-philadelphia-water-department-announce-northeast-water-pollution-control-plant.
[11] "Phila. Water Department's Biogas Electric Plant Goes on Line," accessed March 23, 2015, http://philadelphia.cbslocal.com/2013/11/22/phila-water-departments-biogas-electric-plant-goes-on-line.
[12] Kohl, "Lessons Learned from a Combined Heat and Power Biogas Capture Project."

Chapter 8
Identifying and Engaging External Stakeholders

Stakeholder engagement is a crucial component of an organization's sustainability strategy. Stakeholders fall into two groups, internal and external stakeholders. Project management professionals are more familiar with engaging with internal stakeholders. External stakeholders, however, can have a significant impact on an organization's long-term viability and operations. Through the process of stakeholder identification, dialogue, and feedback, management gains insight into the material issues to consider in their sustainability vision. Organizations at the beginning of their sustainability journey focus on stakeholder identification, needs assessment, and requirements gathering. As the data are incorporated into the strategic process, these requirements inform sustainability goals and targets as well as programs and projects designed to support these goals and targets. The stakeholder engagement process is iterative, with feedback being provided to stakeholders on the impact of their recommendations on organizational strategies, policies, and programs. As an organization's experience and maturity with sustainability grows, stakeholder engagement leads to an ongoing relationship between the organization and its stakeholders. Trust between the organization and its stakeholders develops over time as management demonstrates that stakeholders' recommendations have been heard, evaluated, and considered seriously for incorporation into their sustainability program.

Incorporating stakeholder recommendations into an organization's sustainability vision and strategy involves changing organizational priorities, processes, and procedures. In order to effect these changes, programs and projects are created to integrate material sustainability goals into core operations. As program and project managers drive these initiatives, they need to understand fully the impact that stakeholders have on the success of sustainability programs and projects. Moving through the stakeholder engagement process for sustainability programs, the variety of stakeholders that need to be considered in the stakeholder engagement plan is greater and more diverse than in typical programs. Sustainability champions and project management professionals must ensure that sufficient time, budget, and resources are being allocated to stakeholder engagement. As an organization becomes more experienced in sustainability, portfolio management criteria incorporate stakeholder requirements, embedding sustainability further into the organization. Stakeholders create both positive and negative impacts on organizations. The key to successful stakeholder engagement lies in the quality and authenticity of the stakeholder management plan.

8.1 Impact of Engagement

Engaging with stakeholders provides a broader perspective to management and often highlights risks and opportunities that have not been previously identified. Visioning with key stakeholders facilitates creating a long-term strategic sustainability and business strategy. The process provides management with a roadmap to develop new business opportunities, improve processes, generate cost savings, and improve risk mitigation. Stakeholder engagement benefits include:

STAKE HOLDER ENGAGEM

1. New partnerships
   a. Joint ventures
   b. Access to new markets
   c. New opportunities
2. Additional products and services
   a. New products, services, and solutions for ESG-related challenges
   b. Solutions for underserved markets
   c. Identification of unique opportunities
3. Risk mitigation
   a. Operational risk
   b. Reputation and brand risk
   c. Compliance risk
   d. Strategic risk
   e. Financial risk

Through stakeholder engagement, management seeks to maximize opportunity and minimize risk. While organizations often focus on the risk mitigation aspect of stakeholder engagement, the greatest organizational value add comes from identifying opportunities for new products, services, and sustainable solutions for both existing and new markets. These include ideas such as new processes that reduce waste and cost, partnerships and joint ventures to tackle global environmental, health, and human rights challenges, and new products and services to solve client-driven sustainability issues or meet the needs of underserved markets. Adopting sustainable practices and building community relations creates business opportunities for all sizes of business.

> **Irv & Shelly's Fresh Picks Engages with Diverse Stakeholders to Create New Opportunities[1]**
>
> Opportunities created through stakeholder engagement aren't just for large multinationals. Small and mid-size businesses can benefit from effective stakeholder engagement as well.
>
> From the very beginning of their business, stakeholder engagement—especially community partnerships—were a priority for Irv and Shelly. Stakeholder engagement is part of their mission statement, which includes supporting local farmers and providing high-quality organic products for customers. Their mission is to make great, healthy food available to everyone through locally sourced farmers and suppliers. They bring their customers the freshest organic and natural foods in the most sustainable way. Their initial inspiration was born when they met at MIT. They dreamed of creating a socially responsible business based on a triple bottom line approach—healthier people and communities, healthier planet, and healthier profits.
>
> The inspiration for Irv & Shelly's Fresh Picks came from the missing link between farmers who wanted to spend their time growing high-quality and healthy organic food

and consumers who needed an easier, more convenient method to get fresh, healthy food year-round. So, Irv & Shelly's Fresh Picks was launched to provide a better food delivery system between urban areas such as Chicago and Milwaukee and the surrounding farms. In addition, their business model promotes a healthier environment by reducing GHG emissions related to food distribution. According to their research, the average travel distance for large factory farm products is 1500 miles. Their home delivery approach further reduces GHG emissions by curtailing grocery store traffic. Their metrics indicate that they save 50–100 car trips a day with each van delivery. They promote eliminating pesticides, antibiotics, and hormones in their products, as these contaminants pollute local habitats and wildlife.

In addition, they have been a strong community partner, creating joint opportunities with community groups, food pantries, and schools to improve access to healthy food and to promote sustainable lifestyles. Their business structure supports the local economy by providing opportunities for farm families and local workers. Through their stakeholder engagement and sustainable policies, they have developed a strong local reputation.

As a result of their reputation, they were approached by Jim LoBianco, Executive Director of StreetWise, and Ken Waagner, founder of e.a.t. (education, agriculture, and technology), with an opportunity to provide organic and locally sourced food products for a new venture in downtown Chicago. StreetWise, a workforce development agency, is best known for assisting men and women at risk for becoming homeless. The business model is based on selling Streetwise Magazine, one of the largest "street papers" in the United States, in the Chicago area. The magazine covers topics that are socially conscious and Chicago-oriented. As a result of their service to the community, Lo Bianco of StreetWise was offered an opportunity to use vacant newsstands to sell their magazine. As a business model, he realized that selling periodicals from a newsstand was obsolete. He began looking for an innovative and creative new venture to house in the newsstands. The goal was to find a venture that would provide stable employment for the StreetWise workforce. His organization had some early success partnering with Ken Waagner, of e.a.t., offering produce via mobile carts. Lo Bianco and Waagner approached the city with the idea of leasing the newsstands in order to offer reasonably priced healthy and nutritious prepared foods to workers and residents in the neighborhoods where the newsstands were located.

Creating a partnership with a diverse group of stakeholders provided economic, environmental, and social value for this community. The venture provided employment for the StreetWise workforce, additional demand for local farmers, and healthy food options for the community in an otherwise urban food desert.[2]

This business model brings diverse stakeholders together to deliver a sustainable product, providing jobs to the local community, and meeting the needs of an underserved population. The e.a.t. organization prepares the food and operates the newsstand locations. StreetWise provides the training and development of the workforce. Irv & Shelly's Fresh Picks handles the food supply and delivery. It is a triple bottom line win for the partners and the stakeholders. The previously vacant newsstands are now serving reasonably priced, healthy lunches in the downtown Loop area of Chicago.

Another area of stakeholder engagement benefit is risk mitigation. These benefits include reducing strategic risk in terms of meeting competitive challenges, industry trends, customer needs, and maintaining a license not only to operate but also to enter new markets. Stakeholders' perceptions impact compliance and operational risk, including community acceptance of facilities, supply chain compliance, and employees' actions and perceptions. Regulators, communities, and customers are examples of groups that make demands on organizations and impact operational and reputational risk management.

These risk components combine to impact financial risk as well. While traditional financial risk includes interest rates, exchange rates, liquidity, and cash flow, stakeholders' actions can impact access to capital from investors and financial institutions. They can also impact project timelines and continued operating privileges. Lapses in stakeholder engagement through lack of dialogue, transparency, and meaningful engagement impacts the success of ongoing operations and new ventures as well as the ability of management to maximize opportunities and minimize risks.

Research conducted by professors from the Wharton School of the University of Pennsylvania, New York University, and the University of South Carolina found empirical evidence that increasing stakeholder engagement improves the financial valuation of an organization. Their analysis of 26 gold mines, which were owned by 19 publicly traded companies, found that the stock market valuation discounted the firm's stock price up to 72% because of lack of transparency about local community and governmental engagement.[3] Uncertainties about the impact of governments, regulators, community leaders, and civil society on mine openings, continued operations, and potential cost overruns were being built into the valuations. By tracking over 50,000 stakeholder engagement events through media sources, they developed an algorithm to measure the degree of stakeholder conflict or cooperation at these mines. Introducing stakeholder engagement measures into a market capitalization analysis, they were able to reduce the discount placed on these mining firms between 13% and 37%. Two mines with the same gold deposits, cost of extraction, and global gold prices varied in valuation because one of the mines had local stakeholder support and the other did not.[4]

Mining operations are long-term, capital-intensive investments. Because of the long-term investment horizon and sizable initial capital required, the risk associated with local stakeholder disruption is significant. At any point in the lengthy process of developing, opening, and operating a mine, concerned local stakeholders can wield their significant power over the continuation and the ultimate success of the mining project. For any capital-intensive project that requires significant up-front investment, project managers should consider project parameters such as capital intensity, timeframe, equipment requirements, and the life cycle of the project in order to better assess the risks and opportunities presented by stakeholders. As part of the risk assessment and response, consider the impact of disruption from local citizens and governments and include a robust stakeholder engagement plan as part of the project planning process.

In order for stakeholder engagement to be effective, it must be transparent and authentic. If stakeholders feel that they are not valued and treated with integrity, they often take action against an organization. This action can take the form of social media campaigns, community activism against projects, or even job-site disruptions. The impact of stakeholder actions can damage a firm's reputation, valuation, and opportunities for continued business success. The negative impacts may be long-lasting and significant, including:

1. Flight of quality investors
2. Allocation of time and resources to investigation and response to an incident
3. Reputation damage to C-Suite leadership
4. Potential for regulatory response toward the organization and the industry
5. Negative customer response and loss of business
6. Negative impact on retention and recruitment of employees
7. Social media impact and media scrutiny
8. Loss of license to operate
9. Loss of stock valuation
10. Shareholder resolutions[5]

Choosing not to engage with stakeholders has risks and repercussions as well. If organizations do not create a structure to engage with key stakeholder groups, these groups may take it upon themselves to engage with the organization's senior leaders at shareholder meetings. "Socially responsible investors" (SRIs) have pursued this route, introducing shareholder resolutions. Their actions have generated significant impacts on global organizations' policies and practices. One SRI group that has had a sizable impact through shareholder resolutions and senior management meetings is the Interfaith Center on Corporate Responsibility (ICCR). Their mission is to build a more just and sustainable world through management of their investments as a tool to create social change. Their staff and members engage multinational leadership on environmental, social, and governance issues to create a better future for investors, employees, and customers.[6] The ICCR has moved beyond simply filing shareholder resolutions and actively seeks to engage with senior management on sustainability issues. Some major accomplishments include ICCR's engagement with Walmart between 1998 and 2007, which led to changes in equal employment opportunity rights. More recently, they have been part of a broad coalition addressing labor and pay practices at Walmart, which announced in 2015 that it would raise base pay rate to $9/hour, and to $10/hour in 2016. Currently, ICCR's priority issues are corporate governance, health care and health insurance, environment especially global warming, the financial services industry in terms of risk management and responsible lending, food access and quality, global health issues, human rights, water rights, and corporate water impacts.

Research suggests that through this process of engagement, which can be contentious, a dialogue emerges that raises awareness, reframes issues, and builds partnerships in order to address significant sustainability challenges. Positive impacts include the role of activists in shifting major corporate policy away from policies that are damaging to sustainability initiatives and toward more favorable policies. Examples include Ford Motor Company's change from supporting climate change propaganda to becoming a champion of sustainability in the automotive sector, and Merck & Co.'s change of policy on generic anti-HIV drugs in South Africa. By raising awareness, building networks and coalitions with an organization, and reframing the issues in terms of business impact, external stakeholders can have a profound effect on organizational strategy and policy.[7]

8.2 Creating an Engagement Plan

Begin the process by creating a stakeholder map and engagement plan. Before considering whom to engage, identify issues on which management wishes to engage stakeholders. What are the environmental, social, and governance challenges and opportunities facing the organization? Use tools such as the SASB materiality matrix discussed in Chapter 4 to identify material issues for your industry. Consider peers' and competitors' sustainability challenges as well as issues and concerns raised by stakeholder groups that have reached out to the organization. While the list of issues is not static, project management professionals need a starting point to begin the engagement process.

Once the issues have been identified, consider what the organization is hoping to achieve through the stakeholder engagement process. Is the objective to gather information? Is management seeking stakeholder recommendations? Is the purpose to receive feedback on existing sustainability programs? Is feedback being solicited on new programs and projects? Once the objective of the engagement has been determined, consider the organizational readiness to hear and respond to stakeholder feedback. Does the organization have a means and structure for the engagement process? Is leadership willing, ready, and able to hear recommendations and incorporate them into organizational strategy? The project management process benefits from taking the time to create a stakeholder engagement plan.

While responses from stakeholders may be different from original expectations, going through the planning process helps project management professionals, sponsors, and internal stakeholders such as

Table 8.1 Stakeholder Engagement Plan

Plan	What are our issues about which we are seeking stakeholder engagement? What groups of stakeholders should be considered? What is our objective for stakeholder engagement? How is it defined? What type of engagement format will best suit our objective? Who will be accountable for the engagement process? What metrics will be used to measure success? How will stakeholders be evaluated for initial and ongoing inclusion? How will we rank stakeholders on impact and influence?
Do	How and when should the invitation be extended to stakeholders for engagement? What are the ground rules for engagement? What issues/objectives might be good starting points? What format should be used for the engagement sessions? What resources, materials, training are needed in this process? Is a third-party facilitator needed? Is third-party verification or audit required?
Check	Did we meet our objectives? Do we need additional sessions? Was the process helpful, or are process changes needed? Do we have the correct stakeholders in the room? Is the right person/group assigned to own key stakeholders?
Act	How will the stakeholder requirements be communicated to key decision makers? How will the impact of their contribution be demonstrated to stakeholders? How will ongoing stakeholder engagements be modified to maximize effectiveness? What is the best frequency and form for ongoing communication with each of the stakeholder groups? Do we need to add any stakeholder groups? Do we need to realign any stakeholder relationships?

the C-suite and management better understand the likely outcomes and expected impacts from the external stakeholder engagement process. If management is not willing to consider external stakeholder input, then entering into an engagement process is ill-advised. Gathering recommendations and requirements but then ignoring them creates stakeholder discontent. The project plan should include a communication vehicle to update both internal and external stakeholders on strategy, development, progress toward goals and targets, and final outcomes. As the stakeholder engagement process unfolds, the organization is developing a chapter in its sustainability story. The shared experiences created through this engagement become stories to share with future stakeholders. The organization's sustainability story should demonstrate how stakeholder feedback was captured and reflected in its sustainable strategy through related project plans and actions. Table 8.1 suggests specific questions to facilitate the stakeholder engagement plan development. Then it poses considerations for execution, and finally, assessment. The plan also includes incorporating lessons learned back into the engagement process, promoting continuous improvement.

Creating a stakeholder engagement plan provides project management professionals with a roadmap to manage a diverse set of external stakeholders effectively. While lessons will be learned and modifications made to the plan, it provides a means to engage authentically with key stakeholder groups and to ensure that they feel they are being heard and that their input is impactful to the organization.

8.3 Identifying Stakeholders and Collecting Requirements

Project management stakeholder engagement usually focuses on internal stakeholders engaging with the sponsor, project team, and business units, with specific client or supplier projects engaging external

Identifying and Engaging External Stakeholders 171

customers or vendors. In the context of sustainability programs and projects, stakeholder management requires engaging with a more diverse and dispersed group. For many organizations, management has never engaged with non-governmental organizations (NGOs), environmental groups, community groups, or even government agencies in this manner. Identification of stakeholders begins with outlining key sustainability issues and considering groups or individuals that are most closely aligned with those issues. Often, management has been approached by some of these groups in the past. The key for consideration is who can impact our industry, and whom do we impact? Figure 8.1 is an example of a Stakeholder Engagement Map. This example includes both internal and external stakeholder examples. Internal stakeholders are the C-suite, management, employees, owners, and investors. External stakeholders are divided into three main groups: regulators, customers, and resources.

Stakeholders vary by organization, industry, and country, but core stakeholder groups include employees, investors, unions, customers, community members, trade groups, regulators, suppliers, and NGOs. Each of these groups has an interest or "stake" in the outcome of a project, program, or organization. As a first step, look at existing stakeholder groups and current engagement activities. How can these relationships be leveraged to develop or improve the organization's sustainable strategy? Who should be added to this group in order to make it a more diverse and effective group of key stakeholders? Stakeholders will have different levels of contributions based on their expertise, experience, and willingness to engage. Certain stakeholders will have a higher degree of influence on your organization and must be engaged to keep programs and projects on track. Organizations may want to engage others

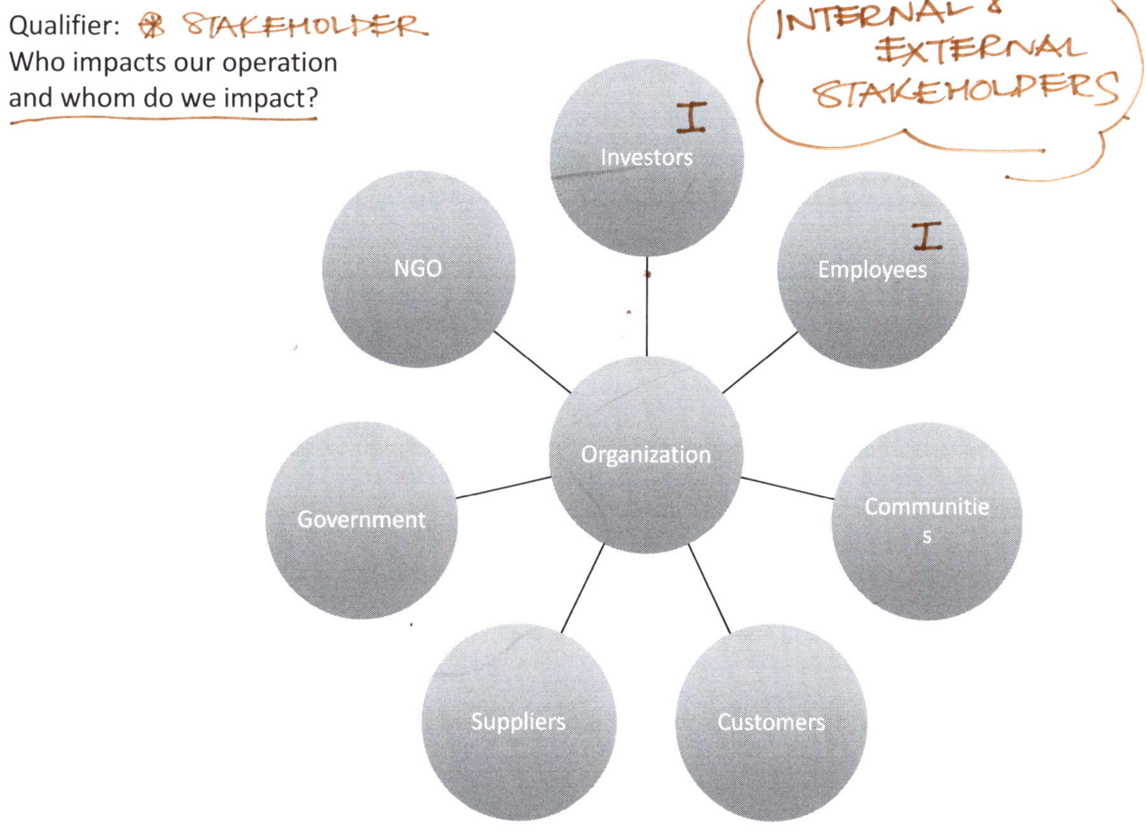

Qualifier: **STAKEHOLDER**
Who impacts our operation
and whom do we impact?

INTERNAL & EXTERNAL STAKEHOLDERS

Figure 8.1 Identifying Stakeholders

because of their unique perspectives related to target markets, technologies, regulations, or potential for partnership. The opinions of the stakeholder groups may vary, and the priorities may not align with each other or with the organization's priorities. A best practice is to be open, honest, and direct with regard to the organization's goals, the intentions for the engagement, and the planned use of the stakeholder feedback.

The Dow Chemical experience discussed in Chapter 3 speaks to some of the challenges of engaging diverse stakeholders. In their Sustainability External Advisory Council (SEAC), Dow provides an opportunity for a diverse group of key stakeholders to interact with leadership. The SEAC group includes academics, NGOs, regulators, and representatives of other businesses. They all have their own requirements and unique perspectives. When management entered their first SEAC meeting thinking that they had done a good job of addressing sustainability challenges facing Dow, some of the SEAC stakeholders had a different perspective. They delivered a challenge to management that necessitated rethinking Dow's strategy about sustainability issues.

The size, location, duration, and nature of a project will impact who should be included as stakeholders. If an individual, group, or community believes that they are impacted, then they see themselves as stakeholders. As such, they need to be considered and engaged as stakeholders. Stakeholders have a significant impact on the success of both sustainability and traditional programs and projects. Stakeholder engagement is an important step in the process of identifying and ranking material issues and establishing an organization's sustainability vision and plan.

8.4 Engaging Stakeholders to Identify Needs and Define Requirements

Engaging stakeholders is about getting the "voice of the customer," or in this case the stakeholder, into the process. In order to accomplish this task, ask stakeholders what issues, concerns, or opportunities need to be addressed by the organization's sustainability program in order to meet their organization's needs. The objective is to identify the stakeholder requirements that your organization's sustainability program should address for that group. While stakeholder groups may have overlapping priorities, each will have some unique requirements. Customer responses may be focused on needs for more environmentally friendly products or services. NGOs may request human rights or environmental standards concerning issues such as a living wage or GHG emission reductions. Internal stakeholders such as employees and management may require that sustainability goals align with key business targets.

Once a list of stakeholder needs and requirements has been formulated, circle back to the stakeholders and verify that your understanding is their intention. Moving through this process, tools such as

Table 8.2 Stakeholder Requirements Matrix[8]

Stakeholder	Group	Requirements	Strategy
Save the Lake	Community	No impact on lake water levels and biodiversity.	Establish net zero water target. Report on water usage versus target.
VP of Sales	Employee	ESG solutions and products for customers.	Report on percent of products meeting ESG criteria.
CFO	Employee	ROI 10%.	Report on business case quarterly.
XYZ Co.	Customer	Develop products with a lower water footprint.	Issue reports on products' water footprints.
Friends of the Fish	Community	Promote indigenous fish in the lake.	Partner with NGO to develop species baseline and ongoing sampling.

interviews and surveys facilitate validation and help to quantify targets and goals with the stakeholder population. Once the stakeholder requirements are understood, align them with organizational strategic initiatives. The degree of alignment determines organizational readiness in terms of mission, scope, and resources to engage with stakeholders on these issues. From a program or project management perspective, requirements that fall outside the current strategic mission are referred to senior management or the executive steering committee overseeing the project. As part of the stakeholder engagement plan, include a mechanism to address stakeholder requirements that do not fall within current sustainable strategy. These items may become portfolio requirements for future project selection, or they may be tabled as not relevant to the organization.

Unexpected issues may arise from this process. The designated leadership group can then identify how this input will be incorporated into the sustainability plan and over what timeframe. Using a matrix to identify key stakeholders groups, their requirements, and the corresponding organizational strategy serves as a useful planning and communication tool. An example matrix is shown in Table 8.2.

Developing a stakeholder requirements matrix facilitates identifying key stakeholders and their priorities as well as providing mapping to the organization's sustainable strategy plan development. It facilitates selecting strategic sustainability targets best aligned with these requirements. It allows for identification of requirements that are not currently being addressed through sustainability programs and projects, providing opportunities to identify how to strategically address these concerns.

8.5 The Stakeholder Engagement Process

Once stakeholders have been identified, needs have been gathered, and requirements have been detailed, it is time to incorporate stakeholder input into organizational vision. The stakeholder engagement process is summarized in Figure 8.2.

One of the keys to building the relationship is to demonstrate that the project team is listening to the stakeholders and incorporating their requirements into the sustainability goals of the organization. If stakeholders spend the time to meet and engage with the project team, they want to see that their key requirements have been heard, understood, and incorporated into the sustainability process. Translating this input into SMART goals (see Chapter 6) for the organization facilitates communication, accountability, measurement, and transparency with both internal and external stakeholders. From a management and employee perspective, these goals provide clear and measurable targets for success. From an external stakeholder perspective, they offer a clear metric on which to assess an organization's progress. While leadership is sometimes hesitant to commit to clearly defined and measurable sustainable goals, doing so increases the credibility of the organization's sustainability program and performance with both internal and external stakeholders. Clarity of goals identifies areas for changes in policies and processes to support sustainability initiatives and strategies. If an organization wants to improve ethics compliance, leadership will need to support this goal with an action plan that includes making sure that all employees have read and are trained in the code of conduct. From a policy perspective, there needs to be a code of conduct and a requirement that all employees be trained on it. In addition, clarity of goals and expectations facilitates development of sustainability programs and projects to create new products and services, conserve resources, and develop new solutions and efficiencies. The process is iterative, with the sustainability team and stakeholder relationship managers reporting back to stakeholders on organizational changes based on stakeholder feedback. The engagement process then begins again, allowing for further refinement and improvement in the organization's sustainable strategy.

EMC Corporation, a technology firm specializing in data storage, information security, and cloud-based solutions, engages with stakeholders in order to encourage dialogue about emerging trends that are of interest to stakeholders, build trust, and gain perspective from their expertise. The purpose is to validate stakeholders' requirements and to communicate changes in EMC's policies and processes

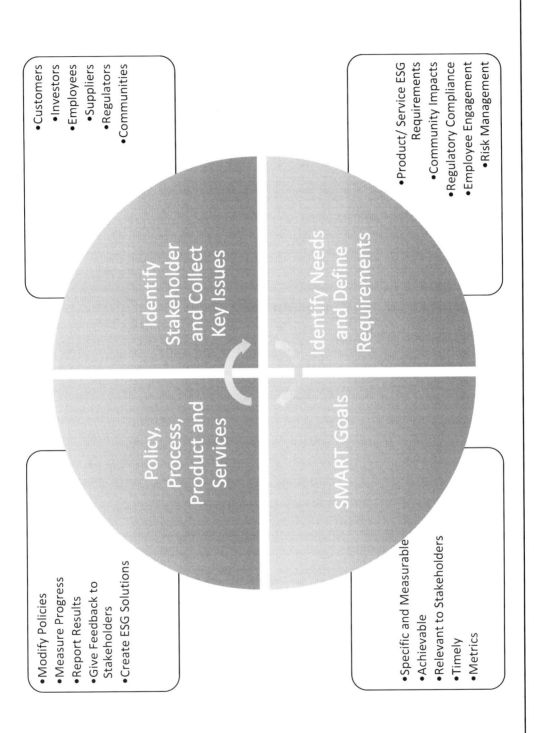

Figure 8.2 Stakeholder Engagement Process

to address previously raised stakeholder issues. In 2013, EMC conducted two stakeholder engagement events in conjunction with a third-party facilitator, Ceres; EMC participants included subject-matter experts, sustainability professionals, managers, and executives. Stakeholder participants included consultants, NGOs, socially responsible investment analysts, and supply chain auditors. Two different events were held. The first was an in-person forum created to address materiality, reporting, eWaste, diversity and inclusion, supply chain, energy usage, and carbon. The second was structured as a teleconference with a concentration on creating employee engagement through sustainability.[9]

Topics covered at the two separate events included:

Forum Topics[10]

Forum I
1. Communication of material issues
2. Incorporating sustainability into risk management
3. Improve transparency by disclosing key suppliers and diversity information
4. Creating productivity metrics for electronic waste
5. Tying executive compensation to sustainability goals

Forum II
1. Goals and metrics to measure employee engagement on sustainability
2. Integration of sustainability into daily operations and incentives to support goals
3. Communication strategy to enhance employee knowledge of sustainability
4. Communication strategy for customers in line with EMC's sustainability goals
5. Communication strategy for value chain to provide leadership and promote collaboration.

Feedback from these engagements provided guidance on creating EMC's 2013 Sustainability Reports and future sustainability practices and priorities. In addition, the Compensation Committee made changes to the compensation program for 2014 based on recommendations from these forums. They simplified the equity program and created a long-term equity incentive plan in lieu of the annual performance equity award. In addition, they included a Total Shareholder Return performance metric to evaluate long-term organizational performance holistically.[11] The EMC example highlights the benefit of listening to stakeholders and demonstrating the impact of their input in terms of changes to the organization's policies and practices.

8.6 Identifying Key Stakeholders

While all stakeholders are important, they fall into different buckets for strategic engagement purposes, based on their relevance to the program and to their potential for engagement. The matrix in Figure 8.3 provides a mechanism to lay out stakeholder groups based on the potential to engage the group and the group's relevance to the organization's sustainability program or project. The categories range from high engagement to low engagement and provide a map for engagement techniques with various stakeholder groups depending on their category.

Stakeholders impact projects by introducing and raising awareness about issues, building either support or opposition groups within or outside the organization, and reframing issues to facilitate communication among diverse parties. In order for trust to be developed and maintained, stakeholders need an ongoing connection to the organization. Whether that connection is simply access to information on the project or includes a voice in the process really depends on the type of stakeholder.

High-engagement (HE) stakeholders are well organized, influential, and well informed; they have a meaningful impact on the organization, program, or project. Engagement activities with these

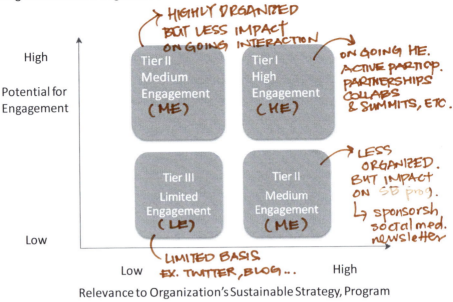

Figure 8.3 Stakeholder Engagement Matrix[12]

stakeholders need to be robust, as they have the greatest impact on organizational initiatives. The process should include regular, open, and honest dialogue about sustainability programs. The goal is active participation from this group. Consider including representatives from the HE group on steering committees, at meetings with senior management, and in facilitated strategy development meetings. Other types of engagement might include joint ventures, partnerships, research collaborations, or summits.[13] Identify a key member of your organization to be the owner of the stakeholder relationship with HE stakeholders in order to develop an ongoing, trust-based relationship. As an organization matures in its sustainability process, HE stakeholders have been asked to join influential groups such as board of director advisory councils in order to provide guidance and insight about sustainability visioning. This group of stakeholders is very impactful on the successful outcome of an organization's sustainable strategy and corresponding programs and projects.

Medium-engagement (ME) stakeholders normally fall into two groups. Those stakeholders in the upper left quadrant of the stakeholder engagement matrix are often highly organized and available for interaction but have less impact on the project outcome. These stakeholders usually want to be kept informed, have access to regular updates, and feel they are part of the process. Ongoing interaction can be achieved with periodic information forums, meetings during which stakeholder feedback is sought, and access to information and feedback portals. Consider making these stakeholders part of an issues-gathering focus group. The second group of ME stakeholders, those in the lower right quadrant of the stakeholder engagement matrix, tends to be less organized or fragmented groups of stakeholders. As a result, it is harder to organize engagement events. While they aren't easily engaged, their actions can impact your sustainability program. These stakeholders need to be engaged with a formal communication plan that promotes two-way communication. Communication strategies can include sponsorship, surveys, newsletters, conferences, and social media. Establishing a quarterly teleconferencing event with a Q&A session is an effective means of updating these stakeholders. Marks & Spencer, a British retailer, is using real-time, mobile survey tools to engage with global factory line workers in their facilities and those of their suppliers. These surveys have given management feedback from 60,000 employees in five countries on working conditions and labor standards.[14]

In the lower left quadrant of Figure 8.3 are limited-engagement (LE) stakeholders with whom you need to maintain a relationship but on only a limited basis. They may require access to information or periodic updates through sustainability reports, publications, social media, or blogs. Social media platforms such as Twitter can be used to inform and track comments from these stakeholders. Members of Intel's corporate responsibility team publish a blog (blogs.intel.com/csr) discussing their views and opinions on sustainability issues and receive and respond to members of the blog community.[15]

Spending the time to identify the broad base of external stakeholders, understanding their concerns, and determining the best approach for engagement is key for successful sustainable strategy development and execution. As illustrated in Figure 8.4, Tier I stakeholders should be fully engaged in ongoing dialogue. Tier II stakeholders need an effective communication plan. Tier III stakeholders should be kept informed through access to information. Dividing stakeholders into these groups allows for better management of stakeholder requirements and more efficient allocation of an organization's resources.

The global digital technology firm, Orange, has 236 million customers in 30 countries with revenues of 41 billion euros. In order to operate effectively around the globe, Orange uses a variety of stakeholder engagement strategies. The following story provides a glimpse at the goals and format of one of their stakeholder engagement events targeting a widely distributed ME group.

Orange has a comprehensive sustainability program and they are recognized globally for their sustainability performance, with inclusion in the FTSE4Good and the NYSE Euronext VIGEO World

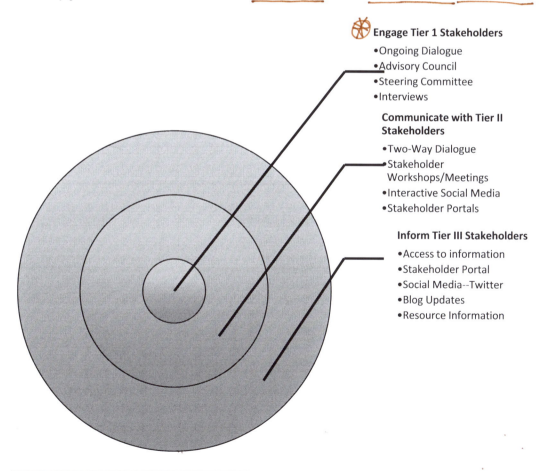

Figure 8.4 Stakeholder Communication Strategies

120 stock indexes.[16] Top sustainability priorities include reducing their carbon footprint by optimizing facilities to reduce energy usage, using renewable energy, and promoting equipment recycling—primarily mobile phones. Spanning the digital divide through technological and innovative solutions and education, especially concerning data security, are also top priorities.

In order to better engage with French sustainability bloggers, a diverse group with far-reaching stakeholder impact, the Orange Corporate Social Responsibility (CSR) team set up a private dinner and workshop for about 30–40 participants. The format for the event was a series of round-table discussions, with the tables divided by the following topics:

1. Freedom of Speech & Respect of Privacy
2. Digital Technologies & Environmental Impacts and Solutions
3. Mobile Recycling: An Example of the Circular Economy
4. CSR, an Obligation or Innovation Opportunity for Big Companies
5. Social Entrepreneurship & Digital Inclusion

A CSR team member as well as an expert on the topic from academia or a business partner hosted each table discussion. The format was an open dialogue, with Orange representatives willing to answer tough questions. While the evening served as an opportunity for Orange to tout its sustainability program and its successes, it also created an environment for a frank, honest, and open exchange with an impactful stakeholder group. The reaction from participants was very positive. Here are some of their best practices that were utilized to make the event a successful engagement:

1. Allowed any questions, setting the tone for open and honest conversation and feedback.
2. Selected a stakeholder group that was receptive to the topic and open to meaningful engagement.
3. Provided a platform for Orange to discuss key sustainability initiatives and an opportunity to highlight achievements.
4. Provided an opportunity to brainstorm solutions to challenging issues with an informed stakeholder group.
5. Provided Orange with a public relations opportunity to promote their sustainability program to a group of influential bloggers.[17]

Engaging with stakeholders may take many different formats. The casual dinner and workshop was a very successful format for Orange. It allowed for engagement with an influential but not highly organized group of stakeholders. (Lower right quadrant ME group in Figure 8.3.) Selecting an appropriate format for engagement with each stakeholder group and using techniques to foster open and honest dialogue goes a long way in promoting successful stakeholder interactions. This approach allowed the Orange CSR team to communicate effectively with the groups about their initiatives as well as to gain insight into the bloggers' areas of interest for existing and future programs.

8.7 Portfolio, Program, and Project Management Impact

In order to maximize the stakeholder engagement experience, keep in mind that some fundamental techniques will improve the overall success of projects and programs. The most important requirement is to foster an environment of inclusion. If you are not sure about a stakeholder group, include them in the process. It is better to overcommunicate than undercommunicate. Once the dialogue begins, ensure that your organizational representatives are listening. As part of the planning process, create a structure to keep senior leadership in the loop and, if possible, engaged in the process. Having the right people in the room with stakeholders helps build credibility and long-term trust. Including senior representatives in the engagement process demonstrates the importance of the stakeholders to the organization. If stakeholder representatives are not senior within an organization, those representatives must have access to decision makers in order to report back on stakeholder recommendations. In order to promote credibility, stakeholder input must be incorporated into the sustainability planning process. Stakeholders expect to receive feedback on their input and to be given demonstrated evidence of impact on strategies, programs, and projects. Without a meaningful link to senior leadership, this process will not succeed.

Table 8.3 lists some guidelines that have been developed by leading global companies to provide best practices in the stakeholder engagement process.

Once organizations and the stakeholders have agreed on stakeholder requirements, incorporate these recommendations into sustainability strategy. Use this information to modify portfolio standards and component selection criteria in order to select programs and projects that address these material stakeholder interests. Then, communicate the organization's actions to stakeholders and establish an ongoing communication platform to keep stakeholders involved and engaged.

Engaging with stakeholders is informative, but it takes time and resource commitments. If your organization interacts with a great many different types of stakeholders, managing the volume and diversity of issues can be challenging. Opening up the flow of information is an eye-opening experience for management, but gathering, maintaining, and analyzing the data requires dedicated resources and new tools and technologies to manage the process. Stakeholder engagement can be expensive and time-consuming, but it significantly improves management's awareness of issues, opportunities, and challenges. In the age of cell phones and 24/7 information, reactions to press releases, reports, and management comments are disseminated instantaneously. Aspects of operations that could be contained in the past now go "viral." Engaging in stakeholder management requires organizational

→ having spread throughout.

Table 8.3 Guidelines for Stakeholder Engagement[18]

Inclusion	Promote broad involvement. Welcome interested parties. Promote tolerance and respect. Invite diverse participants to gain a variety of perspectives.
Build Trust	Candidly share information, goals, limitations, and strategies. Consider all issues. Create an environment for open exchange of differing ideas.
Relevance	Identify and focus on material issues. Engage on issues in a timely manner in order to maximize impact on decisions and actions.
Listen	Learn new perspectives. Seek mutually beneficial solutions. Focus on the future. Focus on areas of impact.
Action	Incorporate recommendations and requirements into sustainable and business strategies. Outline how stakeholder impact will be utilized. Demonstrate impact of previous recommendations to stakeholders. Invite to events such as LEED building openings. Include in stakeholder engagement forums. Invite to industry-related sustainability events.
Communicate	Maintain an ongoing dialogue. Provide push and pull information sources. Alternate mode. Determine frequency. Include a feedback loop. Provide timely stakeholder response.

commitment from the board and senior management as well as from the project team, owners of stakeholder relationships, and business function leaders.

Stakeholder management is a complex but crucial part of developing an effective sustainable strategy. The process begins with creating a strategy for engagement, identifying key stakeholders, and preparing engagement and communication plans. Executing the plan, gathering stakeholder requirements, communicating findings internally, and incorporating feedback into strategy are all important steps in a stakeholder engagement program. The final step is completing the loop and reporting back to stakeholders on your organizational progress.

Information gathered through this engagement process helps to direct strategic focus to areas of materiality for an organization. It also facilitates prioritization of sustainability projects and programs. If a project is important to an influential group of stakeholders, it is more likely to get the green light ahead of other proposals. Stakeholder feedback also leads to discovering opportunities for partnership and joint venture solutions. Often, stakeholder engagement is focused on threats or concerns; it is important to remember that stakeholder engagement can also lead to new products, services, and markets.

After project teams have gathered stakeholder requirements, they can begin to incorporate priority requirements into their organizational strategy. In order to address these requirements long-term, portfolio managers must incorporate them into sustainability component selection criteria to ensure that resources are being allocated toward projects that support strategic sustainability vision. As an organization's sustainability program matures, stakeholder requirements become embedded into programs and projects across the organization. With maturation in the sustainability process, senior leaders realize that stakeholder requirements impact not only sustainability portfolio components but also traditional portfolio components. As the process evolves, points of stakeholder engagement expand horizontally and vertically across the organization, moving from the sustainability champion to other business leaders, such as the CFO addressing investor concerns, the VP Sales addressing customer challenges and opportunities, and the CEO addressing employee engagement. The most effective stakeholder engagement programs reflect multiple levels of organizational engagement and relationship management. The relationship management selection process is a function of both the stakeholder group and their level of impact.

Adding these roles to the stakeholder engagement matrix identifies the best internal person to own the stakeholder relationship based on the strength of the relationship between the owner and the stakeholder group. While not every relationship can begin as a strong one, leverage strong existing relationships in order to manage the process more effectively. Select stakeholder owners based on the best fit with the stakeholder group in order to facilitate building a long-term relationship with the group. An example stakeholder engagement matrix is shown in Table 8.4.

Understanding the role of the stakeholder group (champion, decision maker, influencer, blocker) as it relates to your program or project informs the approach that is taken by the owner in terms of developing and maintaining a relationship. In the case of the Save the Lake Group illustrated in Table 8.4, the CSO may want to work on developing a stronger personal relationship with the group in order to have more credibility in the relationship. As blockers of the project, it will be difficult to overcome the group's objections without developing a trust-based relationship. All new relationships are weak, but with consistent credible action they can become strong relationships over time. Each of these stakeholders has a different level of influence on the organization and vice versa. Engagement techniques need to be adjusted to address these varying levels of impact. In addition, stakeholder relationships are impacted by the level of influence of the person assigned to interact with the group. Typically, high-engagement stakeholders receive the most senior attention, often being part of a C-suite steering committee as in the case of Dow's SEAC. Stakeholder groups may move between categories, and a best practice includes an annual stakeholders category reassessment process. If a junior person is assigned to a stakeholder group, the group may feel their concerns are not being taken seriously. An annual meeting with senior leadership for these stakeholder groups may help alleviate their concerns about having a

Table 8.4 Stakeholder Engagement Matrix[19]

Stakeholder	Group	Require	Power*	Owner	Relationship	Strategy
Save the Lake	Community	Seeking zero impact on lake water	B	CSO	Weak	Establish net zero water target. Report on water usage versus target.
VP Sales	Employee	ESG solutions and products for customers	C	CSO	Strong	Report on percentage of products meeting ESG criteria.
CFO	Employee	ROI 10%	DM	CSO	Average	Report on business case quarterly.
XYZ Co.	Customer	Develop products with a lower water footprint	I	VP Sales	Average	Issue reports on products' water footprint.

* Power: C = Champion, DM = Decision Maker, I = Influencer, B = Blocker

more junior contact person. The quality and strength of the relationship between the project team, especially the designated owner of the relationship, and the stakeholder group has a significant impact on the anticipated level of alignment and support of sustainability programs and projects.

The benefits of engaging in the stakeholder management process include identifying risks and managing them proactively, having a seat at the table for regulatory and policy development, identifying opportunities through joint ventures, leading industry events, creating goodwill, and saving resources. Most important, establishing a relationship with key stakeholders provides a forum to discuss ongoing challenges, concerns, and opportunities. Developing relationships with stakeholders and addressing their concerns early on in the life of a program or project improves the likelihood of a positive outcome. Programs are often multiyear in duration, with significant up-front capital. Project management professionals who include internal and external stakeholder management as part of the portfolio component management process will be more likely to achieve success.

> ### Managing Stakeholder Engagement and Reputational Risk
>
> Stakeholder engagement has a significant impact on reputational risk and an organization's right to operate. The gold mining case discussed earlier in the chapter addressed the stock valuation impact. This case discusses incorporating stakeholder engagement into the project planning process to ensure that a long-term capital-intensive project is viable from a stakeholder management perspective, prior to investing sizable organizational assets.
>
> One of the most hotly contested issues involving shale resource development using hydraulic fracturing has been the impact on freshwater. In the United States, there has been a significant backlash from communities about the safety of drinking water in hydraulic fracturing areas. The innovative techniques used in hydraulic fracturing have allowed for development of shale gas, natural gas liquids, and oil across the United States and Canada. The economic impact and energy independence that this source of fuel has generated for the United States and Canada is being watched around the world. Because of this new technology, the world's natural gas and oil reserves have grown significantly. However, local and regional concerns persist about the impact of this process on local water supplies.

The World Resource Institute (WRI) is a global research organization focusing on environmental, social, and economic issues. Their work focuses on seven areas: climate, energy, food, forests, water, cities, and transportation. According to the WRI, more than a billion people live in water-scarce regions, with as many as 3.5 billion anticipated to face water shortages by 2025.[20] The WRI works with businesses, governments, and society to promote future water security. Their Aqueduct project uses the most accurate data to create global water risk maps so that organizations can better understand and assess current and future water challenges. WRI research has taken some of the lessons learned from this project and created a recommended protocol for dealing with stakeholder concerns on the impact of local water supply.

One of the major obstacles to shale exploration is the availability and use of freshwater in the extraction process. In the United States, there has been community pushback against "fracking" because of concerns about the availability of and the impact on the freshwater supply. Stakeholder concerns about water quality have been given an international voice and global awareness by celebrity spokespersons supporting their efforts. As a result, any hydraulic fracturing project must have a proactive stakeholder engagement plan to move forward.

In order to better assess the impact of shale resource extraction on freshwater, WRI prepared a global and country-specific analysis providing a tool to evaluate the availability of freshwater in shale reserve areas.[21] (Map: Shale Resources and Water Risks, *Interactive Maps Showing the Overlap of Shale Oil and Gas Resources and Water Stress*, http://www.wri.org/our-work/project/aqueduct.) As many countries around the globe consider developing their shale gas reserves, the WRI mapping tool facilitates identifying areas where freshwater management will be most important if shale resources are developed. WRI's analysis provides a foundation for organizations, governments, and local stakeholders to better understand the business risks, the advantages of water stewardship, and the need for government oversight and corporate policies for proper water oversight.

The purpose of the report is to increase the dialogue among the stakeholders, including industry, government, and civil society. In order to engage all the stakeholders in this meaningful conversation, the WRI has created metrics with which to measure freshwater availability and business risks. The purpose of these indicators, listed here in Table 8.5, is to provide a common reference point and language in order to discuss the impact of water-related issues with both internal and external stakeholders. Promoting common language

Table 8.5 Indicators Selected to Evaluate Freshwater Availability and Related Business Risks[22]

Indicator	Definition
Baseline Water Stress	Ratio of total water withdrawals to available renewable supply.
Seasonal Variability	Variation in water supply based on months of year.
Drought Severity	Average length of droughts times the severity of droughts over past +100 years. Higher values indicate areas of more severe drought.
Groundwater Stress	Ratio of groundwater withdrawal to aquifer recharge rate. Values over 1 indicate unsustainable groundwater usage.
Dominate Water Usage	Sector with largest annual water usage.
Population Density	Average number of people per square kilometer.
Reserve Depth Interval	Range of depths of shale area. Deeper formations usually need a higher volume of water for drilling.

and definitions is crucial to communicating sustainability issues effectively, both internally and externally.

Based on their research, the WRI was able to identify the location of the world's technically recoverable shale gas and the corresponding level of baseline water stress. Their findings indicate that, for the most part, large shale reserves are not located where freshwater is abundant. Countries such as China, Mexico, and South Africa have large shale gas reserves but also have high water stress. In fact, eight of the top twenty countries with shale gas reserves have arid conditions or extremely high baseline water stress. As a result, a shale extraction company may have difficulty meeting its demand for freshwater for the extractive process in these areas. Firms need to enter exploration projects understanding that local and regional stakeholder concerns over new competition for water resources can negatively impact their project. In some cases, the reaction affects a company's social license to operate, inviting local protests and restrictive government regulation. As a result, companies entering these areas for extractive exploration must consider the local water impact and have a viable plan for water management and stewardship. In addition, they must engage local, regional, and national stakeholders to better understand their needs and to ensure minimizing environmental impacts and water resource depletion, while maintaining positive stakeholder relationships..

Best Practice to Engage Stakeholders on the Issues of Fresh Water Impact of Shale Resource Development[23]

1. Perform local water risk assessments to determine resource availability in order to minimize the business risk of launching a project in a water-scarce area.
2. Engage with communities, regulators, and industries in a transparent manner by disclosing and communicating water usage. Take the time to understand their issues and concerns.
3. Support water governance and environmental standards to ensure water security, minimize the development of new regulations, and protect an organization's reputation and license to operate.
4. Minimize freshwater use through conservation, water reclamation, and technology.
5. Develop a water stewardship program to reduce the impacts on water availability.

By adopting these best practices, an organization gains tools to better gather priorities from local stakeholders such as communities, regulators, other businesses, and governments. Developing robust baseline information and detailed estimates of future water supply and demand provides a shared knowledge base around which governments and communities can develop effective water policies.

Building a strong foundation requires transparency about actual and projected water use and about water management and stewardship policies. Ongoing operational disclosure is necessary to build trust and to protect an organization's reputation. Collaboration with local and regional governments, communities, agriculture, and other industry allows for better understanding of the hydrological conditions and regulatory framework within a water basin. Gaining this level of collaboration improves the likelihood of project success through more accurate cost estimates, technology requirements, better understanding of baseline and seasonal water requirements, and environmental issues and concerns.

It is in the company's best interest to promote water governance and environmental protection standards that are supported by effective enforcement to minimize environmental

> degradation by all users. Understanding and working within a known regulatory structure allows companies and investors to feel more confident in the long-term value of this investment. Working with stakeholders, companies can promote shared water sourcing, recycling infrastructure, and better management of watersheds and aquifers.
>
> Current technology allows for the use of non-freshwater sources such as brackish water, or recycled or reused water, in order to minimize freshwater usage. Adopting a goal of limiting freshwater usage as part of an organization's sustainable strategy and making water management and targets part of organizational benchmarks will promote judicious usage of freshwater and help ensure long-term water availability for all stakeholders. Incorporating these strategies into the project planning process provides for more realistic projections of project scope, time, and costs, ensuring that stakeholder expectations are well managed.
>
> Engaging stakeholders to better understand local, regional, and national issues is an important step in planning and executing global projects. It can mean the difference between a successful project and an expensive failure. Transparency and authenticity of information are crucial to the stakeholder engagement process. As this case exemplifies, stakeholder engagement contributes to a triple bottom line win for organizations and communities.

8.8 Conclusion

Engaging with external stakeholders at the beginning of a project brings issues to the forefront, which reduces the chance of an event occurring that impacts strategic, operational, and reputational risk. Ignoring stakeholders concerns significantly increases the risk associated with programs and projects. Stakeholder engagement is a complex, multilevel, and interwoven process that requires planning, organization, and resources dedicated to keeping the lines of communication and information open. Stakeholders lose trust in an organization when information is not transparent, reliable, or clearly and efficiently communicated. Once an organization engages stakeholders in the process of sustainable strategy development and, more specifically, sustainability partnerships and joint ventures, senior leadership must be committed to an ongoing stakeholder engagement. They must support this strategy with both their own time and organizational resources. Stakeholders require evidence of their impact on an organization's sustainability decisions, priorities, and strategies. Managing the feedback loop is just as important as the initial requirements gathering. Stakeholders have a significant impact on organizations and play a crucial role in developing and effectively implementing an organization's sustainable strategy.

Notes

[1] Irvin Cernauskas and Shelly Herman, "Irv & Shelly's Fresh Picks New Joint Venture," Interview with Irvin Cernauskas, Owner, Irv & Shelly's Fresh Picks, September 6, 2014.

[2] "Old Newsstands Now Serving Lunch in the Loop as E.A.T. Spots—Downtown—DNAinfo.com Chicago," *DNAinfo Chicago*, accessed September 6, 2014, http://www.dnainfo.com/chicago/20140819/downtown/vacant-newsstands-now-serving-lunch-loop-as-eat-spots.

[3] J. Henisz Witold, Sinziana Dorobantu, and Lite J. Nartey, "Spinning Gold: The Financial Returns to Stakeholder Engagement," *Strategic Management Journal* 35, no. 12 (December 2014): 1727–1748, doi:10.1002/smj.2180.

[4] Ibid.

[5] Andrea Bonime-Blanc, "Risk and Opportunity—the Role of Stakeholder Trust," *Ethical Corporation*, CR Reporting and Communications Mini Report 2014 (November 13, 2014): 6–7.

[6] ICCR, "About ICCR," accessed April 6, 2015, http://www.iccr.org/about-iccr.
[7] Fabrizio Ferraro and Daniel Beunza, "Why Talk? A Process of Model of Dialogue in Shareholder Engagement," *Social Science Electronic Publishing*, March 26, 2014, http://ssrn.com/abstract=2419571.
[8] Thomas Mc Carty, Michael Jordan, and Daniel Probst, *Six Sigma for Sustainability,* New York: McGraw Hill, 2011.
[9] EMC, "Governance/Stakeholder Engagement," accessed April 6, 2015, http://www.emc.com/corporate/sustainability/delivering-value/stakeholder.htm.
[10] Ibid.
[11] Ibid.
[12] Jonathan Morris and Baddache, "Back to Basics: How to Make Stakeholder Engagement Meaningful for Your Company," BSR, January 2012, www.bsr.org.
[13] Ibid.
[14] Laura Gitman and Sara Enright, "Breakthroughs in Stakeholder Engagement," accessed April 8, 2015, http://www.bsr.org/en/our-insights/blog-view/breakthroughs-in-stakeholder-engagement.
[15] Jonathan Ballantine, "Building Trust: Approaching Stakeholder Engagement," *Environmental Management & Sustainable Development News*, accessed November 6, 2014, http://www.environmentalleader.com/2009/07/21/building-trust-approaching-stakeholder-engagement.
[16] Orange, "Corporate Social Responsibility 2013 Complete Report," 2013, http://www.orange.com/en/content/download/23330/480379/version/3/file/Orange_2013_CSR_report.pdf.
[17] Marion Dupont, "Best Practices in Stakeholders Engagement: How Orange Dialogues with Sustainability Bloggers," *Sustainability Reporting & Digital Communications*, accessed January 7, 2015, http://blog.wizness.com/best-practices-in-stakeholders-engagement-how-orange-dialogues-with-sustainability-bloggers.
[18] BSR and Altria, Inc., "Stakeholder Engagement Planning Overview," December 4, 2003, http://www.forumstrategies.com/content/pdf/stakeholder_engagement.pdf.
[19] Jordan McCarty and Daniel Probst, *Six Sigma for Sustainability*.
[20] World Resources Institute, "Water: Mapping, Measuring and Mitigating Global Water Challenges," accessed April 13, 2015, http://www.wri.org/our-work/topics/water.
[21] World Resources Institute, "Map: Shale Resources and Water Risks," accessed April 13, 2015, http://www.wri.org/our-work/project/aqueduct.
[22] Paul Reig, Tianyi Luo, and Jonathan Proctor, "Global Shale Gas Development: Water Availability and Business Risks, Executive Summary," World Resources Institute, n.d.
[23] Ibid.

Chapter 9

Leveraging Organizational Relationships and Assets

We have discussed managing external stakeholders; now, let's take a deeper look at managing internal stakeholders, specifically, business function leaders and those who hold power within organizations. As the sustainability champion, chief sustainability officer (CSO), or project manager (referring to this functional role as CSO), it is important to recognize that sustainability doesn't function well in a silo. Isolating sustainability in the Corporate Social Responsibility or Environmental Stewardship offices will limit the growth and adoption of sustainable practices within your organization, curtailing its sustainability journey. While a dedicated sustainability champion facilitates initiating and managing the process, a cross-functional team approach is significantly more impactful in terms of creating a culture of sustainability. The key to success is getting business function leaders to embrace sustainability by demonstrating how sustainability programs, projects, and processes align with their core business goals. Sustainable strategy provides new tools and techniques for business leaders to improve performance in terms of engaging with stakeholders, creating new products and solutions, minimizing risks, engaging employees, and generating operating efficiencies. Adopting a sustainable approach provides meaningful business impact in terms of resource savings, operating efficiencies, new business opportunities, employee engagement, and organizational transformation.

Once the C-suite has been convinced, engaging other internal stakeholders in the process, especially functional business leaders and managers, expands the depth and breadth of impact throughout the organization. Creating a network of support is key to bridging the gap between sustainability vision and organizational adoption and integration. While the CSO educates, champions, and coordinates, the entire organization needs to be part of the sustainability mission and process in order to create transformation. Issues such as culture and readiness for change significantly impact the speed and effectiveness of sustainable transformation.

If sustainability efforts are limited to the CSO, an organization cannot engage effectively with external stakeholders nor make meaningful changes across the organization that external stakeholders expect. One person and one office is not sufficient for this level of transformation. Leveraging the skills and expertise of functional leaders across the organization results in improved strategic alignment between sustainability vision and implementation. In order to create this type of holistic change, sustainable strategy must include engaging leaders and influencers who impact core projects, processes, and policies

188 Becoming a Sustainable Organization

across the organization. Leveraging key internal relationships and gaining leadership support and buy-in drives sustainability program and project selection, approval, funding, and resource allocation.

9.1 Impact of an Organization's Sustainability Maturity on Stakeholder Engagement

An organization's experience, or maturity, with the sustainability process impacts its overall approach to sustainable strategy. (See Figure 5.4, the Sustainability Continuum.) Drivers for business leaders to engage in sustainable strategy vary depending on the organization's maturity in its own sustainability journey. The more experience management has with sustainable strategy, the higher the frequency of demonstrated success and the more ingrained sustainability becomes in the operating culture. As a result, the levers that CSOs choose vary significantly across the stages of development. Once the C-suite has bought into the value of a sustainable strategy, the CSO still needs to get the functional business leaders on board. Even with C-suite support, it can be a challenge to get support from the next tier of management. Expect responses to range from cautious collaboration to total unavailability. Some managers may throw up barriers to projects, others may express interest but lack resources. One or two brave souls may be willing to pilot a project.

Business leaders have short-term goals because, for the most part, their compensation is tied closely to achieving those goals. They are not interested in allocating resources to new programs unless they are convinced of both C-suite support and the business value creation. Exposure-stage levers for engagement include information, education, crisis management, and pilot projects. Many sustainability practitioners will tell you not to underestimate the benefits of providing a sustainability solution to address an environmental, regulatory, customer, or community crisis. This crisis is often the launching platform for a much more expansive sustainability program. Once organization leaders have experience with a successful project, they become more open to the concepts of sustainability. Pilot projects work in a similar fashion by building a comfort level with sustainability projects within the management team by demonstrating cost savings, resource minimization, and improved operations. As the pilot project delivers on its goals, management becomes more engaged and willing to scale the project across the organization.

As a result, the levers for sustainability engagement expand within the Integration Stage. Business leaders begin to see the value in terms of risk mitigation, process improvement, and improved community relations. The stakeholder engagement focus expands to incorporate a broad range of internal and external stakeholders. At this stage, organization leaders begin to see the potential for meaningful long-term business impact by adopting sustainability. Identifying key business influencers and developing joint projects with those leaders has a significant impact on the adoption of sustainability projects and programs, both across the organization and by other business unit leaders. Enlisting a champion to promote sustainability within the peer group of business function leaders is a powerful lever.

As sustainability projects and programs become more accepted within the business units, incentives and compensation formulas begin to change to incorporate sustainability performance metrics. Tying sustainability activities and performance to compensation is a powerful change management tool. Ultimately, sustainability becomes ingrained within the culture of the organization, with policies, projects, and processes aligned to support the overall sustainable strategic vision of the organization. The outgrowth of this stage is full transformation of the organization's strategy for creating and delivering business products and services to its customers. The Transformative Stage delivers the most competitive advantage as sustainability moves from supporting business strategy to becoming a pillar of business strategy by promoting new platforms for customer sustainability solutions and expanding business opportunities.

In order to better refine drivers and levers for sustainable engagement by an organization's stage on the sustainability continuum, Table 9.1 outlines activities and impacts by stage. It provides details

Table 9.1 Internal Stakeholder Engagement Techniques and Impacts

	Exposure	Integration	Transformation
Drivers	Regulation Compliance Cost Savings Personal Interest Competitive Pressure	Risk Mitigation Operating Efficiency Stakeholder Requirements Value Creation Sustainability Reporting	Strategic Alignment New Markets New Products & Services Standardized Sustainability Reporting Global Image
Levers	Education Information Crisis Response Pilot Programs C-Suite Support Industry Rankings Competitive Rankings	Demonstrate alignment with business function. Scale pilot projects. Celebrate wins. Develop alliances. Business function sustainability rankings. Incentives tied to sustainability. Partner with business unit heads. Adopt lessons learned. Transfer project/program sponsorship to business units. Publish rankings and results. Formalize stakeholder engagement.	Sustainability is included in the strategy development process. Align business and sustainable strategic vision. Promote cross-functional engagement. Hold long-term visioning sessions. Sustainability goals become part of management performance scorecard. Build stakeholder alliances. Support industry or global sustainability initiatives.
Actions	Pilot Projects Energy Efficiency Community Volunteerism	Product Life-Cycle Assessment Supply Chain Management Engagement with External Stakeholders Sustainability Reporting Scaling Pilot Projects	Fundamental Changes in Business Process, Policy, & People Incorporation of Circular Economy Approach Transparent Engagement Sustainability Reporting with Recognized Standards and Third-Party Verification Business Unit Empowerment Active Participation in UN Global Initiatives Senior Leaders Speak about Sustainable Strategy
Projects	Energy Focused One-Off Silo Pilots Sustainability Champion Internal Operational	Programs Cross-Functional Projects Integration with Sustainability Champion & Business Unit Head Resource Minimization External Value Chain	Portfolio Standards Embedded Cross-Functional Programs & Projects New Platforms to Launch Sustainable Solutions Academic & Industry Partnerships to Promote Solutions to Global Challenges New Technologies

(Continued on the following page)

Table 9.1 Internal Stakeholder Engagement Techniques and Impacts (Continued)

Engagement Focus	Exposure	Integration	Transformation
	Limited	Expanded	Embedded
Stakeholders	C-suite Government Customers	C-suite Business Unit Leaders Employees Government Customers Suppliers Trade Groups Community Groups NGOs Academia	C-suite Business Unit Leaders Employees Government Customers Suppliers Trade Groups Community Groups NGOs Academia Industry UN Global Initiatives
Business Impact	License to Operate Fine Avoidance Cost Savings	License to Operate Fine Avoidance Cost Savings Resource Conservation More Sustainable Products/Services Improved Efficiencies Improved Risk Management Improved Employee Engagement Attract SRI Supply Chain Management	License to Operate Fine Avoidance Cost Savings Resource Conservation More Sustainable Products/Services Improved Efficiencies Improved Risk Management Improved Employee Engagement Attract SRI New Product and Service Platforms New Markets & Populations to Serve Sustainable Solutions for Customers Re-engineer Work Flow & Processes Value Chain Enhancement
Budget	Limited Project Based	Sustainability Office Based. Convince Business Units to Underwrite Sustainability Initiatives	Cross-Functional, Fully Funded Sustainability Programs Part of Annual Budgeting Process Business Units Fund Projects
People	Project Team	Small Sustainability Office Convince Business Units to Allocate Resources	Structured Cross-Functional Programs with Dedicated Resources from Business Units
Impact	Positive	Significant	Transformative

on recommended stakeholder engagement, sustainability project types, and business impact. It also provides a framework for expectations about budget allocation, project team structure, and internal stakeholder engagement based on the organization's sustainability maturity.

Engaging internal stakeholders moves organizational sustainability from one-off projects to cross-functional projects and programs and ultimately to incorporating sustainability portfolio criteria in all projects, programs, and processes. Sustainability and business strategy merge into sustainable strategy, transforming the culture of the organization into a sustainable organization.

9.2 Managing Internal Stakeholders

Engaging with internal stakeholders generates a range of responses. Some are advocates, some ignore attempts at engagement, and others block forward progress. The roles that business function leaders adopt has a significant impact on the sustainable transformation process. While a few may be champions helping to move the sustainability initiatives forward, others will act as blockers trying to stop the forward momentum of sustainable programs and projects within the organization. Although all business leaders are important, some have more influence on the success of sustainability programs and projects than others. Some, such as the Chief Financial Officer (CFO) have decision-making authority that impacts the funding and ultimate success of sustainability programs and projects. Others aren't decision makers but do have the ability to influence those who make the decisions. In order to develop a successful sustainability program, it is important to engage both decision makers and influencers. As the CSO, understanding the various positions of the key internal stakeholders and the levers available to help move the sustainability agenda forward improves outcomes. Table 9.2 identifies key stakeholders, their roles, and their potential impact in promoting or blocking sustainable strategy.

While internal stakeholders begin with certain roles and levels of support, these roles evolve and change as they become better informed about the benefits of a sustainable approach. Strategic sustainability vision cascades from senior management to business function leaders, then to managers, and down to employees. As organizational acceptance of the sustainable value proposition grows, it becomes more difficult for internal stakeholders to act as blockers of sustainability programs and projects. Once enough influential organizational leaders have bought into the value proposition of sustainability, a "tipping point" is reached within management ranks, so there are more supporters than detractors.

The importance of both internal and external stakeholder engagement increases as management's sustainability maturity evolves, because the scope and complexity of programs and projects increases. As key internal stakeholders, employees are crucial to a successful sustainability transformation. Ensuring that employees understand the organization's sustainability vision and their role in the process opens the door for employee action, engagement, and feedback. Encouraging employees to suggest new ideas to enhance sustainability performance within their function changes the way in which work is performed. As internal stakeholder engagement progresses, managers identify and better understand the

Table 9.2 Internal Stakeholder Management

Stakeholder	Role	Business Impact
VP Sales	Champion	New Products & Solutions to Meet Customer Demand
CFO	Decision Maker	ROI Thresholds Project Funding
VP Product Development	Influencer	Product Reformulated to Include More Sustainable Ingredients
VP Operations	Blocker	Assessment of Manufacturing Facilities to Better Mitigate Climate Change Risk

Table 9.3 External Stakeholder Management Plan

Stakeholder	Group	Issues	Requirements	Owner	Strategy
Advisory Council	Multiple	Business Strategy Guidance	Sustainability Standards Environmental, Social, & Governance Policies Global Guidelines Local Actions	CEO CSO	Annual Long-Term Strategy Focus Group Annual Status Meeting Access to Senior Management
Socially Responsible Investors	Investors	Risk Management Social, Environmental, & Governance Policies ROI	Transparency of Reporting Industry-Recognized Sustainability Reporting Standards Specific Environmental & Social Issues	CFO Investor Relations CSO	GRI or CDP Reporting Standards Address Issues at Annual Meeting & Quarterly Calls Access to Senior Management
River Watershed Protectors	Community	Protect & Improve Watershed Biodiversity Maintain Quality & Quantity of Water	Zero Impact on River Watershed Water Quantity & Quality	VP Operations CSO	Monitor Water Outflows and Inflows to Ensure Zero Impact Treat All Water to Ensure Quality of the Water Is Returned to River as Fresh Water Meeting Quality Standards Promote Volunteer Efforts to Clean up the Watershed and to Plant Native Plant Species

impactful nature of sustainability, including the benefits of external stakeholder engagement. As they begin to understand the advantages of direct engagement with external stakeholders, they become more involved and invested in stakeholder relationships. Stakeholder relationships and engagement programs move from the CSO to other appropriate business leaders including the C-suite and board of directors. Ultimately, ownership of impactful external stakeholder relationships is spread across the organization, matching stakeholder group to the owner with the best strategic fit. As the impact of these engagements cascades into programs, projects, and processes, project managers need to consider appropriate stakeholders groups in their project plans. Project teams should include diverse cross-functional members in order to facilitate strong relationships with stakeholder groups. Table 9.3 outlines an example plan for engaging external stakeholders based on the group type, their key issues, and requirements. Based on this information, organizational owners are identified to best match the needs of the group as well as to identify an effective strategy for engagement.

Demonstrating that sustainability-based business solutions address challenges and create opportunities effectively is key for engaging business function managers. Once business leaders see evidence of impact, conversations about allocation of limited budgets and resources to sustainability programs becomes easier. Moving along the sustainability continuum is evidenced by greater buy-in of key internal stakeholders to the process. As acceptance multiplies, so does the impact of a sustainable approach. It becomes exponential. Once the organizational "tipping point" is reached, business sustainability champions, leaders, and influencers become more prevalent. As a result, competition heats up among business functions to find sustainable solutions in order to meet their sustainability goals and to differentiate their area's performance. Internal stakeholder engagement is iterative, with program and project feedback generating improvements to future component selection, planning, and implementation. This continuous improvement cycle creates broader organizational understanding and acceptance of sustainability.

9.3 Engaging Function Business Leaders

Start with sharing the corporate strategic vision of sustainability, highlighting the beneficial impact on specific business functions. Explaining "why" sustainability is important to both the organization and a business function's mission facilitates the acceptance process.

The U.S. Army embraces sustainability and has a unique message about its impact. The mission of the U.S. Army is to protect and defend the United States using its strategic assets, which are primarily soldiers. The Army leadership's focus is on protecting their soldiers' lives while they are engaged in very risky activities. Their sustainability message is that resource conservation can build resiliency and save lives. As an army moves into battle, its supply lines need to be restocked. By embracing renewable energy, alternative power sources, and water conservation techniques, transport conveys can be sent into dangerous areas less frequently. As a result, lives are saved. This sustainability message has resonated with Army leadership. As a result, they support making decisions and taking actions that align with the Army's net-zero goals for water, energy, and waste. In addition, the Army is actively seeking sustainable solutions for energy generation, battery technology, and risk-mitigation strategies to improve their operational resiliency in the field.[1]

Identifying the messages that resonate with organizational business leaders is a crucial aspect of leveraging leadership assets within an organization. Taking the time to understand business function challenges and goals facilitates the process of aligning sustainability business strategy. Creating an engagement map like the one shown in Figure 9.1 highlights the benefits of sustainable strategy on functional areas.

In order to engage with each of the business function leaders, take the time to gain an understanding of the material issues that motivate them. Then, select points of alignment between their goals and

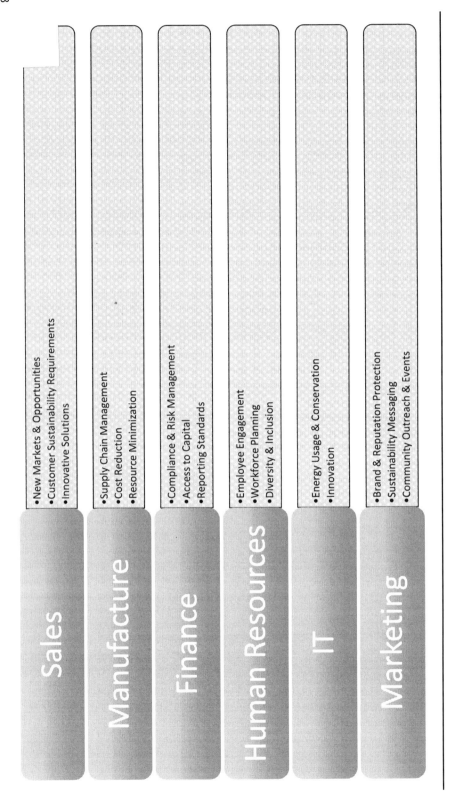

Figure 9.1 Engagement Map[2]

sustainability goals. It is not unusual for business leaders to be motivated by different aspects of sustainability. The VP of Sales may be engaged through examples and stories that demonstrate increased market share and market penetration gained from product differentiation through more sustainable raw material sourcing and increased customer demand for sustainable solutions. From the perspective of the VP of Sales, the goal is to meet clients' needs with a better solution than the competition. Offering a sustainable product or service can be a product differentiator, especially for those customers that have sustainability guidelines for purchasing. In order to redesign a product to improve its sustainability rating, the sustainability champion needs to engage the VP of Manufacturing. Mutual benefits include reputation protection, improved source of supply, lower cost, and reduced waste. Meaningful sustainability programs and projects require a cross-functional approach, as new platforms and solutions require programs and projects that reach across the organization's value chain.

In order to reduce greenhouse gas (GHG) emissions in a service-based organization, a project may focus on redesigning the sales process to minimize business travel. An effective project will engage multiple functions, including IT, HR, and Sales, to provide the best solution. This approach provides a solution that includes supporting sales representatives with technology and training to facilitate online customer meetings. Sustainable solutions need functional expertise from all of these areas in order to realize benefits such as increased productivity, lower emissions, attentive customer service, and improved work–life balance for the sales group. The Engagement Map can be adapted to reflect an organization's functional areas, highlighting the impact of a sustainable approach on functional business goals and targets.

The more embedded sustainability becomes in an organization, the more entwined it becomes in the decision-making process. Once sustainability takes root in an organization, it fundamentally changes the approach that business leaders adopt in evaluating challenges and opportunities. In some industries, such as the personal computer, tablet, and laptop market, industry sustainability benchmark standards are a threshold for sales to a growing number of client categories. If an organization does not meet these standards, then certain public and private organizations cannot include that firm on their approved vendor lists.

In 2012, Apple requested that the Electronic Product Environmental Assessment Tool (EPEAT) registry pull all of its desktop computers, laptops, and monitors from its green products list. While Apple had adopted sustainability product standards of its own, their design direction was no longer in line with EPEAT requirements. EPEAT awards a seal to computers, displays, and laptops to certify that they are recyclable, designed for energy efficiency and minimal environmental impact. From Apple's perspective, they published their environmental story and felt that EPEAT standards did not sufficiently reflect current technological advancements and trends. Many of Apple's customers, including federal, state, and local governments, corporations, and higher education institutions, all had procurement standards requiring that desktop and laptop purchases be from EPEAT-approved sources. Once the announcement was made, the city of San Francisco notified its various agencies that Apple desktops and laptops were no longer eligible under procurement protocols. Within days of the initial announcement, Apple's management reversed their position and reinstated the use of the EPEAT registry for product certification.[3] Management realized that EPEAT registration had become a threshold standard for doing business and that its certification as an environmentally sound product had been fully incorporated into their clients' business processes. One can only imagine the internal conversations between the SVP of Sales and the SVP of Engineering. The end result was an apology issued by Bob Mansfield, SVP of Hardware Engineering, reversing Apple's decision and deciding to re-list Apple's products with EPEAT based on the input from many loyal customers. At Apple. a clear relationship exists between sustainable strategy and sales performance and stakeholder input.[4]

This example highlights the interconnectivity of sustainability both within the organization and externally with stakeholders such as customers. It also demonstrates the impact that sustainability standards have on clients' purchasing decisions. Increasingly, these standards are being pushed further down the supply chain to mid-size and smaller organizations. While Apple prides itself on taking a holistic approach to its environmental impact, including requirements that were not considered by EPEAT at the time, such as the removal of toxic materials from their products, they forget a crucial component of sustainable-strategy stakeholder engagement. Even though EPEAT was a voluntary program, it had become an industry standard for computer manufacturers. As such, their customers had adopted the EPEAT certification as a threshold requirement for their purchasing departments. Apple's decision to withdraw from EPEAT may have been sound from an engineering and environmental policy perspective, but from a sales perspective it was a disaster. Engaging with both internal and external stakeholders as part of the policy review process helps to avoid costly missteps such as this example. As sustainability grows within an organization, the levels of engagement become more complex and intertwined, requiring a cross-functional approach.

As an organization moves forward in its sustainability journey, the integration of sustainable standards, expectations of customers, and product and solution offerings are all impacted by stakeholder perceptions. This case reminds us that client requirements for sustainable products drive business, and that business leaders must consider stakeholders' requirements in their decision-making process. Even though Apple's management felt they were exceeding EPEAT standards, the decision to no longer have products listed as EPEAT certified had a significant impact in their marketplace. Understanding the relationship between sustainable strategy and client requirements is vital for effective stakeholder engagement and successful business operations.

Environmental, health, and safety (EHS) facilities and operations are often where organizations begin their sustainability journey, with compliance reporting, energy-savings, or resource reduction projects. EHS drives the reporting functions to comply with government regulations and provides a resource for data and information to support program and project proposals. Facilities managers can be engaged with projects that generate utilities savings through process changes, education, and behavior changes. Other savings comes from evaluating waste streams to identify opportunities for reusing and recycling. In 2012, Ford recycled 586,000 tons of scrap metal in its North American operations, adding $225 million in revenues from previously discarded waste.[5] We refer to these projects as the "low-hanging fruit" because they are relatively straightforward and low in cost. Once the dual benefits of cost savings and preservation of natural capital are demonstrated to business leaders, the door begins to open for more complex programs and projects.

Next-level projects include engaging the manufacturing function and focus on designing products for energy efficiency, utilizing energy star–rated equipment, reconfiguring product formulations to incorporate sustainable materials, and rethinking the layout and space requirements of facilities. Conducting a product or service life-cycle assessment in order to fully understand resource consumption, as well as environmental and social impacts, is a great way to engage with the manufacturing leadership team. Looking into design and manufacturing processes is a starting point for understanding how products and services are designed, developed, and created. Evaluating an organization's supply chain and gathering ESG information from suppliers through surveys and/or scorecards provides management with a broader perspective on the global ESG impacts of their products and services. As a result, they have information with which to make better-informed vendor selection decisions—which vendors have programs that align with their organization's sustainability agenda and which vendors leave them open to previously unconsidered risks. With this information, programs and projects can be selected to improve supply chain sustainability performance and outcomes.

9.4 Internal Stakeholder Drivers for Sustainability

In order to demonstrate the value of a sustainable approach, consider an organization's full value-creation process, "from cradle to grave." Most people are familiar with the linear model of value creation, in which an organization starts with raw materials and other resources, manufactures a product, sells the product to an intermediate, who sells to the end user. Management doesn't really think about what happens to its product at the end of the product's useful life. This model relies on readily available resources and energy to thrive. As we see from the outlook for our global future, inexpensive and abundant resources will not be continuously available to meet the needs of a growing world population. In order to create a more resource-sensitive model, the concept of the circular economy was developed. It is based on the study of living systems in nature, which seek to optimize systems rather than components. Taking a deeper dive into the benefits of the circular economy, which was introduced in Chapter 4 with the Caterpillar case, identifies specific impacts on business functions. It begins to paint a picture of how sustainable strategy helps business leaders achieve their own goals.

As shown in Figure 9.2, the process begins with the design phase, during which the full life cycle of the product is considered. Products are designed with end of life in mind so that they can be deconstructed into nontoxic biological materials for composting and man-made materials designed for reuse in the manufacturing process.[6] The concept is to design out waste in the process. The design stage creates parts of a product as components that can be reused in a new product, reducing the time and costs to manufacture a product. From a manufacturing viewpoint, the circular approach reduces raw materials

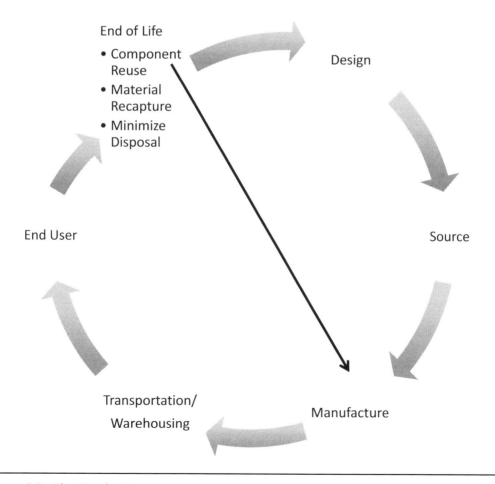

Figure 9.2 The Circular Economy

purchase and storage, energy usage, production time, waste, and disposal costs. From a finance perspective, it lowers resource requirements and results in a shorter manufacturing cycle. The contraction in production time lowers working-capital financing requirements. From a sales perspective, the new product has a great sustainability story to tell to clients in support of their own sustainability initiatives. In addition, product prices may be lower with faster delivery time, because the product creation process begins further along in the traditional manufacturing process. The circular economy is an alternative approach to creating value as it considers the production process impacts on other systems and has multiple entry points for creating value in the process. From a business perspective, the circular economy:

1. Improves efficiency
2. Reduces costs
3. Enhances cash flow
4. Addresses customers' ESG requirements for products and services
5. Proactively addresses ESG concerns; minimizing regulatory impact
6. Provides new opportunities
7. Reduces risks

Mud Jeans is leading the way in developing a new model for its customers to enjoy wearing its organic cotton jeans. One of the challenges faced by Mud Jeans's management is the volatility of cotton prices. The product is made from 30% recycled content and 70% organic cotton, thereby leaving the company significantly exposed to virgin raw material price volatility and potential supply disruption. Textile recovery rates in the European Union are relatively low, with 25% of textiles recovered per year. In order to address this challenge, Mud Jeans's CEO, Bert van Son, considered a business model in which manufacturers retain ownership of their own products and consumers pay to lease the product. As a result, a new product offering was introduced whereby customers can either purchase or lease jeans for 5 euros per month. After a year, the consumers have several options. They can:

1. Swap their jeans for a new pair and continue their leasing agreement
2. Pay a deposit and continue to wear the jeans
3. Return the jeans to Mud

The returned products can be reused through three processes:

1. Clean and re-use
2. Repair and re-use
3. Return to manufacturer for textile recycling

Bert van Son believes that this business model provides endless marketing opportunities and that it addresses the significant challenge of raw material price volatility and scarcity.[7]

The Mud case provides a glimpse of the circular economy in action. Rather than source all materials from virgin organic cotton, Mud has created a circular value chain by taking back used product and providing recycled cotton to the manufacturing facilities. Engaging a circular process has numerous benefits for Mud, including reducing its operational risk toward cotton price volatility, providing a product-recapture program to keep old jeans out of the waste stream, meeting customers' demands for a more sustainable product, reducing working capital tied up in purchases of virgin cotton, and improving operating efficiency and time to produce the product. Engaging business function leaders to reevaluate organizational processes through a circular rather than a linear lens has reduced Mud's environmental footprint while improving their operating performance. Business leaders are motivated by risk reduction, cost savings, and new product delivery structures.

The benefits of sustainability include protecting brands, image, and license to operate in a world that is constantly connected with stakeholders demanding transparency. Well-known organizations such as Nike have had their reputation damaged because of actions taken by their suppliers. While Nike outsources the manufacture of its clothing and shoe lines, its reputation and image are on the line when subcontracting factories breach human and labor rights. In 1998, Nike became the "poster child" for what not to do with your supply chain. Then-CEO Phil Knight admitted, "The Nike product has become synonymous with slave wages, forced overtime, and arbitrary abuse."[8] Sales were falling, and he believed that the American consumer didn't want to purchase products from a company that was using factories where human and labor rights were abused. In order to address the issue, Nike sought to change conditions for factory workers around the globe. They became a founding member of the Fair Labor Association (FLA), a not-for-profit collaboration of companies, universities, and civil society organizations to create solutions to end abusive labor practices. The FLA established a code of conduct for factories, created tools and resources for companies, developed training for factory management and workers, and conducted independent audits of factories. The FLA continues to advocate for greater transparency and accountability from the global supply chain. Nike undertook 600 factory audits between 2002 and 2004 and became the first in its industry to publish its list of contract factories.[9] By taking responsibility as a global brand leader for its global supply chain, Nike set a great example of how sustainability can protect and enhance a company's reputation and global brand image.

Many organizations find themselves on both sides of the sustainable supply chain, both asking their suppliers to complete surveys or scorecards and also completing them for their own business clients. A few years ago, Walmart created significant awareness and action by requiring completion of sustainability surveys from its suppliers. For many companies, this was their first foray into the field of sustainability. Walmart has moved its procurement process from requesting surveys, and later scorecards, to ranking suppliers based on Sustainability Index performance. In 2012, CEO Mike Duke announced that by 2017, Walmart would buy 70% of its products for its U.S. operations from suppliers listed on its Sustainability Index. The Sustainability Index requests information about the sustainability of a supplier's processes and products.[10] Business function leaders become more interested in adopting sustainable strategies when significant clients such as Walmart require information and compliance with sustainability standards.

Information Technology (IT) has a significant impact on an organization's carbon footprint. Data centers are large users of energy, not only for processing information but also for keeping equipment at optimal temperatures for functioning. Through working with the CIO, a sustainability champion helps rethink strategies and opportunities for IT to become part of a sustainable solution. Some IT groups are using the cloud to reduce the need for data centers. By sharing server capacity, the servers function more efficiently. Others have redesigned data centers, creating warm and cool aisles, thereby reducing energy requirements for equipment cooling.

For some firms, IT has a crucial customer service delivery and support function. New products and services are developed to reduce the product line's environmental impact, such as offering software as a service delivered via the Web rather than requiring clients to purchase hardware and software to use the product. As a result, IT facility footprints shrink further, reducing carbon emissions. Additional sustainability impacts come from developing new software solutions to help clients manage their own sustainability challenges.

Through IBM's Smarter Cities platform, a city can gather data on its waste streams in order to optimize diversion and to monetize valuable waste streams such as aluminum. Ultimately, the goal of the Smarter Cities platform is to facilitate cities moving to zero waste. Technology is offering a solution to a centuries-old problem of what to do with the human waste stream. The Rubicon Global case in Chapter 6 discusses the impact of technology as well.

Data is an increasingly important driver of management's decisions across the entire production process, from sourcing to waste. According to Thomas Odenwald, SVP of Sustainability at SAP, they purchased Ariba, a business-to-business (B2B) network where buyers and sellers can exchange data, in order to help address this challenge. The Product Stewardship Network portion of Ariba allows suppliers and product manufactures to disclose their products' sustainability information. From these data, buyers can create scorecards based on algorithms allowing for product sustainability comparisons across suppliers.[11] Management teams are very interested in their competitive ranking against key competitors.

Electronic waste is another great opportunity for IT to have an impact on sustainability. E-waste is the fastest-growing municipal waste stream in the United States. Cell phones and other electronic devices contain valuable metals as well as toxic materials. In the United States, we throw away over $60 million in gold and silver each year.[12] In addition, e-waste represents an information security risk. IT professionals need to ensure that vital data are not being compromised because of inappropriate disposal of e-waste.

From a finance perspective, issues of engagement for the CFO include risk mitigation, access to capital, operating margin, and financial reporting. From a risk mitigation perspective, examples from competitors in the industry about the potential risks and business impacts offer a meaningful topic for engagement. The CDP (formerly the Carbon Disclosure Project) is a good source for industry-specific examples of how organizations are tackling the risks of climate change and water scarcity. If the organization is a publicly traded U.S. company, the CFO should be familiar with SEC requirements for reporting on climate-related financial risk in the company's 10-K reports. According to a report prepared by Ceres, 41% of the companies listed in the S&P 500 did not include any climate-related information in their 2013 10-K filings. As a result, firms are at risk for not only the impact of climate change on their operations but also SEC noncompliance. As investors demand more disclosure of ESG information, the SEC will need to evaluate its approach to enforcing its disclosure requirement. In approaching the CFO about sustainability reporting requirements, think about what is important from the CFO's perspective:

CFO Requirement Questions

1. Is the organization launching a new product, division, or making an acquisition that will be taking the organization into the capital markets?
2. Is the firm's stock price at the level it should be relative to its peer group? Do investors discount the organization's stock price because of inadequate disclosure of ESG risks? Do existing programs and policies address environmental, social, and governance (ESG) risks?
3. Would engaging in sustainability reporting give the organization increased access to capital?
4. Are insurance underwriters asking questions about organizational response to climate-change risk?
5. Are climate-change risks disclosed on the 10-K filing with the SEC?
6. Are investors or bankers requesting more information on ESG issues?

The list could go on, but the point is that there are a significant number of issues on which to engage your organization's financial team. Select those issues that are priorities to them and demonstrate the impact that sustainable strategy can have on helping them achieve the organization's finance-related goals.

When speaking to a CFO about sustainability, use business language. Many CFOs don't think in terms of GHG emissions, alternative energy, or social capital. They think about business value and economic opportunity, using concepts such as cost, risk, profit, margins, ROI, and payback periods. Tailor the sustainability conversation to their areas of interest, speak in their language, and convert environmental and social impacts into dollar values. Do foundation work ahead of time to demonstrate how projects and programs create business value and environmental or social impact.

When speaking to the CFO about a sustainability program or project, be prepared to discuss facts and figures. If the proposal does not meet project portfolio standards in terms of ROI and payback periods, anticipate pushback and questions. Come prepared with a compelling business case. For example, the organization may not generate a significant carbon footprint, but it is in the business of helping clients manage their carbon footprints, which drives revenue creation. An example is Jones Lang LaSalle (JLL), whose business model is based on helping clients manage and operate significant real estate holdings. Lauralee Martin, CFO of JLL, shares that combining the environmental and value-creation aspects of sustainability gave her a great message to send out to investors. Through energy and other resource efficiency efforts, JLL helped clients save $125 million. Their efforts not only reduced the client's carbon footprint, it also increased the value of the client's property by $2 billion through operating costs savings.[13]

CFOs are looking for real value that can be projected, documented, and reported. They are looking for sustainable solutions that create new opportunities, generate financial returns, and reduce risk. Once CFOs understand that environmental and social stewardship drives business value creation, they become sustainability champions.

Engaging Human Resource (HR) professionals on sustainability has a significant impact on creating a culture of sustainability. From research and best practices, we know that one-off projects do not make a sustainable organization. Sustainability needs to be a core value for an organization, reinforced through practice and policy, for it to become part of the culture. As the keepers of the culture, who better to engage than your HR professionals. (In Chapter 10, we will take a deeper dive into leveraging this relationship.)

In making the case to HR, consider their drivers for talent acquisition, workforce planning, retention, compensation, and engagement. Demonstrate how sustainability can positively impact employee turnover, absenteeism, employee engagement, employee satisfaction, talent acquisition, and retention. HR professionals are uniquely positioned within the organization with a strong network of cross-functional relationships. If an organization is functional in structure, HR can be a crucial link in helping embed sustainability strategy into the organizational culture by helping to bridge functional silos.

While technology, supply chain management, and energy conservation equipment all have an impact in reaching an organization's sustainability goals, getting your people involved is the real key to transformative change. Engaging the workforce unleashes a massive resource to generate environmental and social change while driving business value. (Chapter 12 delves into this topic in greater detail.) Embedding sustainability into an organization begins with formulating a sustainable strategy that aligns with business strategy. Drilling down within each business function, work with business leaders to demonstrate alignment between sustainable strategy the function's business, goals, and targets. Laying out the business area's opportunities for sustainable impact initiates identification and selection of programs and projects. Further portfolio component refinement comes from engaging with stakeholders. Once sustainability challenges and opportunities are identified, they can be incorporated into the organizational portfolio component planning process, helping business function leaders improve their own triple-bottom-line performance: people, planet, and profit. See Figure 9.3.

Engaging business leaders is an iterative process that begins with creating awareness about sustainability and identifying connection points with functional managers. Sustainability drivers include regulatory or client compliance, new products or services, innovation, competitive advantage, brand and image, cost savings, industry standards, personal interest, and past experience. This engagement approach is tailored to the drivers or group of drivers for that manager. Engagement issues for the sales group might include a strategy to help the organization meet a client's sustainability certification threshold or complete a client's sustainability scorecard more accurately in order to improve the organization's place on the client's vendor registry. Preparing a detailed presentation on energy savings for the sales department will not garner the same level of interest and support as a plan to address client requirements for sustainable solutions. Each business manager has a connection point to sustainable strategy. Successful adoption of sustainable strategy relies on finding the most relevant connection point and creating a plan to maximize it. As internal stakeholders become more engaged in sustainable

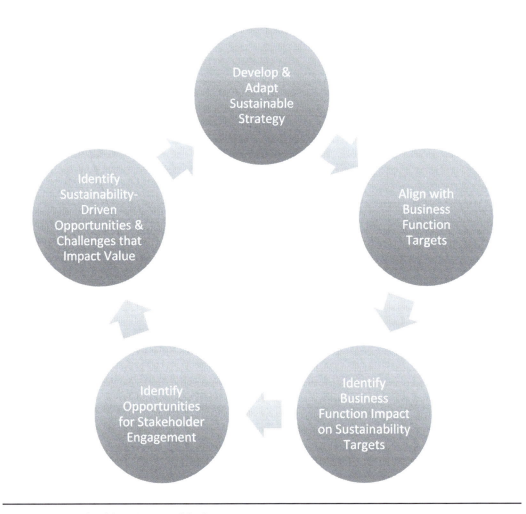

Figure 9.3 Embedding Sustainable Strategy

strategy, the more sustainability becomes a part of core business strategy. As the sustainability process matures, the more interconnected business functions become in order to address challenges and maximize opportunities. Once an organization transforms into a sustainable organization, it can develop platforms to better leverage cross-functional collaboration to provide sustainable solutions.

9.5 Best Practices for Engagement

In order to engage key internal stakeholders and bridge the gap between sustainability vision and organizational action, create clear alignment between sustainability and business value creation. The sustainability message should be positive and focused on how sustainability can help business leaders realize their organizational goals. While communicating the sustainability message can be complex, taking the time to define the benefits clearly will pay dividends in higher rates of acceptance and improved outcomes.

The following practices facilitate engagement and help to bridge the gap between sustainability vision and performance:

1. Create an organizational structure that provides decision-making and resource support for a sustainable strategy.
2. Develop a body of knowledge that provides a resource for project managers to plan, manage, implement, and control their sustainability-related projects effectively.
3. Select material industry issues and recognized standards in order to develop a common language of sustainability with clear goals and defined metrics.
4. Develop a stakeholder map and an effective engagement strategy for each stakeholder group.
5. Create training programs to educate the workforce on the organization's sustainability vision and mission, including their role in the process.
6. Develop an effective communication plan that delivers consistent messaging to both internal and external stakeholders.

While crisis management and risk avoidance strategies may get management's attention at first, the development of sustainable strategy creates long-term, business value. The key to a successful plan revolves around managing the triple constraints of time, cost, and scope of sustainability programs.

Engaging Business Leaders Across Your Organization on Sustainability Opportunities and Challenges

Kristina Kohl interview with Trudy Heller, M.S., Ph.D.

Trudy Heller, Founder and President of Executive Education for the Environment, has a proven track record of demonstrating how to integrate environmental stewardship into an organization's business model. She has worked extensively with leadership at a number of well-known organizations such as Campbell Soup, Lockheed Martin, and Merck. Trudy is an adjunct professor at both the University of Pennsylvania and Rutgers University. She has traveled the world to teach, lecture, and train on the topic of sustainability, especially about opportunities to reinvent or create new business models for a green economy.

Trudy is a Wharton School graduate, a researcher of innovation, and a well-known author who has published numerous articles. Trudy presents at conferences both in the United States and around the world.

Kristina Kohl: In your opinion, what are the top three most influential drivers of sustainability within an organization?

Trudy Heller: In my experience, senior leaders need to develop an organizational culture that empowers people to unleash their greenness within the organization. So often, organizations have employees who are very interested in environmental stewardship and are willing and able to pick up the green mantle, but their interests aren't cultivated in the work environment. They reserve their green activities for home, as they believe environmental stewardship isn't valued at work.

Customer demands are increasingly driving sustainability in organizations. It comes from a variety of avenues, both from an environmental compliance side and from a customer problem solution side. Several years ago, Chrysler approached their paint supplier, PPG, with a request to develop a solution to help reduce their environmental footprint by using less paint on their vehicles. PPG worked with Chrysler to help find a solution, thereby solidifying their position as a key supplier partner who was willing to support environmental stewardship.

Many are motivated by the cost savings impact of environmental programs such as lighting retrofits and other energy saving initiatives. In my experience, many of these projects tend to be stand-alone projects and they are often not incorporated into the business processes or culture. As a result, they have a one-time impact on cost savings and limited value for environmental stewardship.

Kristina: What do you find to be the motivation for most line business leaders to engage in sustainability?

Trudy: I find that most people are engaged through a positive message about sustainability. I use examples of innovations and creative solutions to inspire and motivate. In my experience, a doom-and-gloom perspective is a turnoff. I share real-life examples that generate creative thinking.

Kristina: What techniques have been most effective in conveying the business value generation of sustainability?

Trudy: I find that sharing a story provides a mechanism for people to relate to the topic. If it worked in someone else's organization, maybe it can work in mine. I am a big believer in interactive learning. One of my favorite exercises is to work with a group to draw out the value chains for a linear versus a circular economy. Very quickly, participants get the picture that in a circular economy there are many more value add points. One example is a brewery started by Gunter Pauli in Namibia that diverts waste by-products from landfills to support worm farms and feed chickens. In turn, the chicken manure is used to power the brewery. It is a positive story for the environment and for the company's bottom line. It is a process of changing the way we think about our business model.

Kristina: What are the most significant barriers to adoption?

Trudy: Green skeptics can be found among both environmental and business professionals. Environmental professionals often suspect that business profit motives preclude adopting environmental stewardship. They question the long-term impact of corporate sustainability without regulatory changes. Business professionals have a very limited understanding of the scope of sustainability. When I begin a conversation on sustainability, many think only of energy. It is more of an issue of education and broadening their understanding of the many aspects and impacts of sustainability.

Kristina: Are you able to share a story of a business leader who represents an effective role model for sustainable strategy adoption?

Trudy: When Volvo first launched their environmental stewardship program, "Dialogue for the Environment," they took employee engagement very seriously. Leadership hired a well-known celebrity speaker and shut down all facilities, including manufacturing, for an hour in order to launch the event. It was a symbolic action by leadership that indelibly inked the concept into the minds of the employees and into the DNA of Volvo's culture.

Several years ago, I had the opportunity to work with a gentleman in the R&D department at Kodak. He was tasked with redesigning single-use cameras so popular at the time for recycling. While his assignment was in the headquarters location, he had the vision to think globally about the issue. In choosing his team, he selected those who had a strong interest in recycling. His team was able to develop and launch an effective global recycling

> process for the single-use camera. It had a tremendous impact on Kodak's environmental footprint and it began a fabulous story to share with others.
>
> **Kristina:** How do business leaders react to your stories about innovation and new opportunity being a benefit of sustainable strategy?
>
> **Trudy:** In my experience, it is hard to get people in the door to hear the message. Once they are there, I plant the seeds for new opportunities through sustainable strategy. Years later, I hear from former students about how the lessons that I taught are impacting their decisions in the workplace.
>
> **Kristina:** Are you able to share any best practices about initiating and managing a sustainability-driven innovation project?
>
> **Trudy:** Several years ago, I worked with two different companies following their environmental stewardship programs. While each of the companies undertook an environmental project, they handled the project totally differently. The first company embraced the environmental project and made it part of their culture through changing processes within their organization. The second company simply complied with the environmental requirement. Ultimately, the first company had a much more positive business outcome.
>
> In order to adopt sustainable strategy across an organization, you need to change the mindset of the organization. I find this is best done in one-to-one conversations. It is important to have a dialogue. While a website is good as a resource for information, it isn't going to drive adoption. You need to make a personal connection.
>
> **Kristina:** Please share with us one of your favorite stories about the positive business impact of sustainable innovation.
>
> **Trudy:** My favorite story is an entrepreneurial, clean tech firm in Finland. They are mining landfill for past trash that is now recyclable. The business model creates two revenue streams. One stream is from the recycling of valuable resources and the second is from creating new space in the existing landfill for additional dumping. It is a story about looking at an old problem with a new mindset.
>
> **Kristina:** Thank you, Trudy, for sharing your valuable insights and experience with us.

Communication techniques have a significant impact on garnering management support. One-to-one meetings promote frank and open discussion. Consider the language being used to discuss sustainability. Use business language and ensure that sustainability terms are clearly defined. Many business leaders think of sustainability only in terms of the environmental impact and aren't thinking about the social, governance, and economic aspects. Discussing the risk mitigation, cost savings, employee engagement, brand and image protection, and the competitive advantages lets business leaders hear that you have an understanding of their pressing issues, and they will be more receptive to the conversation. Whenever possible, use real-life examples from the organization's industry and immediate peer group. Business leaders are very interested in the competition and how their organization ranks relative to the competition. A crucial part of this conversation is to help leadership understand that sustainable strategy goes beyond programs and projects that seek energy-efficient equipment or technology as a solution. The real impact of sustainable strategy comes from programs and projects that promote behavior change within the organization. Changing people, processes, and policies within an organization is a complex portfolio component challenge which requires senior management support as well as support throughout the organization.

In order to drive this change, ask business leader who have piloted a sustainability project to share their personal experiences with their peers. If possible, review the story ahead of time and encourage the individual to focus on behavior change and business benefits. Encouraging a peer to act as a role model for other leaders is one of the most effective ways to sell sustainable strategy to this group. One of their own is vouching for the program and backing up the sustainability presentation with real-life examples. Story telling is one of the most effective forms of communication. Through the sustainability story, demonstrate understanding of the audience by addressing issues and concerns that are relevant to them. Use the sustainability story to create awareness about sustainability within your organization and to address areas where change is needed to processes, behaviors, policies, and incentives. Leadership needs to understand that transforming into a sustainable organization is crucial for the long-term viability of their organization. They need to begin to take ownership of the sustainability program to drive this change.

Encourage senior management to share and promote success stories both internally and externally. Share not only the results but also effective strategies and techniques. Create an organizational structure to promote sustainable decision. Form a cross-functional steering committee to evaluate and recommend sustainability projects and programs. Members should include senior-level representatives from each of the functional areas to serve on this committee. Projects and programs approved by the steering committee should include commitments of resources such as people, time, and budget. As the process unfolds, the business unit leaders rather than the CSO become the drivers of sustainability programs and projects as they realize the beneficial impacts of sustainability on their operation.

9.6 Measuring Engagement

How can sustainability champions measure the impact that they are having on internal stakeholder engagement? Here are some indicators:

1. Business groups include sustainable strategy in their planning process.
2. Organizational sustainability goals have been established and cascaded through divisions and departments to the individual level.
3. Sustainability is part of programs and projects. Tasks that promote sustainability are included in work streams.
4. Presentations from senior and middle management include tracking their sustainability results against targets.
5. Management is evaluating and changing their sustainable strategy and process to improve results.
6. Sustainability terms and goals become part of the regular business conversation with business leaders and their teams, promoting sustainable strategy as part of their function.
7. Function leaders and their teams embrace sustainability, coming up with their own ideas and projects.
8. Sustainability projects come from across the organization and not just from the Chief Sustainability Officer.
9. Rewards and compensation include sustainability goals at both the employee and managerial levels.
10. Metrics and analytics include tracking results against sustainability targets.
11. Internal and external messaging about sustainability is consistent.
12. Employees can identify the sustainability vision of the organization and articulate their role in the process.

9.7 Project, Program, and Portfolio Management Impact

Managing internal stakeholders is a familiar process to project management professionals; however, a sustainable approach provides a different lens through which to consider these relationships. In order to move pilot sustainability projects into organization-wide projects, sponsors and project managers need to consider their organization's culture, readiness for change, and sustainability maturity in selecting appropriate levers and drivers for a successful business outcome. As an organization's sustainability journey evolves, the internal stakeholder management process evolves as well. Moving into cross-functional project and program management requires support from a broad base of business leaders, including access to their budget, time, and personnel. Promoting success and celebrating business leaders who support sustainability is a useful tool to motivate internal stakeholders. The goal of program and project managers is to engage with stakeholders to be better informed about issues, concerns, and requirements. As internal stakeholders become more engaged, the value proposition of sustainability becomes clearer. Through this process of engagement, stakeholders move from blockers and bystanders to champions and influencers. As an organization matures in its sustainability journey, the deeper the concepts of sustainability are embedded into the program and project management processes.

9.8 Conclusion

Internal stakeholder management is a vital link to embedding sustainability into an organization. The more broad-based is acceptance and adoption of sustainable strategy by business leaders, the greater is the impact on organization policies, processes, and people. The key to leveraging internal relationships is to demonstrate the specific value creation capabilities of sustainable strategy by business function. Once a sustainability champion is able to garner support, leverage those relationships to encourage greater leadership participation in specific sustainability programs and projects. As the sustainability wave moves through management, it carries change to all levels of the organization. In order to gain budget, time, and other resources for sustainability programs and projects, business function leaders must be educated and informed about the value proposition of sustainable strategy. Once these key alliances are formed, an organization can progress on its sustainability journey.

Notes

[1] Katherine Hammack, "Business Takes the Lead: How Innovation Will Drive Our Mitigation and Adaptation to Climate Change," Wharton School, University of Pennsylvania, April 22, 2015.

[2] Peter Graf, "Sustainability and Business Innovation," openSAP Course, Week 5, Unit 1, "Engaging Line of Business Leaders," April 29, 2014–June17,2014.

[3] Jason D. O'Grady, "Apple Shoots Itself in the Foot with EPEAT Withdrawal (Updated)," *ZDNet*, accessed November 24, 2014, http://www.zdnet.com/apple-shoots-itself-in-the-foot-with-epeat-withdrawal-updated-7000000658.

[4] Ibid.

[5] Initiative for Global Environmental Leadership, "Disrupting the World's Oldest Industry," March 2014, http://d1c25a6gwz7q5e.cloudfront.net/reports/2014-03-06-Disrupting%20the%20World's%20Oldest%20Industry.pdf.

[6] Ellen MacArthur Foundation, "The Circular Model—An Overview," *Ellen MacArthur Foundation*, accessed November 24, 2014, http://www.ellenmacarthurfoundation.org/francais/leconomie-circulaire/the-circular-model-an-overview.

[7] "Mud Jeans—Case Studies," *Ellen MacArthur Foundation*, accessed November 24, 2014, http://www.ellenmacarthurfoundation.org/case_studies/mud-jeans.

[8] "How Nike Solved Its Sweatshop Problem," *Business Insider*, accessed December 1, 2014, http://www.businessinsider.com/how-nike-solved-its-sweatshop-problem-2013-5.
[9] Ibid.
[10] Initiative for Global Environmental Leadership, "Disrupting the World's Oldest Industry."
[11] Ibid.
[12] "11 Facts About E-Waste | DoSomething.org | America's Largest Organization for Youth Volunteering Opportunities, with 2,700,000 Members and Counting," accessed December 1, 2014, https://www.dosomething.org/facts/11-facts-about-e-waste.
[13] Don Bray, President, AltaTerra Research, "CFOs Seek Boldness and Shareholder Value in Sustainability," *Environmental Management & Sustainable Development News*, accessed November 28, 2014, http://www.environmentalleader.com/2012/06/13/cfos-seek-boldness-and-shareholder-value-in-sustainability.

Chapter 10

Leveraging Human Capital Professionals

An often overlooked internal stakeholder group during the sustainability transformation process is the Human Resources (HR) department. When I mention the value that human capital professionals contribute to the sustainability process, many colleagues give me a strange look. Some ask, "What do they have to do with carbon emissions or environmental stewardship?" The answer is "A lot." The reality is that over 30% of the measured criteria for industry standard sustainability compliance reporting falls within the areas of expertise of human capital professionals. In order to move an organization's sustainability program from one-off projects to a fully integrated program, HR leadership needs to be involved in creating and executing sustainable strategy. People are the key to your success in becoming a sustainable organization. If you consider the sustainability journey of an organization, the growth of the organization during the journey is based on how well the triple bottom line concepts (people, planet, profit) of value creation are embedded into the organization. One-off projects such as installing a building management system to reduce energy usage can help an organization begin its sustainability journey; however, in order to truly transform into a sustainable organization, sustainability needs to be integrated into all functions of the organization.

In order to transform an organization, the culture must be changed to make sustainability part of organizational core values. As an organization moves forward on its sustainability journey (see Figure 3.2, Sustainability Journey), each of the steps depicts sustainability becoming more embedded within the organization. The differentiators between these steps deal with leadership, policies, processes, and people. In order to reach the transformation phase, an organization's employees must be fully engaged in the sustainability process and aligned around its core business mission. Reaching this phase is not easily achieved and requires fundamental change in policies, people, and process. According to a Society for Human Resource Management (SHRM) survey, less than 10% of respondents are really able to achieve transformation.[1] Stakeholder engagement, change management, and communication planning are integral to transforming an organization. In order to succeed, project and program managers responsible for sustainability projects need human capital professionals as part of their team. HR professionals are experts in cross-functional communication, change management, collaboration, and human capital policies and procedures. Leveraging the relationship between the Sustainability Office and Human Resources has a significant impact on the overall adoption and success of the sustainability program.

10.1 Human Capital Relevance in Reporting Frameworks

We dive deeper into sustainability reporting paradigms in Chapter 14, but it is important for chief sustainability officers (CSOs) and sustainability champions to note that sustainability reporting frameworks such as the Global Reporting Initiative (GRI) include a significant number of questions about issues related to organizational human capital policies and procedures. Figure 10.1 shows GRI G4 categories and guidelines.

Approximately a third of the performance indicators included in the GRI G4 Guidelines for Sustainability Reporting are within HR's areas of expertise. In order to produce a GRI-compliant report, HR leadership needs to be part of the strategic development process for creating and implementing an operating and reporting structure to support this level of sustainability compliance. If HR leadership is simply asked for data to complete the report, the functional and behavior changes required to meet these standards will not be achieved. While GRI is a reporting framework, fundamental changes must happen within the policies, processes, and daily behaviors of employees in order to achieve GRI performance standards. Human capital professionals need to be part of the sustainability team in order to effect the type of organizational and culture change needed to achieve GRI reporting standards.

Looking more closely at the GRI standards, Figure 10.2 shows the GRI Social Indicator Subcategories. The reporting requirements are broken into Labor Practices and Decent Work, Human Rights, Society, and Product Responsibility. Clearly within HR's wheelhouse are labor and work structure issues such as employment, labor relations, health and safety, training and education, and equal pay by gender. Human rights components such as nondiscrimination, eliminating child labor, freedom of association and collective bargaining, and policies to avoid using forced labor also fall under HR's area of expertise. In addition, HR professionals can be called on to collaborate on guidelines and best practices to avoid unfair labor practices within the supply chain. While it is common to think about the material, energy, water, and waste components of sustainability reporting, compliance with GRI reporting requires significant disclosure of human capital management, policies, and practices. In order to have a meaningful sustainability program that addresses industry standards, HR professionals must be engaged and part of the sustainability team.

The Dow Jones Sustainability Index, which is produced by RobecoSam, also requires detailed information on human capital management practices. (See Figure 1.2, Dow Jones Sustainability

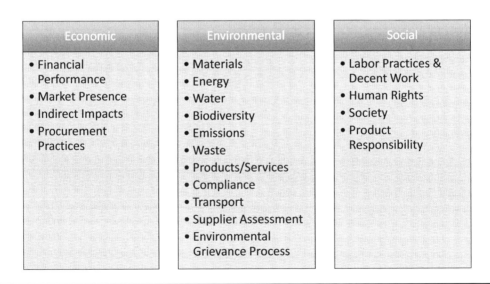

Figure 10.1 GRI G4 Categories and Guidelines[2]

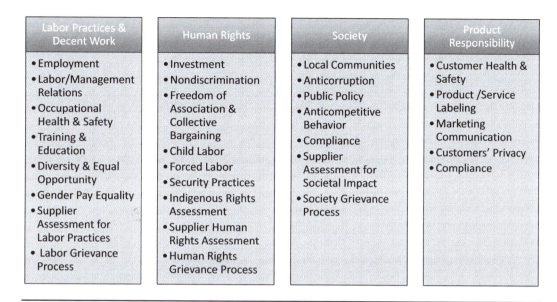

Figure 10.2   GRI Social Subcategories[3]

Assessment Criteria.) According to RobecoSam, key changes in the 2014 criteria included increased focus on materiality in Occupational Health & Safety, Employee Turnover Rate, and Transparency of Senior Management Remuneration.[4] Human capital functions play a significant role in the practice of sustainable strategy, and recognized industry standards for reporting and ranking agencies focus on the human capital impact of sustainability as part of assessments. Developing and leveraging a relationship with HR leadership on a strategic level is crucial to creating an organization's sustainability program. As with other business functions, partnering with HR is best approached through a strategy of mutual benefit. Offering HR leadership a seat at the sustainable strategy table provides them with a voice in the organization's sustainability vision and a strong motivation to provide valuable resources, tools, knowledge, and solutions. Including HR leadership as part of the Sustainable Strategy Steering Committee is an effective way to include senior HR leadership in the strategic development process. Leveraging this relationship provides the CSO or sustainability champion with project team members who are experienced in HR compliance, employee engagement, change management, cross-functional networking, and various channels of internal and external communications. All of these skills are extremely valuable when creating a culture of sustainability within an organization.

10.2   Human Resource Alignment

How do you convince HR leadership that they should be interested in sustainability? Start by creating the business case for HR; demonstrate the impact of adopting sustainable strategy on their key goals such as attracting and retaining top talent. In a recent survey by Ethical Corporation of 1500 sustainability professionals, 53% of corporate respondents indicated that sustainability directly impacts HR functionality.[5] In fact, sustainability professionals' opinions are that human capital management (HCM) is the third most impacted business function by sustainability, behind marketing/communications and supply chain/procurement. As shown in Figure 10.3, sustainability drives HCM benefit.

As an organization moves from making a sustainability commitment into creating an organizational structure that supports stakeholder engagement, internal and external stakeholders communications, and sustainable operations, human capital professionals have a significant role to play. Through

Figure 10.3 Sustainability Drives HCM Benefit

developing projects, programs, and processes to facilitate alignment between sustainability goals and business goals, HCM professionals begin to create a sustainable culture. Tools used in the process include training employees on the benefits of focusing on the environment, society, and governance considerations while performing their core function. Strategic alignment is further reinforced through incentives to support sustainable behavior and to promote innovation around a sustainable approach. An organization promotes its image as an employer of choice by offering employee programs for health and wellness, and promoting a culture of inclusion through recruiting practices, compensation equality, and flexible work arrangements. As HCM professionals develop, communicate, and facilitate the change management process in order to promote sustainability programs throughout the organization, sustainability becomes adopted as a core value of the organization. This approach impacts both internal and external stakeholder perceptions. As external stakeholders begin to see that the organization has truly embraced sustainable strategy as a core mission, other human capital benefits accrue to the organization, such as attracting top talent, engaging employees, employees acting as ambassadors, innovation, and higher profitability for the organization. Surveyed sustainability professionals reported that from their experience, sustainability is important in engaging employees (86%) and stakeholders (90%).[6] When sustainability professionals were asked about 2015 sustainability opportunities for their organizations, the top three responses were

1. "Embedding Sustainability"
2. "Sustainability Innovation"
3. "Culture of Sustainability"[7]

Each of these opportunities requires relying on core HR strengths, including engaging stakeholders, effective communication, and change management. In sustainable organizations, senior leadership is focusing on sustainable strategy as more than just regulatory compliance and cost savings. They view sustainability as a means to create value, with sustainable strategy being a source of competitive advantage, an avenue for innovation, an opportunity for new products and services, and risk mitigation. HCM professionals are a key resource to effect the organizational change needed to transform into a sustainable organization.

Engaging HR leadership through sustainable strategy in order to align capabilities and functionality around core business goals gives HR a "seat at the table." So often, HR is perceived as transactional rather than strategic in focus. Facilitating the process for HR leadership to have a more strategic role in developing a business vision for sustainability creates a strong alliance for the CSO or sustainability champion. As we saw in Chapter 9, one of the key ways to demonstrate the impact of sustainability strategy to internal stakeholders is through identifying the benefits to that functional area with an alignment matrix. Figure 10.4 provides a more granular look into the benefits of sustainable strategy to human resource management.

Focusing on the benefits to HR leadership, such as workforce planning and development, employee engagement, work/life balance, diversity and inclusion, total rewards, and incentives designed to attract, motivate, and retain the best employees, creates alignment with core HR strategic mission. Research indicates that potential employees, especially the Millennial generation, are looking at an organization's corporate social responsibility (CSR) plan and performance, and environmental policies and outcomes, as part of their decision-making process. According to a 2012 Net Impact study, 53% of all respondents felt a corporation's social responsibility performance was an important factor in selecting an employer, and 72% of college students (the Millennials) said that they were looking for a job in which they can make a difference.[8] At a job fair hosted by the Initiative for Global Environmental Leadership at the Wharton School, University of Pennsylvania, students were very interested in opportunities to work for a company where sustainability was a core business pillar. One student shared her story of accepting a position with a European firm that seemed to embrace sustainability based on its public messaging and sustainability reporting. Once the student joined the firm, it became apparent that sustainability issues were not really part of the firm's core business values. In practice, the sustainability report was done to be in compliance with regulations. While the organization appeared to have a culture of sustainability

Figure 10.4 Alignment Matrix: Engaging HR Leadership

from an external view, the actual culture was not in alignment with a mission of sustainability. Needless to say, the person was seeking an opportunity to find another employer better aligned with her view on the importance and value of sustainability in an organization.

> **Greenwashing's Impact on Employees**
>
> Firms are often accused of "greenwashing" when they make environmental or social claims about their products or services that are not demonstrable or entirely accurate. While we think of greenwashing as being an issue for external stakeholders such as customers or consumer watchdog groups, it can have an impact on internal stakeholders such as employees as well. A core foundation of sustainable strategy is belief by employees in the organization's sustainability mission. Management that presents one message to external stakeholders and another to internal stakeholders runs the risk of being perceived as greenwashing their own employees. If management represents to external stakeholders that they have sustainability programs and goals but they don't incorporate them into their policies and processes for employees, there is an inherent mismatch between the external messaging and internal actions. As a result, management runs the risk of disenfranchising their employees, which can lead to disengagement, turnover, poor customer service, and bad public relations. In order to be effective, an organization's sustainability message must be authentic and align with internal policies and processes. Otherwise, the positive impact of a sustainability story to external stakeholders, including potential new recruits, will turn into a negative experience. Given the ease of communication and 24/7 access to information, an organization's disconnect between sustainability messaging and actions will not remain a secret for long. Greenwashing damages brands, reputations, and employee relations.
>
> In early 2002, for example, Nike was sued for misrepresentation of working conditions and workers' rights in its factories (*Kasky v. Nike, Inc.,* 45 P. 3d 243, 119 Cal. Rptr. 2d 296, 27 Cal. 4th … - Cal: Supreme …, 2002). The suit arose from Nike's statements in press releases, in letters to university presidents and athletic directors, and in other publicly distributed documents about rights and benefits afforded workers at factories, such as wages equal to double the local minimum wage, free meals, health care, and compliance with local laws. This image was shattered when an internal report documenting work-rule violations and unsafe working conditions at a Vietnamese factory was leaked by an internal source to the public. While Nike has made tremendous strides in correcting these issues in the ensuing years, the example is a good one for demonstrating the negative impact of greenwashing on brand, reputation, and employees.[9]

Results from a survey conducted by the Society for Human Resource Management (SHRM) of senior human resource leaders in the United States confirms that they find sustainability important for attracting employees. Almost 90% of respondents indicated that sustainability creates a positive employer brand that assists in attracting top talent.[10] In order to thrive, organizations must build a workforce that has the depth and breadth of skills to address the challenges and opportunities presented by global megatrends facing governments, society, and private enterprise. (See Chapter 1 for a detailed list of megatrends.). A broad set of talents is needed to engage global stakeholders, lead diverse companies, identify new opportunities, and innovate creative solutions to address the world's challenges. Highlighting an organization's sustainability story in a transparent and authentic manner helps HCM professionals keep the talent pipeline full of high-quality candidates who possess the skills needed to address these challenges.

10.3 Impacting Global Challenges

HCM professionals are part of the solution for globally challenging threats and opportunities. As population rates in developing countries accelerate relative to more stable growth rates in the developed world, we will see an increase in diversity of the global population and greater pressure for both public and private institutions to manage diverse workforces. In order to take advantage of the business opportunities created in developing countries, organizations need to create a culture of inclusion that respects cultural differences. Attracting and managing a diverse workforce including race, gender identification, ethnic identification, cultural beliefs and norms, age, and education, to name a few, is a significant management challenge. Engaging with HCM professionals to better address these issues within an organization's workforce creates a talent pool that is better prepared to address the challenges and opportunities of a changing world.

Gender equality in terms of rights, pay, and access to opportunity continues to be a significant challenge globally for organizations. In order to highlight global gender diversity challenges, the Millennium Development Goals includes them alongside other global development challenges such as poverty and disease. Numerous governments, international organizations, civil society groups, and academic institutions have issued statements of support for the Millennium Project.

> **The Millennium Development Goals**
>
> United Nations members adopted the Millennium Development Goals (MDG) as a result of the Millennium Summit in 2000. The eight global development goals address issues including halving extreme poverty, halting the spread of malaria and HIV/AIDS, providing universal primary education, empowering females and gender equality, reducing infant mortality rate, improving maternal health, ensuring environmental sustainability, and building global partnerships to support continued development. While strides have been made in reducing extreme poverty, improving drinking water, reducing malaria rates, and improving maternal and infant health, challenges remain, including the acceleration of global greenhouse gas (GHG) emissions, uneven access to education, and the persistence of gender-based inequality. Globally, gender inequality remains an issue, with women unable to gain equal access to education, jobs, and economic opportunities.[11]

In both developed and developing countries, women continue to be relegated to less secure jobs and are paid less than men performing similar jobs. Research by Michelle Budig, a sociology professor at the University of Massachusetts, found that even in the United States, a "motherhood penalty" exists for working mothers, while a "fatherhood bonus" is given to men who have families. On average, men's income increased 6% when they had children. For working mothers, earnings decreased 4% for each child they had in their household.[12] Even in a developed country such as the United States, we have a cultural bias against working mothers. Savvy business leaders are aware of the importance of empowering women and grooming them as leaders, in order to build new markets, support innovation, and improve productivity within their organizations. It is no surprise that the "Best Place to Work" companies have programs in place to empower their workforce, including women and working mothers. If you are questioning the financial benefits of these types of programs, consider these results from Russell Investment Group research on the performance of publicly traded "Best Companies," compared with the S&P 500 and the Russell 3000 indexes, which indicates that returns for the "Best Companies" from the period 1998 to 2009 were over 224% compared to 42% (S&P 500) and 47% (Russell 3000).[13] Collaborating with HCM professionals to develop employee engagement programs for a diverse workforce clearly adds value to an organization's financial bottom line while improving its reputation and image.

10.4 Employee Engagement

Employee engagement is another crucial Human Resource function. According to Gallup, the cost of employee disengagement in the United States is between $450 and $500 billion[14] (see Figure 10.5). Approximately 30% of the workforce is engaged, 20% of the workforce is disengaged, and 50% is merely present.[15] Employees who are engaged are more productive and have better customer satisfaction ratings, less absenteeism, and lower rates of work-related accidents. Organizations with higher levels of employee engagement have significantly better financial performance than their peers. Organizations with world-class engagement rates generate earnings per share that approach 1.5 times greater than their competition.

Employee engagement is a definable and measurable metric. While we all know intuitively that we do a better job when we are interested and excited about a project, it is possible to provide tangible metrics about employee engagement to support your business case with HR professionals. Table 10.1 lists some examples of human capital metrics that can be used to evaluate employee engagement.

In the UNGC/Accenture survey discussed in Chapter 2 (Figure 2.2, CEO Top Drivers of Sustainability), CEOs reported that one of the key drivers for their investment in sustainability programs is the impact on employee engagement and recruitment.[17] The employee engagement category has risen in importance to CEOs since the previous 2010 survey as management's sustainability maturity has increased with their experience level. This viewpoint is shared by senior HR leaders, with 85% of those surveyed indicating that they believe sustainability is an important factor in attracting and retaining employees.[18] These surveys indicate a strong linkage between sustainable strategy and employee attraction, engagement, and retention. Using defined metrics, management is able to measure the impact of sustainability efforts on employee engagement. One of the most commonly monitored metrics in HR is the turnover ratio. Demonstrating to HR leadership that sustainable strategy reduces employee turnover and the associated costs, disruption, and training of new people is a hot button for enlisting HR support for sustainable strategy.

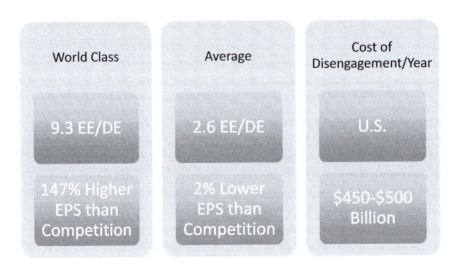

Figure 10.5 Impact of Employee Engagement[16]

Table 10.1 Examples of Human Capital Metrics for Engagement

Metric	Measurement
Employee Turnover Ratio	Employee Separations/Total Active Employees
Productivity	Output/Total Employees
Absenteeism	Average Number of Days Absent per Annum
Employee Retention Rate	Percentage of Employees Retained Annually
Employee Satisfaction	Annual Employee Survey Results

Employee engagement and the return on investment on work/life balance and on health and wellness programs are also very convincing. Firms with flexible work arrangements were found to have 3.5% higher market value than their peers.[19] These arrangements include flexible hours, job sharing, flexible schedules, and working from home. Companies ranked as "Best Places to Work," such as Intel, go even further, providing sabbaticals for employees. In 2013, a total of 4,067 employees took advantage of Intel's 8-week sabbatical, which is offered to employees every 7 years.[20] Companies that offer this level of employee benefit cite the positive impact the program has on attracting and retaining the best and the brightest employees.

Another sustainability lever for HR professionals is offering health and wellness and work/life balance programs to employees in order to improve engagement and overall satisfaction as well as to promote a healthier lifestyle. Examples of some of these programs are given in Table 10.2.

Table 10.2 Health and Wellness Examples[21]

Organization	Program	Return	Benefits
Aetna	Extending Job Guarantees for New Mothers	$1,000,000	Retained Qualified Employees
KPMG	Emergency BackUp Child Care	521% ROI	Reduce Absenteeism
Johnson & Johnson	Wellness Program	$225 per employee per year	Improved Health & Reduced Health Care Expense

Research indicates that these programs can save organizations significantly more than they cost—Medical costs fall $3.27 for every dollar spent on wellness programs, which is an ROI of 327%.[22] In addition, absenteeism expenses fall by $2.73 for every $1 spent on wellness programs.[23] These figures are impressive in their impact and provide meaningful support for advocating for human capital sustainability programs for employees. While the results are impactful, offering programs to engage employees on lifestyle risk issues such as stress, obesity, and exercise is challenging. The following are best practices from U.S. companies that have achieved a higher rate of employee engagement in their own health and wellness programs.

Best Practices to Engage Employees[24]

1. Senior Management Support for Programs
2. Develop a Strategy and Plan for Execution
3. Enlist Managers as Role Models
4. Utilize Employee Engagement Strategies Such as Incentives
5. Develop a Communication Plan and Communicate Frequently
6. Make the Process Easy
7. Establish Metrics and Measure Outcomes

Project management professionals are familiar with many of these best practices. They provide a useful framework on which to collaborate with HCM professionals to engage employees in sustainable strategy. Following these best practices ensures that the programs launched generate the positive impacts anticipated.

10.5 Creating Business Value

One of the challenges of programs and initiatives that impact employees' engagement is measuring the impact of this intangible relative to the organization's financial performance. HR professionals consistently struggle to provide metrics that demonstrate the financial impact of their programs. Often, HR's role is perceived as transactional rather than strategic in nature, partially because of the challenges involved in measuring the impact of intangibles, such as the financial impact of employee engagement. From an HCM's professional perspective, "not having a seat at the table" is one of the top mentioned challenges in the human capital management field. This comment refers to HR leadership not being included in the strategic decision-making process of the organization. Embracing sustainable strategy is a means to getting a seat at leadership's strategy table. In order to engage HR, CSOs and sustainability champions have research and case studies that substantiate the beneficial impact on human capital management. In addition, they offer frameworks that provide a set of tools and metrics that connect the impact of improved employee engagement and retention to financial performance.

An example of a firm demonstrating the link between sustainable strategy, employee engagement, and financial impact is SAP. The SAP Integrated Report 2013 (http://www.sapintegratedreport.com/2013/en) provides a comprehensive assessment of their operation, including the social, environmental, and economic impacts on the firm's performance. The Integrated Performance Analysis, Strategy, and Business Model is especially useful in understanding the interconnectivity of sustainability strategies and their impacts on business operations. Tracing through the model, it becomes clear that actions performed in an area such as environmental stewardship impacts outcomes in other areas such as employee engagement. To promote environmental stewardship, management established GHG emission-reduction targets and reports on the firm's progress relative to the targets. Initiating programs and tracking environmental performance encourages employees to take pride in SAP for addressing air quality and pollution in their local and global communities. As a result, employees act as ambassadors championing their firm's environmental achievements. Because the experience is positive for the employees, they become invested in furthering the environmental success of the firm. As a result, they begin to identify new opportunities to save energy and reduce GHG emissions. Further promoting environmental stewardship through collaborating with HR to develop recognition and incentive programs results in additional cost and resource savings, generating significant bottom-line impact.

Another organization, Alcoa, has received over 1000 ideas to promote sustainability submitted by employees and turned them into over $100 million in savings.[25] Developing an engagement program for employees to contribute ideas to promote sustainability goals provides significant economic as well as environmental and social benefits.

In order to better understand the impact of the Integrated Performance Analysis, Strategy, and Business Model on SAP's human capital, Figure 10.6 portrays a subsection of the model that traces the impact of human resource activities on employee engagement and retention.

The Cause-and-Effect Chain for Business Health Culture Index (BHCI) explains the interconnectivity between nonfinancial and financial indicators, specifically, how improved employee engagement and retention creates business value. The SAP model provides a process to link activities that impact an intangible such as a health and wellness campaign through the BHCI, which measures the health of employees and the health of the organizational culture, to impacts on revenue and operating margin. The cause-and-effect chain begins with activities on the left side of the diagram designed to

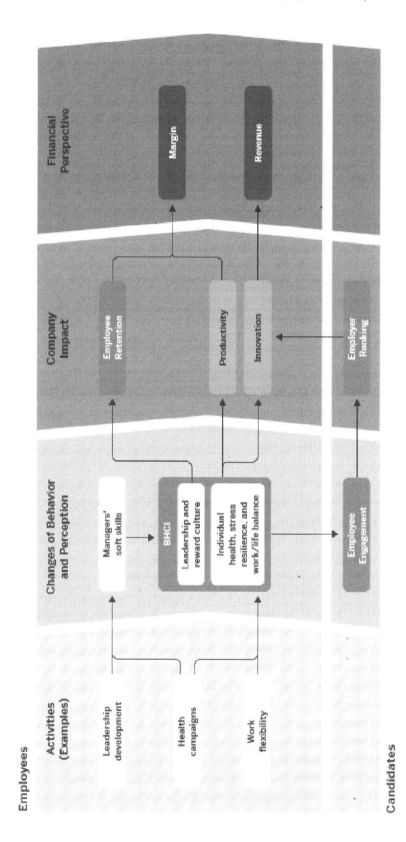

Figure 10.6 SAP's Cause-and-Effect Chain for Business Health Culture Index. (© 2013 SAP SE. All rights reserved. Reproduced with permission of SAP SE.)[26]

support improved leadership, better employee health, and flexible work programs. SAP management finds that these activities improve their organizational culture and employees' morale through better work/life balance, more empowerment, and lower stress. Ultimately, these activities improve employee engagement, productivity, and retention. Strong employee engagement improves SAP's employer ranking, making them an employer of choice and helping them attract the best and brightest personnel. Attracting a top talent pool drives innovation, creating new products and services for clients, which enhances revenues.

Leadership development improves mangers' skills and supports building a sustainable organizational culture, which drives productivity and employee retention, leading to improve operating margins. SAP reports that for every percentage point change in employee retention, operating profit is impacted by 62 million euros.[27] This impact is bi-directional in that improved employee engagement adds to profits and employee disengagement detracts from profits. Promoting engagement in order to minimize employee turnover and the associated costs of recruitment, training, productivity loss, and client disruption makes a compelling case for the value of human capital programs that align with sustainable strategy. Providing HR leadership with tools and financial metrics to build the business case for human capital programs designed to build and support a sustainable culture completes a crucial piece of the puzzle to get HCM professionals onboard with sustainability initiatives.

10.6 Integrating Sustainability into Human Capital Management

Once senior management supports the value added by HR involvement in the sustainability process, project management tools such as the Deming cycle, the continuous improvement process based on plan, do, check, and act, are useful for planning and integrating sustainability into human capital functions and processes. Begin with the organizational vision statement about sustainability. Engage with key internal stakeholders, gather requirements, and identify materials issues on which HR can be impactful. Develop a plan that focuses on the organization's unique skills and opportunities and the role that HCM professionals can play in promoting and engaging internal stakeholders to maximize the beneficial impact of sustainability on the organizational culture. The cycle is illustrated in Figure 10.7.

Next, consider the types of information and programs that need to be communicated and implemented in order to get your people interested in the organization's vision of sustainability. What types of initiatives have an impact on employees? Are changes needed to the organizational structure, work arrangements, training, and incentives programs? Understanding the employee population and their needs in the process has a big impact. How can we encourage our people to think differently about their jobs, the environment, their health and wellness, their communities?

When Walmart thought about how best to adopt a sustainable strategy, management knew that they needed to get their employees to understand sustainability on a personal level before sustainable strategy could have an impact at the organizational level. They developed a program called "My Sustainability Plan" (MSP), designed to help their associates select and track personal sustainability goals in three major categories, as indicated in Table 10.3.

Over 100,000 Walmart associates have selected personal MSP goals, tracked their progress, and inspired others to undertake their own sustainability journey.[29] Engaging in the MSP program helps employees understand that their individual actions can make a big difference not only to themselves but also around the globe. They are not just improving their own lives but are also acting as ambassadors spreading the message to fellow associates, family members, friends, and community members.

Brooke Buchanan, Communications Director at Walmart, made the following comment about the program: "What makes MSP so successful is that anyone can get involved. It's not simply about working at Walmart, but making a personal commitment to be healthier, greener, or more active in the community. We are living better and making the world a better place—one associate at a time."[30]

Leveraging Human Capital Professionals   221

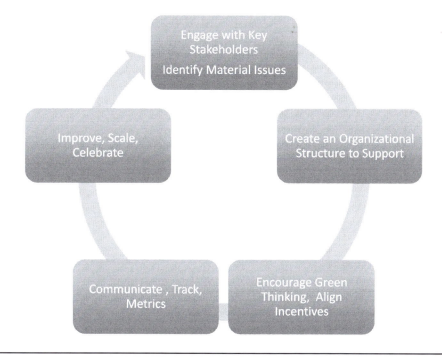

Figure 10.7 Integrating Sustainability into HR Planning and Processes

One of the primary goals of MSP is to develop personal foundational steps on which Walmart can engage its workforce and then use that base to address broader organizational sustainability issues. The MSP program has been a success with employees, especially Millennials, who are seeking employment with organizations that demonstrate a dedication to core values that include stewardship of the environment and contribution to local communities and global society.

As part of their commitment to sustainability, Walmart shares this best practice under a royalty-free MSP license arrangement. More than 100 organizations have taken advantage of this resource to build engagement within their own employee base.

Realigning the incentive system to reward behaviors that support sustainability goals facilitates employee behavior change. HCM professionals play a crucial role in establishing and implementing compensation and total rewards for employees. Incorporating sustainability criteria as part of a performance scorecard across all levels of the organization has a significant impact on employees' interest and willingness to focus on sustainability concerns such as energy savings, waste reduction, sustainable solutions, and innovative tools and techniques. Aligning incentives with sustainability targets and goals is a meaningful change management tool to produce desired behaviors.

Table 10.3 Walmart's My Sustainability Plan[28]

My Health	My Planner	My Life
Stress Reduction	Water Conservation	Learning New Skills
Healthy Eating	Waste Reduction	Money Management
Increased Exercise	Energy Savings	Voluntarism
Quitting Smoking	Biodiversity Preservation	Quality Time

The right communication plan with appropriate communication channels improves engaging employees in sustainability programs. In the Walmart example, associates' use of a social media platform to discuss their MSP actions, best practices, and success stories with each other promoted widespread adoption of the MSP program. Consider how your organization shares information. Is it a formal or an informal process? Are your people centrally located, or are they spread around the world? All of these factors should be considered in developing the communication plan. (We will discuss communication plans in detail in Chapter 11.)

As with any project, metrics matter, because success can only be achieved when a successful outcome has been defined and agreed on in advance. Engage with senior business leaders, HR management, and the sustainability champion and agree on which human capital metrics measure employee adoption of sustainable strategy. Establish a mechanism to track and gather data to measure the agreed-on metrics. Without meaningful metrics, the human capital sustainability program is measured on antidotal rather than tangible results. Gather your lessons learned from projects, scale successful pilots, and celebrate successes with the team. Then, begin the planning process again.

10.7 Relevance to Project Management Professionals

Successful projects rely on collaboration. Project management professionals need to understand who within the organization holds the roles of champion, decision maker, facilitator, or blocker relative to their project. Because of HR's unique cross-functional position within an organization, HCM professionals offer considerable expertise in managing complex internal relationships. They know everyone's name and phone number. They can make introductions and get you on the schedule of someone who has been unavailable to you. They are well versed in the culture of the organization as well as the unique cultures of various divisions. Many organizations are a sum of very distinctive parts, and getting collaboration across divisional management is extremely challenging.

As a program manager tasked with reducing an organization's GHG footprint, consider the stakeholders in a solar energy project for one of your factories. As indicated in Table 10.4, each of these stakeholders has different issues and concerns. Many of them don't know each other very well. The one relationship constant is that each of these stakeholders has a relationship with HR.

Table 10.4 Internal Stakeholder Requirements

Internal Stakeholder	Role	Issues/Concerns	Relationships
Factory Manager	Blocker	Energy Projects & Production Upgrades Not Compete for Capital Funding	VP/Manufacturing Controller HR
VP of Sales	Champion	Prospect Requesting Information on Factories Use of Alternative Energy	VP/Manufacturing HR
Controller	Influencer	Alternative Energy Proposal Meeting ROI Hurdles	Factory Manager VP/Manufacturing HR
VP of Manufacturing	Decision Maker	Quality Product & GHG Emission Targets	Factory Manager Controller HR
HR Generalist	Influencer	Employee Engagement	Factory Manager VP of Sales Controller VP/Manufacturing

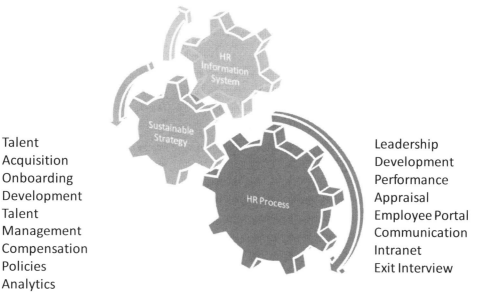

Talent Acquisition
Onboarding
Development
Talent Management
Compensation
Policies
Analytics

Leadership Development
Performance Appraisal
Employee Portal
Communication
Intranet
Exit Interview

Figure 10.8 Integrating Sustainability into Core HR Functionality

Successful sustainability program and project implementation depends on internal stakeholder engagement and collaboration. Sustainability initiatives leverage interconnectivity between functions and departments to offer better, more impactful solutions. HCM professionals have cross-functional networks and relationships that can be leveraged to help project management professionals effectively manage cross-functional portfolios, programs, and projects. Developing an alliance with HR leadership and including HCM professionals on project teams facilitates effective communication and change management across the organization. HR professionals' communication and collaboration skills facilitate engagement with both internal and external stakeholders, helping to move projects toward successful completion. HR's collaboration skills promote project ownership, stakeholder trust and commitment, and project team interest and excitement. Leveraging the skill set of HR professionals to create an organizational culture that holds sustainability as a shared core value improves the likelihood of successful portfolio component adoption and sustainability transformation.

Figure 10.8 highlights positive impacts of integrating sustainability into core HR functionality. Beginning with strategy and workforce planning, HR professionals help identify, attract, and develop the talent pool that is needed to lead and manage an organization that is committed to sustainable strategy.

Incorporating an organization's sustainability message and story into the talent acquisition process helps to attract top talent that is looking for a strong fit between an employer's values and their own values. Promoting an organization's sustainability story and developing a corporate image and brand in line with management's sustainability vision helps to attract the best and the brightest employees, especially within the Millennial generation. The HR team further promotes the sustainability message by incorporating sustainability materials and training into the onboarding process. Onboarding is an ideal time to explain and demonstrate the culture of an organization. Sustainability champions and CSOs who leverage this opportunity to engage new recruits are building sustainability into the foundation of the organization.

Expand this concept of sharing information on the organization's sustainability story to include training and development for existing employees. Through the process of workforce gap analysis, identify skill sets that are needed to support the organization's sustainability vision. Through collaborating with HR, create training and development to develop leaders and managers to meet these challenges.

Developing goals and performance scorecards that include sustainability targets is crucial to creating a culture of sustainability. Human capital professionals develop and manage the process of compensation and incentives, so they are a natural fit for a sustainability project seeking to realign work functions and remuneration to incorporate sustainability goals. HCM professionals manage internal communications, employee portals, and employee social media sites. HR professionals add value through their internal stakeholder communications and change management expertise. Employee surveys and exit interviews are resources to measure employee engagement and the impact that a sustainability program is having on an organization's employees and culture. Incorporating questions about the organization's sustainability mission and vision is a great way to measure the progress that is being made in terms of engaging employees in the sustainability process. Whether portraying a corporate brand to prospective employees, developing employees' skill sets, engaging employees, or creating a culture of sustainability, HCM professionals have a significant role to play in furthering an organization's sustainability journey.

Through their "HP Living Progress" global citizenship program, Hewlett Packard (HP) uses their people and technology to help solve some of the world's greatest sustainability challenges. One of management's areas of sustainability focus is supply chain responsibility. HP has a complex supply chain operating on 6 continents and in 45 countries, each with a different culture and legal requirements. Since 2007, HP has worked to develop its vendors' capacity to address sustainability issues. Initiatives include the Supply Chain Responsibility program, which requires suppliers to meet strict social and economic requirements verified through factory audits. The HP team believes that significant social and environmental progress relies on the ability of their supply chain to improve their policies, processes, and operations. Working collaboratively with their suppliers, HP invests in developing supply chain capacity to address areas such as improving health and safety conditions, eliminating discrimination, enhancing social management systems, and raising awareness of women's health issues. The following is a list of examples of supplier capacity building in the areas of social and environmental responsibility.

1. Implemented 26 programs in 12 countries on topics such as worker empowerment, antidiscrimination, energy efficiency, labor rights, and women's health
2. Trained more than 460,000 workers and managers beyond their walls
3. Impacted 133,000 students through pre-departure training
4. Provided training for 277 second-tier suppliers
5. Audited suppliers engaging in capability-building programs to verify improved performance[31]

John Frey, Sustainable Relations Manager, HP, shared the following story that highlights the interrelationships among employee engagement, product quality, and supply chain management. In reviewing their supply chain in China, HP questioned the logic behind having all of the factories located in the south of China near Hong Kong, when the factory workers came from the north. Because of the distance, employees were able to return home only once a year to see their families. While these factory workers were not their own employees, HP reached out to its supply chain and began to discuss the possibility of building people-centric factories with improved facilities in locations that were more convenient for the workers. Collaborative partnerships were formed with cities and local governments to address the issue. As a result, a modern factory with worker housing and facilities was constructed. The end result was happier, healthier, and more engaged workers because they were able to visit their families more frequently. In addition, the new facilities offered improved housing and recreation facilities, enhancing the workers' quality of life. HP has found that the product quality from these factories has improved significantly. In addition, the more northern location provides the opportunity to use rail transportation instead of airplanes to ship goods to the European market, thereby reducing their GHG footprint. Collaborating with their supply chain, cities, and local governments, HP was able to focus on human capital issues to address both social and environmental challenges.

Table 10.5 Ways for HR to Add Value to the Sustainability Project Team

Sustainability Project	HR Contribution
GHG Emission Reduction	Reengineer Work to Reduce Emissions Through Energy Conservation, Travel Reduction, Functional Changes.
Supply Chain Labor & Human Rights Disclosure & Improvements	Design and Deliver Training on Labor & Human Rights Best Practices.
Increased Representation of Women & Minorities at the Board of Director Level	Develop a Program to Promote Diversity & Inclusion within the Organization. Develop a Pool of Qualified Candidates That Includes Women & Minorities for Board Consideration.
Cross-functional Sustainable Product Development & Launch	Facilitate Internal & External Stakeholder Management. Facilitate Internal Networking & Communications.
Development of a Leadership Team to Drive Sustainability	Skills Gap Analysis. Training & Development. Embed Sustainability Story into Recruitment & Onboarding

The HP example highlights the impact of collaboration with both internal and external stakeholders. Many of HP's efforts have been focused on core HR functionality such as training and development designed to change behaviors, policies, and processes. Programs and projects created to effect change on sustainability issues require a multistakeholder approach and soft skills in building consensus, trust, and communicating goals, desired actions, and outcomes effectively. Including HR professionals on the project team provides expertise and skills to address these challenges.

Table 10.5 highlights several examples of the ways in which HR adds value to sustainability projects. In order to maximize impact, their involvement should extend beyond traditional HR functionality. HCM professionals add value to sustainability project teams because they bring a skill set that includes managing complex, interpersonal, cross-functional, and diverse stakeholder relationships.

While a project to reduce energy dependence on fossil fuels through installation of a solar array can be a one-off facilities project, project management professionals who incorporate HR professionals into their teams are able to consider the impact on employees and highlight this project as part of the change management process. What information should employees receive about the conversion to solar? How will this change impact them? Engaging HR professionals on these issues begins to lay the foundation for an organizational transformation toward sustainability. In fact, not giving consideration to the impact of energy conservation programs on employees can cause the best intentions to become a source of disruption.

As an example, a global company had entered into demand response contract agreeing to reduce energy usage during peak demand times in return for a financial incentive. In the Northeastern region of the United States, demand reduction requests are frequently activated during July and August, when air conditioning requirements push the electrical grid to peak capacity. In this facility, the building lighting and HVAC systems began to cycle down to reduce load as per their contract. The employees were taken totally by surprise. Work stopped, and many employees phoned Facilities to ask what was happening. The employees thought they were experiencing a power outage. While Facilities had pursued the organization's sustainability vision by entering into a demand response contract to reduce peak energy demand load and to help protect the grid, the employees were not aware of the program or the impact that it would have on them. Rather than being a positive sustainability action and message—"We helped our local community by going into demand response mode to preserve power for the community"—it was a negative employee message that caused work disruption and annoyance. Sustainability programs are as much about the stakeholder management and communication as they are about the energy-saving action. Including HCM professionals as part of the cross-functional team to facilitate internal stakeholder communications could have prevented this problem and would have added value for the project team.

10.8 Conclusion

Sustainable strategy focuses on the triple bottom line, incorporating people, planet, and profit. Organizations that fully incorporate the people aspect of sustainability through internal and external stakeholder engagement reap the greatest benefit. Engaging with an organization's human capital management team and demonstrating the alignment between sustainable strategy and HCM's core strategic and functional goals lays a solid foundation on which to create a meaningful stakeholder engagement plan. Sustainability impacts recruitment, engagement, and innovation, all of which are core functional areas for HCM. In turn, HCM professionals add value to sustainability programs and projects by leveraging their relationships and skills to improve stakeholder management and communication, and to facilitate acceptance of the sustainability change process. Including HCM professionals in the sustainability transformation process facilitates internal and external buy-in and promotes a culture of sustainability within the organization.

Notes

[1] Society for Human Resource Management, BSR, and Aurosoorya, *Advancing Sustainability: HR's Role,* http://www.shrm.org/research/surveyfindings/articles/pages/advancingsustainabilityhr%E2%80%99srole.aspx, April 10, 2011.

[2] "G4 Sustainability Reporting Guidelines," accessed May 5, 2015., https://www.globalreporting.org/resourcelibrary/GRIG4-Part1-Reporting-Principles-and-Standard-Disclosures.pdf.

[3] Ibid.

[4] RobecoSAM, "2014 Methodology Update: RobecoSAM Corporate Sustainability Assessment 2014," *Robecosam.com,* March 2014, https://assessments.robecosam.com/documents/Methodology_Changes_2014.pdf.

[5] Stephen Gardner, *The State of Sustainability 2015,* Ethical Corporation, April 2015, http://www.ethicalcorp.com/search?search_key=The%20State%20of%20Sustainability, Accessed May 6, 2015.

[6] Ibid.

[7] Ibid.

[8] Jeanne Meister"Corporate Social Responsibility: A Lever for Employee Attraction & Engagement," *Forbes,* accessed August 14, 2014, http://www.forbes.com/sites/jeannemeister/2012/06/07/corporate-social-responsibility-a-lever-for-employee-attraction-engagement.

[9] Jacob Vos, "Actions Speak Louder than Words: Greenwashing in Corporate America," *Notre Dame Journal of Law, Ethics & Public Policy* 23, no. 2, Symposium on the Environment, February 2014, http://scholarship.law.nd.edu/cgi/viewcontent.cgi?article=1111&context=ndjlepp.

[10] Society for Human Resource Management, BSR, and Aurosoorya, *Advancing Sustainability: HR's Role,* http://www.shrm.org/research/surveyfindings/articles/pages/advancingsustainabilityhr%E2%80%99srole.aspx, April 10, 2011.

[11] "United Nations Millennium Development Goals," accessed June 10, 2014, http://www.un.org/millenniumgoals.

[12] Claire Miller, "For Working Mothers, a Price to Pay," *The New York Times,* September 7, 2014, sec. Business.

[13] Michael Burchell, "Great Places to Work: How Company Culture Affects Company Success," accessed May 4, 2015, http://hiring.monster.com/hr/hr-best-practices/workforce-management/employee-performance-management/great-places-to-work.aspx.

[14] Gallup, Inc., *State of The American Workplace: Employee Engagement Insights for U.S. Business Leaders,* 2013, http://www.gallup.com/strategicconsulting/163007/state-american-workplace.aspx.

[15] Ibid., 12.

[16] Ibid.

[17] Accenture and United Nations Global Compact, *The Accenture-UN Global Compact, CEO Study on Sustainability (2013)*, 2013, www.unglobalcompact.org.
[18] Society for Human Resource Management, *Advancing Sustainability: HR's Role*, 40.
[19] "AWLP—Alliance for Work-Life Progress," *AWLP*, accessed August 20, 2014, http://www.awlp.org/awlp/home/html/homepage.jsp.
[20] "100 Best Companies to Work For 2013—Intel Corporation," *Fortune*, accessed August 29, 2014, http://archive.money.com/magazines/fortune/best-companies/2013/snapshots/68.html.
[21] Elaine Cohen, *CSR for HR,* Green Leaf Publishing Limited, 2010.
[22] Katherine Baicker, David Cutler, and Zirui Song, "Workplace Wellness Programs Can Generate Savings," *Health Affairs*, January 14, 2010, doi:10.1377/hlthaff.2009.0626.
[23] Ibid.
[24] "Employers to Boost Success of Health and Productivity Programs," *Towers Watson*, accessed May 5, 2015, http://www.towerswatson.com/en-US/Press/2013/09/employers-taking-steps-to-boost-success-of-health-and-productivity-programs.
[25] Matthew Wheeland, "Hitting the Sweet Spot of Employees, Innovation & Sustainability," Text, *GreenBiz*, January 30, 2012, http://www.greenbiz.com/blog/2012/01/30/hitting-sweet-spot-innovation-employees-sustainability.
[26] SAP, "SAP Integrated Report 2013—Integrated Performance Analysis," accessed August 25, 2014, http://www.sapintegratedreport.com/2013/index.php?id=354&L=1.
[27] Ibid.
[28] Walmart, "My Sustainability Plan," May 11, 2015, http://corporate.walmart.com/microsites/global-responsibility-report-2013/msp.html.
[29] Ibid.
[30] Ellen Weinreb, "How Walmart Associates Put the 'U' and 'I' into Sustainability," *GreenBiz.com*, accessed September 12, 2014, http://www.greenbiz.com/blog/2013/01/09/walmart-associates-u-i-sustainability.
[31] HP, "Living Progress," accessed May 12, 2015, http://www8.hp.com/us/en/hp-information/global-citizenship/data-and-goals.html.

Chapter 11

Adopting a Culture of Change to Unlock the Benefits of Sustainable Strategy

Creating a sustainable culture and one that is ready to accept and embrace change is a significant challenge for sustainability champions. Creating an organizational structure that supports collaboration, agility, and reallocation of resources is crucial for long-term change. Unlocking the benefits of sustainable strategy requires a collaborative team approach. Senior leadership must understand its role relative to developing strategy in order to deliver triple bottom line value to stakeholders. Their approach must be collaborative, seeking input from both external and internal stakeholders, because the global challenges and opportunities facing organizations are interconnected and complex. Effective solutions come from creating cross-functional teams with a variety of expertise to collaborate and develop solutions. In order to support this collaborative approach, new business platforms need to be developed to create an adaptive and responsive organization. Budgets and resources need to be reallocated in order to support a more adaptive structure. Communication and change management play a crucial role. As discussed in Chapter 10, human resource professionals are valuable team members, assisting with delivering key messages to internal stakeholders and helping to manage change.

For some organizations, this type of changes means that a fundamental change in organizational structure must occur in order to begin to break down fiefdoms that keep talent and resources siloed. According to Arulf Grubler, a Yale professor who studies patterns and drivers of technological change and their impact on the environment, "Changes in technologies and social techniques are not one-time, discrete events but rather a process characterized by time lags and often lengthy periods of diffusion."[1] In other words, complex organizational change to technologies and systems takes time and requires an understanding of enablers and obstacles to the systems. These types of change do not happen overnight. The most successful project management professionals consider this complexity and manage change as part of a scripted, planned process of inclusion, orientation, and communication. As sustainability champions drive forward with their sustainability initiatives, they must consider the change management process and its importance for successful project completion and change adoption.

Understanding an organization's culture, structure, and communication channels has a significant impact on the planning, implementing, and ultimately the success of a project. In Chapter 6 we discussed creating a culture of sustainability, which includes assessing your current culture and laying out what a sustainable culture means to your organization. The path between the two points requires managing a change process in order to create a culture that embraces sustainability. The sustainability journey is really a series of targets and goals that change through an iterative process of continuous improvement. Organizations that have made significant strides forward in their sustainability journey don't rest when they have achieved their initial goals. They adopt a culture of change and strive for improved results and set more challenging goals because future success builds on current success. Initial success gives both management and employees an incentive to continue on the path of sustainability. In addition, developing a culture of change within an organization helps management and employees be better prepared to respond quickly to evolving competitive, environmental, and social factors.

11.1 Organizational Structure for Change

In order to manage the change process, an organization needs to be structured to address and adapt to change. Many organizations are structured around core functions such as sales, manufacturing, distribution, research & development (R&D), technology, finance, and human resources. The functional reporting structure is linear, with everyone reporting to a supervisor within the same functional group. Most organizations begin with charismatic and autocratic leaders who established operations based on functional expertise. Within those areas of expertise, or silos, the leaders command and control their process, resources, and budget. As a result, it is difficult to integrate sustainability into these cultural silos. Some function leaders, such as those in manufacturing or R&D, might be more interested in adopting sustainable practices because of drivers such as regulatory or customer requirements. However, convincing other departments to engage in sustainability programs can be difficult under a siloed structure. Chief Sustainability Officers (CSOs) working in a functional organizational structure often hear objections such as "I don't have people or budget for those types of non-core projects," or "I thought that was your area of responsibility," or "We just don't have the expertise to address these issues." In

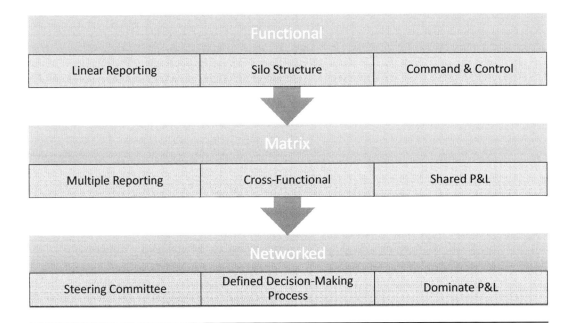

Figure 11.1 Organizational Structure Supports Sustainable Culture

order to effect change in a functionally structured organization, the sustainability champion must build cross-functional strategic alliances supported by senior management. Creating a cross-functional sustainability steering committee that is comprised of a diverse group of leaders and empowered to effect change helps to move sustainability programs forward. This approach simulates a matrix structure that encourages collaboration and resource sharing in order to achieve sustainability goals.

Matrix organizational structure is often referred to as "projectized" organizational structure, as it allows project team members to report to both a project lead and their current manager. This structure allows resources to be better aligned to respond quickly and effectively in a project environment. Moving from a functional organization into a matrix organization is often fraught with challenges because most organizations making this change are complex, with multiple boundaries such as functions, geography, processes, products, channels, and customers. The purpose of a matrix structure is to leverage resources across the organization, offering agile, rapid, and improved responses to sustainability challenges and opportunities. Figure 11.1 highlights the different organizational structures as well as key differentiators.

Integrating sustainability change management under a matrix approach unlocks greater business value because collaborative relationships and communication networks are part of the organizational structure. There is much less work to do in terms of breaking down traditional functional silos. Even though a matrix structure supports collaboration, communication, and response to change, obstacles to change remain. The business value alignment and the decision-making process can become misaligned. Without a clear decision-making process, the intended goal of greater collaboration and flexibility can result in too many decision makers with conflicting priorities.

Some organizations have further transformed from a matrix structure into a networked structure to deliver business services across business units and functions. The networked structure approach is cross-functional and collaborative and is often used in global organizations that are diverse in products, services, and geographic scope. A networked approach is similar to a matrix, but it seeks proactively to create the decision-making hierarchy and to assign profit and loss responsibility based on impact to the business value chain. In order to utilize this type of structure effectively for a sustainable transformation and the corresponding portfolio components, consider the best practices listed in Table 11.1 for making a networked structure effective.

Table 11.1 Keys to Creating an Organization to Support Sustainability[2]

Promote Leadership & Sustainable Culture	Promote a culture of sustainability. Demonstrate leadership action in support of sustainable vision. Leaders act as role models.
Clarify Vision & Alignment	Identify business value drivers. Demonstrate alignment between sustainability & business missions. Engage key decision makers.
Create Structure & Roles	Align people to support priorities & sustainable values. Allocate resources. Identify decision processes. Determine P&L responsibility.
Design Processes & Share Information	Design decision-making responsibility based on sustainable business vision. Design cross-functional processes. Define information-sharing tools. Create a change management plan.
Develop People & Align Incentives	Develop sustainability knowledge & skill sets. Align incentives in support of cross-functional collaboration.

For adoption of sustainable strategy to be successful, senior leadership must set the example by accepting and demonstrating a cross-functional and collaborative management approach. Cascading this management style throughout the organization starts with a good example from senior leadership. Clearly defining the organization's sustainability vision and then aligning resources and processes to support that vision is an important first step. Utilizing a networked management approach and relying on decision-making matrices rather than a linear decision process focuses on sustainable value creation. Identifying the decision-making protocol, streamlining the process, and avoiding unnecessary bottlenecks speeds the process and reduces confusion. Focusing on the sustainability vision provides a set of beliefs and values to guide management's decision-making process in setting goals, gathering data, and determining program and project priorities.

While resource and budget allocations are complex, an effective networked structure aligns revenue and cost allocation to create maximum business value. Nestlé accounts for water using a global Profit & Loss (P&L), as customer requirements and business strategy are similar around the world. With food, however, the approach is to utilize a regional P&L, because tailoring products to regional local requirements drives the value creation process for food products.[3] The key is to select the most meaningful approach for capturing value creation while effectively measuring management's meaningful impact on sustainability programs.

One of the greatest challenges with a networked structure is that people often have multiple bosses. In order to make the decision-making process work, it needs to be clearly defined. Projects and programs can become bogged down if the decision process is not clearly understood. Identify and assign the decision-making process and roles for projects so that reporting and decision matrices are not confused. The timeline is significantly impacted if the decision-making process requires multiple levels as well as cross-functional approvals. Taking the time to agree on those who are given decision-making roles versus those with whom one must consult or inform provides a more streamlined process. Committees and meetings are the norm in a networked decision process. A communication plan that provides a clear understanding of the purpose of the meeting and those who should attend helps to facilitate the decision process. Information needs to be clear, consistent, and transparent. All team members must have access to the same information. In addition, they need to agree on who will be gathering and presenting data as well as the definition of metrics to measure impact.

In the last chapter, we discussed the value that human capital management (HCM) professionals bring to the sustainability process. Managing in a networked structure requires managers to collaborate and to work cross-functionally on programs and projects. They need to model behavior for employees, who as team players must also be willing to collaborate and to accept responsibility for their piece of a project. An organization that is flexible and ready for change requires a workforce with a diverse set of skills to approach business challenges and opportunities through a sustainability lens. Rather than being experts in one functional area, employees need to be exposed to different areas of the organization in order to gain a variety of skills and perspectives so as to facilitate collaboration. Leaders need to focus on creating shared value rather than constructing fiefdoms. This approach means bringing resources from across the organization to develop sustainable solutions. Human resources (HR) professionals provide support for a networked organizational structure by structuring compensation and incentive packages that focus employee attention on strategic issues for the organization. The key objective in creating a networked structure is to build an organizational structure that supports development of a sustainable culture through alignment of resources and processes in order to create triple bottom line value.

The following case from Panera Bread focuses on how their leadership team reorganized in order to better align operations with sustainable strategy. While Panera's leadership team had a vision of sustainable strategy for their organization, roadblocks created by functional silos were making implementation difficult. Their organizational structure was limiting the anticipated value creation from management's strategic sustainability vision. By shifting Panera's strategic focus into three verticals—people, planet, and community—and restructuring its organization, supply chain, and new location construction around these pillars, leadership has been better able to leverage sustainable strategy for improved value creation.

While speaking at the Fast Casual Summit, Scott Davis, EVP and Chief Concept Officer of Panera, shared that reorganizing their company into three verticals has been key to creating a holistic sustainability strategy.[4] In the past, operational silos had made it difficult to embed sustainability across the organization and to generate the value add from sustainability initiatives. Recycling and packaging became top priorities from the planet vertical. Recycling was "low-hanging fruit" for the firm, creating quick and impactful results with an 86% increase in café participation in recycling. Packaging was redesigned and is now 90% compostable.[5] Other gains were made in energy usage and LEED-certified café construction. In the community vertical, they created new partnerships. One initiative is "Soup's On" (https://www.panerabread.com/en-us/articles/fighting-hunger-a-bowl-of-soup-at-a-time-video.html). Panera donates 50% of the money collected through their "Panera Cares Community Breadbox" program to purchasing and distributing their soup to food banks.[6] According to Davis, the people vertical has generated the most impact for Panera in terms of effecting change. They chose one person per café to be the environmental captain, who is charged with rallying the team and driving success. Their Green Warmth Survey score increased 13% over three years as a result of favorable survey responses.[7] Leadership's decision to reorganize the organizational structure into three verticals based around the concepts of the triple bottom line has resulted in breaking down silos and effecting real and impactful change. Management's key to success has been getting the right people involved in order to motivate others and drive change. The process all began with redesigning their structure to focus on sustainable strategy and core values for the organization.

While this level of organizational change is outside the scope of most project management professionals, it is an important lesson for CSOs and sustainability champions who engage with senior management to develop an organization's sustainability vision. Organizational structure impacts the long-term value creation of sustainable strategy. In a recent survey of project management professionals, weak organizational support was the second most frequently cited barrier to working on sustainability projects.[8] Creating an organizational structure that supports a sustainable approach is crucial for embedding sustainability into an organization.

11.2 Change Management Plan

Managing change begins with a clear vision of the organizational mission. Start by crafting a sustainability vision that has senior management buy-in and support. It should reflect the organization's values and goals and be forward-thinking in order to provide a vision of sustainability for the future. Novo Nordisk, a leader in the sustainability field, has the following mission statement: "We believe that a healthy economy, environment and society is fundamental to long-term business success. This is why we manage our business in accordance with the Triple Bottom Line (TBL) business principle and pursue business solutions that maximize value to our stakeholders as well as our shareholders."[9] This statement lays out Novo Nordisk's values and core sustainability vision, and it provides a platform from which to engage both internal and external stakeholders. The mission statement gives employees a clear understanding of their organization's vision. Management has leveraged this understanding to encourage employees' behaviors and actions that align with this core mission. In the United States, this vision includes addressing barriers to quality care for patients and those at risk for Type 2 diabetes, creating a positive social impact in their communities, and supporting stakeholders in attaining their goals. The impact of this approach includes improved quality of life for patients, lower healthcare costs, and jobs and resources for the community. In addition, their workforce is proud of Novo Nordisk's corporate responsibility programs. Their pride is demonstrated through employees' increased commitment to building the Novo business. In the United States, 80% of employees surveyed felt that the firm's TBL strategy had a significant positive impact on its reputation and that it translated into stronger stakeholder support. Novo Nordisk's corporate sustainability vision provides a good example of a clear mission statement that leads to employee engagement and transformational change.

Planning for change is an integral part of the process. Does your organizational sustainability vision span 1–3 years, 5 years, or 10 years? In the early stages of the sustainability journey, the time horizons tend to be on the lower end of the range. However, management teams that are experienced with sustainability lay out transformational goals over longer time periods, such as the next decade or longer. They understand that significant organizational change must occur in order to reach the organization's vision of sustainability. An important step in the change management process is assessing organizational readiness for change in terms of organizational structure, maturity in the sustainability process, sustainability knowledge base, leadership support, and organizational appetite for change. Figure 11.2 details important steps in the change process.

As exemplified by Novo Nordisk, a clear sustainability vision facilitates the change management process. Having a clearly defined mission statement that reflects organizational vision promotes communicating the purpose and the impact of the sustainability program across the organization. In order to have organization-wide buy-in, senior management must understand and support the sustainable strategy. Their actions and participation create roles models for employees and facilitate cascading the message about the importance of sustainable strategy throughout the organization. Inattention from senior management is one of the most common barriers to the change process. Senior leaders need to buy in to the organization's sustainability strategy and understand the value this approach creates for all stakeholders. In addition, the change management plan should lay out their role in the process in order to convey the transformation message to their next level of management effectively. The senior management role might include sitting on a steering committee to review sustainability portfolio management components, sponsoring a cross-functional pilot project, or serving on a committee to address governance changes that are needed to support sustainable strategy.

An organization's readiness for change depends on a wide array of factors. These include the organizational structure, the generational makeup of the workforce, the size of the organization, acceptance of technology as a tool and the frequency with which change has been introduced to the organization. Consider these factors as the change management plan is developed. The change process is entirely different for a start-up staffed by Millennials than it is for a long-established family-run business. Assessing the organization's readiness for change includes understanding how change is managed

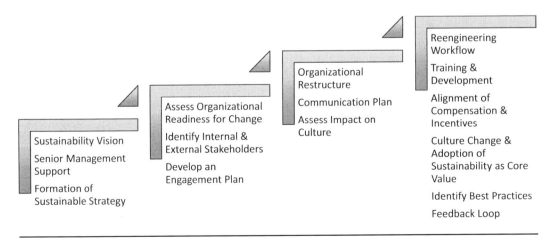

Figure 11.2 Sustainable Strategy Change Management

within the organization. Is there a formal process, or is it handled informally? Is the organization always responding to change, or have operations been static for many years? What is the scope of the change? Who needs to change? What behaviors need to change? Who has the functional expertise? Early in the sustainability journey, the change may come through creation of a "green team." This approach allows employees time and resources to pursue sustainability, usually green, projects identified by the team. As an organization's experience with the benefits of sustainable strategy increase, a project may involve creating a steering committee to review, approve, and monitor sustainability programs. The greater the organization's maturity in the sustainability process, the more complex will be the change process. In order to reach the transformational phase, an organization must undergo significant change in strategy, culture, and organizational philosophy. It may require restructuring the organization, and it definitely includes realigning budgets and resources. Project management professionals who are dealing with large, complex, and static organizations are faced with developing multifaceted and complex change management plans in order to transform their organizations.

In order to effect this type of change, it is important to understand the playing field. Of the stakeholders, who are the influencers, early adopters, and change agents? Who are the resistors and blockers? The first step is to identify internal and external stakeholders who will be impacted by adopting sustainability. Most organizations pilot the process within a particular area and then roll out the program to the entire organization. Many of the internal and external stakeholder engagement techniques discussed in Chapters 8 and 9 are helpful in devising an engagement strategy. If your organization is not ready to adapt to change, consider the changes that need to take place in order to make it ready. These may include education of the C-suite, organizational realignment, competitive pressure, regulatory requirements, risk management, and incentive compensation changes. Panera's organizational restructuring around the three verticals of people, planet, and community is an example of making an organization more adaptable to address the changes required to fully integrate and reap the benefits of a sustainable strategy. Some of these changes require realigning reporting structures, decision-making processes, and budget allocations.. Transparency of process and consistency of messaging both internally and externally provides a credible foundation on which to build the change process.

The key to explaining change in support of a more sustainable organization is to address why the change is important to internal and external stakeholders. Table 11.2 provides a framework for the change management process and some best practices for each of the stages in the cycle. As an organization moves along the sustainability continuum, management engages in a continuous cycle of improvement, with strategy modified as lessons are learned. An important first step is to ask the question: "What does sustainability means to us?" The question is answered by engaging with key internal and external stakeholders and gathering requirements in order to develop a plan.

Once the plan has been created, implement the changes using best practices such as ongoing C-suite engagement, recruiting senior-level sustainability champions, and empowering teams to identify and implement changes. Support the process through effective communication of desired sustainability goals and targets, and reinforce desired behaviors using incentive programs. Once the initial programs and projects have been launched, report on progress and results. Share what works and what doesn't. Be authentic in the reporting, and share both successes and failures. Management learns from both successes and failures. Adopt best practices and revise strategy, programs, and projects based on lessons learned. Encourage feedback from all levels within the organization and from key external stakeholders. Modify portfolios, programs, and projects and repeat the process.

The following case exemplifies leadership engaging in this change process, building a project team, and fundamentally changing the way work is performed within the organization. When Alcatel Lucent's then-CEO, Ben Verwaayen, formed his vision of sustainability, which called for significant greenhouse gas (GHG) emission reduction targets for the firm, he knew that transformation of this magnitude would require fundamental change in how work was performed within the organization. In order to effect this change in management's and the workforce's actions and behaviors, HCM professionals

Table 11.2 Framework for Sustainable Culture Change Management

Cycle	Process	Best Practice
Plan	Define strategic sustainability vision. Identify material issues. Demonstrate alignment with core business mission. Create organizational framework. Identify stakeholders and plan engagement. Select impactful programs & projects. Select a place to begin.	Engage the C-suite. Engage stakeholders in the planning process. Develop long-term goals. Incorporate sustainability goals into core business goals. Create a cross-functional steering committee. Identify sustainability projects. Develop a sustainability project plan. Select pilot programs.
Do	Engage leadership & employees. Communicate the sustainability program. Align corporate incentives. Encourage sustainable thinking. Encourage discussion with stakeholders. Align internal & external messaging.	Reorganize to support sustainability change. Host senior management roundtables. Communicate impacts on products, processes, & employees. Create & empower teams. Utilize champions to act as influencers on managers & employees. Use targeted & frequent communications. Incorporate sustainability goals into performance scorecard. Align compensation & incentives. Create contests, awards, & recognition programs. Encourage personal sustainability plans. Create affinity groups to build employee morale, collaboration, & communication.
Check	Establish a baseline. Engage in internal stakeholder dialogue. Measure & report on metrics and key performance indicators. Conduct tollgate reviews.	Make sustainability progress reports & data available to all stakeholders. Establish steering committee review of program & projects. Conduct employee survey on sustainability culture. Hold cross-functional stakeholder roundtables. Encourage employee feedback through your intranet via chat forum & electronic suggestion box.
Act	Gather feedback to improve program & project performance. Scale change. Identify barriers. Assess need for organizational change. Determine development requirements for knowledge and skills.	Encourage feedback through surveys, suggestion boxes, & intranet polling. Monitor chat forums—Twitter, Facebook, Yammer. Debrief cross-functional teams for lessons learned. Gather lessons learned. Steering committee review for portfolio standards.
Repeat	Share lessons learned & best practices. Modify portfolio component plans. Expand scope & scale.	Scale pilot programs. Restructure organization. Develop education & training programs. Engage more leaders, employees, & other stakeholders. Communicate best practices.

were asked to join the transformation team. They collaborated with functional areas to reengineer job functions, tasks, and role descriptions to reduce the GHG impact of every job. Training programs to develop new skills and knowledge were created for both management and the workforce. The process was iterative, gathering feedback and applying lessons learned to the next series of sustainability projects and programs.

Table 11.3 Sustainability Change Management Plan

ID#	Deliverable	Owner	Key Message	Stakeholder; Cascading Stakeholders	Due Date	Dependency	Work Stream
2	Sustainable Strategy Roadshow	CSO	Adoption of Formal Sustainability Targets	Senior Leadership Team; Business Unit Leaders; Employees	1/31/15	None	Q1 Senior Leadership Meeting

ID#	Delivery	Content Owner	Approver	Consult/Inform	Status
2	In person	CSO	CEO	Business Unit Leaders	Complete

Addressing the global challenge of reducing GHG emissions requires a diverse skill set, one that includes collaboration and communication skills as solutions come through a cross-functional approach. Creating realistic expectations and moving an organization forward requires feedback from employees to verify that the message has been received and understood as intended. It also provides a mechanism for employee input and involvement in the process. Novo Nordisk achieved significant alignment between the employees' and management's perspectives of the mission and goals of the organization. This alignment was achieved over time as the organization refined its programs, projects, and policies to incorporate employees' and other stakeholders' feedback. Further support came from aligning the compensation and incentive structure with sustainability goals and targets encouraging desired behaviors and leading to desired outcomes. That which gets rewarded gets attention.

Adopting a culture of change is a complex process, but with planning, communication, and management of expectations, it can be achieved. Table 11.3 provides a useful layout for managing the change management process. Focusing on the key message for stakeholders allows for a targeted approach to managing change. Identifying deliverables, assigning responsibility, and establishing due dates facilitates accountability for action.

Table 11.3 addresses the multiple reporting requirements of a matrix organizational structure by delineating the decision maker (Approver) from those with whom you may wish to consult or advise (Consult/Inform). Identifying who will be involved in the decision-making process is important to keep projects moving forward. The owner of the deliverable should be the one to obtain the necessary approvals and keep other stakeholders in the loop. This approach allows for each step of the change management plan to be broken down into work streams with designated responsibilities. Incorporating this approach with a communication plan, which will be discussed later in the chapter, provides a solid framework to drive change within an organization.

Effectively managing the change management plan includes managing stakeholder expectations. While a plan facilitates the process, changing people, process, and policies is a complex and time-consuming process. Understanding that there will be resistance and anticipating the problems by incorporating these risks into the change management plan makes sustainability transformation more likely to succeed. Clear definition of roles and responsibilities facilitates the process.

11.3 Barriers to Project Success

Emotions about programs and projects tend to go through a U-shaped life cycle known as the "Valley of Despair." Project teams begin with optimism and enthusiasm, but they move into a despair phase because of barriers, misalignment, resource allocation challenges, and other hurdles. An experienced

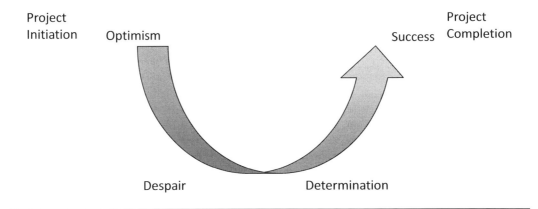

Figure 11.3 Project Life Cycle: The Valley of Despair

project manager plans for the despair phase and addresses barriers to success, rallies the team, and gets the project back on track. A significant role for project management professionals is managing the change process effectively to minimize the negative impacts of this cycle on sustainability projects.

As illustrated in Figure 11.3, project teams begin sustainability projects with optimism and energy. As the project rolls out, clear alignment between the sustainability project and the organization's goals helps to motivate project team members and engage stakeholders. Sustainability projects have the advantage of being good for the environment and/or society, which provides inherent inspiration to team members. However, motivation fizzles if there is lack of mission clarity and leadership interest. Senior sponsor involvement and senior management support in terms of clear messaging, demonstrated actions, and involvement promotes change adoption and helps to minimize project team frustration. Anticipating the Valley of Despair in the project planning process in terms of time frame, resources, and motivational tools improves the likelihood of project success. Consider factors such as how conducive the organization's environment is to accepting change. Can you change the behavior of management and the workforce under the current organizational structure, or will changing the organizational structure need to be part of the transformation process? Enlisting senior management support and active sponsorship keeps the project an organizational priority. One of the most common barriers to success is senior management's lack of attention over the full project life cycle. Once a project is approved, senior management often moves on to other pressing matters. When management is undertaking a cultural transformation, constant reinforcement from senior management in terms of modeling desired behavior, encouraging participation, allocating resources, and spotlighting desired behaviors is required for success. If sustainability is on the lips of senior management, it will be recognized as being important by middle managers and employees.

Resistance to change needs to be anticipated and incorporated into the change management plan. Consider how your organization's compensation structure supports success. Is it based on individual or collective results? How are existing workloads being redistributed to accommodate project responsibilities? Are other employees, who may not be directly involved in the project, required to increase their own workloads in support of the project? Ensure that the communication plan considers these stakeholders and that the compensation structure recognizes and rewards their actions and contributions. Adopting a sustainable culture requires rethinking workflow within the organization, and that process takes time. Ensure that the plan has a realistic time frame with meaningful metrics by which to gauge progress as well as ultimate success. Considering these key issues helps to minimize the Valley of Despair and focus project team members on the benefits of completing the project successfully. The following are some best practices for effecting change from Chip and Dan Heath, authors of *Switch*, a best seller on how to change things.[10]

1. Sell the vision.
2. Outline specific actions and script crucial steps.
3. Encourage stakeholders to understand the issues and to take responsibility for change.
4. Empower the team to recommend and adopt changes.
5. Create a path to make change easier.
6. Break change into more manageable pieces.
7. Create good habits.
8. Rally the team.

In terms of making changes through sustainable strategy, senior management vision becomes workforce action by communicating clearly the sustainable strategy vision and the benefits to the organization, stakeholders, and environment. Establishing a clear plan of action with identified roles and responsibilities for all parties sends a clear message on desired actions and behaviors. Empowering employees through green teams, personal sustainability plans, and "chairman club" contests are all effective means of engaging employees and motivating change. However, clear alignment with core compensation and bonus incentives really drives the message home. Providing a means for stakeholder input and feedback is crucial to the change process. Listen to stakeholders and incorporate their feedback into future programs and projects.

Cisco's HR team has seen success in managing change by creating an online community to share thoughts and ideas through project/team blogs, forums for C-suite executives, and portals for key customers. The forum for C-suite executives provides more than 74,000 employees with a voice in the company's direction as well as their individual work experience.[11] Using social media to give employees a voice and to provide a forum for executive feedback helps to promote change and mange reputation risk. Knowing that the executive team hears their ideas and that they have input into decisions impacting their functional roles empowers employees.

11.4 Communication Plan

Once an organization's sustainability vision has been determined, consider how best to convey the message to both internal and external stakeholders. Focus on the benefits created by implementing the sustainability plan and the alignment between sustainability and core business strategy. In developing the communication plan, consider the following:

1. What is our goal/message?
2. What outcome are we seeking?
3. Who can help deliver the message?
4. What is our budget?
5. What is our action plan, and which communication channels will be most effective?
6. How are we going to measure progress and ultimately success?
7. How will we gather feedback from our stakeholders/audience?

Focus the communication plan on organizational purpose and highlight the benefits driving the action. Link the program or project back to the larger organizational mission. Define the project goals clearly. Examples include promoting energy efficient behaviors or resource conservation from employees, building brand loyalty with customers by offering sustainable solutions, or managing risk by incorporating climate-change risk as part of the operating risk profile for business units.

Think about internal resources from marketing, public relations, public affairs, and human resources to help with the communication plan. After considering the budget, think about external resources such as advertising, marketing, and public relations agencies. Will the mission and message

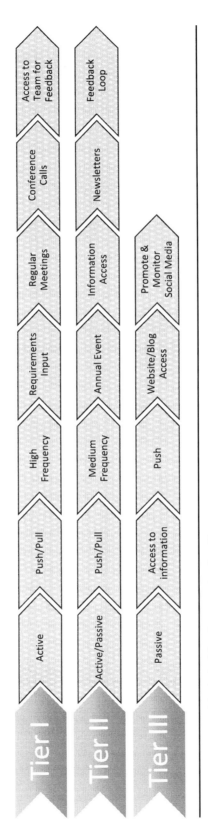

Figure 11.4 Tiered Stakeholder Communication Plan

be enhanced through partnering with community, academic, or nonprofit agencies? If the issues are relevant to other members of the industry, there may be an opportunity to partner with an industry association to promote this aspect of the sustainability agenda. Think globally and act locally. In other words, develop a message that reflects organizational sustainability vision but empower locations to take actions at the local level that collectively create the desired global impact. This approach empowers managers and employees to select programs and projects that create the desired organizational benefit but reflect their personal interests and passions.

In Chapter 8, the idea of stakeholder differentiation for engagement was introduced. (See Figure 8.3, Stakeholder Engagement Matrix.) The basic concept is that a communication plan considers where a stakeholder group falls on the engagement matrix in order to engage the group more effectively. Figure 11.4 is a stakeholder communication plan that considers different communication techniques based on a stakeholder-groups engagement level.

Tier I stakeholders need to be actively included in the process, with open access to leadership. Their communication plan often includes being included on a steering committee or an advisory council with regularly scheduled meetings and routine follow-up conference calls. These stakeholders need senior management attention and a high-touch communication plan. Information is being both given to and received from this impactful group. Tier II stakeholders' communication requirements vary depending on where they fall on the matrix. Some well-organized groups are easily reached and included in formal meeting and outreach events. Other stakeholders are less well organized and harder to reach but still expect to be well informed. Offering a well-publicized and regularly scheduled annual engagement event helps to keep these stakeholders informed. Communication strategies should include both push and pull formats, but the frequency of involvement is lower than that of Tier I. Communication techniques for this group include annual conferences with some access to key decision makers. They require sustainability programs and project status updates on previous discussed stakeholder requirements and on organizational progress and key metrics. Dedicated intranet portals, newsletters, and quarterly emails provide ongoing communication options for this group. Tier III stakeholders are usually satisfied with access to information through tools such as websites, blog access, and social media updates. Feedback monitoring for this tier comes from following website posts and social media comments.

Drilling further into the communication plan, Figure 11.5 outlines a program for internal stakeholder communications to educate and inform them about significant organizational change, which in

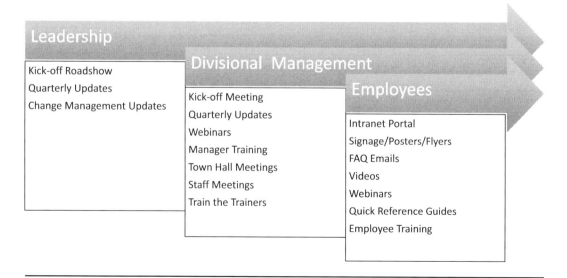

Figure 11.5 Internal Stakeholder Communication Tools

this example is the rolling out of a flexible work program. The communication plan focuses on informing multiple organizational levels about the benefits of adopting flexible work arrangements across the organization, providing information on tools and resources, offering training, and gathering feedback.

Changing how work is managed, performed, and evaluated under a flexible work structure requires a communication strategy that engages a broad range of internal stakeholders, from senior leadership through divisional management and down to the employee level. Because each of these audiences is different, the communication techniques will vary. Roadshows are created and presented to senior leaders to explain the flexible work program, resource requirements, timeline, and business benefits of the program. Details of the change management and communication plans are also shared so that senior people understand the impact of flexible work arrangements on their subordinates as well as how the implementation process will be handled.

The communication focus with divisional management centers on details of the program, the technology to support tracking and adoption, and the details of the change management process to promote manager and employee acceptance. These include a kick-off meeting to explain the program and how it will be rolled out and implemented. Updates are handled through quarterly calls. Several webinar time slots are scheduled to allow managers to see the process and to demonstrate the technology-based tool being used to support the program. Management training is available to promote widespread understanding of the program and the technology-based tool. A "Train the Trainer" series is created to build a network of trained individuals who then educate employees on the program and the technology in their respective areas.

Employee communication is focused on creating awareness of a flexible work option through signs and flyers. The employee kick-off includes tables in common areas such as the cafeteria, offering promotional swag, quick reference guides, and FAQ flyers. In order to support employees who are interested in participating in flexible work arrangements, resources such as a dedicated space on the intranet portal featuring videos, webinars, and program details are made available.

The following case study of Levi focuses on a communication plan designed to generate global impact through local action by employees, customers, and other stakeholders. Levi developed a sustainability campaign to encourage stakeholders to use less water. Levi's approach gives customers, employees, and other stakeholders an issue—water scarcity—on which to focus. The campaign empowers them to have a global impact by making small personal changes in water usage decisions. The campaign has a multipronged approach, encouraging consumers of Levi's to extend the wear of jeans between washings, launching a new line of jeans—Water<Less jeans—and partnering with a non-governmental organization to promote water stewardship. This new line of jeans uses up to 96% less water in the manufacturing process, saving over 172 million liters of water.[12] In addition, Levi teamed with Water.org to launch a campaign involving thousands of people in 1300 cities across the globe in the "Go Water<Less" challenge. Through relatively small actions such as skipping a shower for a day, washing jeans less frequently, using the faucet less, and refilling water bottles, participants significantly reduced water usage. The message is that everyone can have an impact and that each of us taking small steps can generate significant global impact. Table 11.4 highlights some key metrics from the campaign.

Levi's message about saving water is impactful, transparent, and consistent to both internal and external stakeholders. Their internal message about saving water is consistent with their public relations and marketing communications to customers. The significant global water savings were created via local action that engaged stakeholders.

Table 11.4 Go Water<Less Challenge Results[13]

People Shower<Less	Jeans Unwashed	Faucets Untouched	Water Bottles Reused	Liters of Water Unused
1207	1850	714	1374	1.82 billion

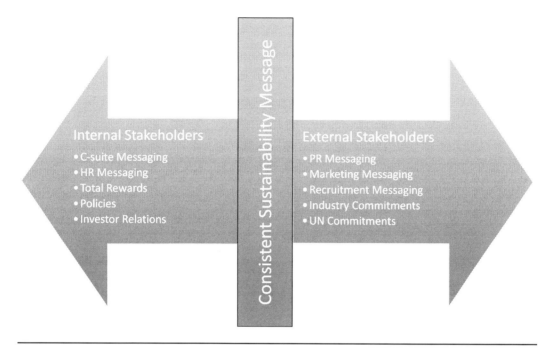

Figure 11.6 Alignment of Internal and External Messaging

In devising the communication strategy, expect employee communications to be made public. It is extremely easy for employee communications to be posted to third-party websites for all to see. Divergence between the external and internal messages might be characterized as greenwashing. The Nike case discussed in Chapter 10 highlights the impact of misaligned messaging. Nike ended up losing a lawsuit because their public image of workers' rights didn't align with their own internal reports. When a report on working conditions in a Vietnamese factory was leaked, the misalignment of their public statements about workers rights and safety and their own internal reporting were made very public. With 24/7 media access and the posting of internal documents one click away, transparent communication of mission, action, and outcome is the only viable option. If an organization is making a sustainability claim, management needs to be able to back it up. When a program falls short of a target, disclose that as well. The sustainability mantra is to disclose the good, the bad, and the ugly. While management naturally wants to promote their successes and minimize their blunders, adopting sustainability means disclosing both successes and shortfalls. It is better to acknowledge that a sustainability program or achieving a target has not gone as planned. Transparency of reporting builds credibility with both internal and external stakeholders. Message misalignment will cause an organization's sustainability journey to derail. Thus, as indicated in Figure 11.6, alignment of internal and external messaging is crucial.

While the overall message needs to be consistent, stakeholders differ in terms of their ability to receive and understand the message. The delivery platform needs to be differentiated based on the group and the stakeholders' readiness to hear and accept the message. Just like consumers, employees are shades of green. The continuum runs from dark green, those employees who are personally interested and fully engaged on environmental and social issues, all the way to brown, those who haven't even heard of sustainability. A dark green employee might be inclined to begin a green team to focus on reducing GHG emissions in her department through organizing ride sharing or biking groups. A light green employee might be more inclined to pull information from an employer portal on the organization's sustainability initiatives because he is personally interested in a specific environmental or social issue. Someone in the middle might be engaged through a contest or an online chat room that focuses on a personal environmental or community "hot button" topic. For the brown employee, education and a linkage between sustainability initiatives and his own function are crucial to getting him engaged.

In order to successfully adopt a sustainable culture, employee participation is crucial. Encourage employee collaboration, communication, and feedback through social networks such as Yammer, Slack, Wiggio, and Jive, among others. Offer incentives for generating ideas. Hold contests and reward high achievers through recognition levels such as a "Chairman's Club" or "Star Teams." These can range from small awards to formalized events such as lunch with the C-suite. Select an incentive that drives the desired behaviors and outcomes in your organization. Empowering and encouraging employees to make suggestions that drive sustainability is the key.

Intel created a program called Intel Environmental Excellence Awards, which recognizes projects that reduce the firm's environmental footprint. In a facility in Ireland, employees found that by changing the frequency of filter changes in scanner tools, the operation saved 350,000 gallons of water and more than $500,000 in waste reduction costs.[14] Intel estimates that projects related to Environmental Excellence Awards have saved the firm over $200 million since 2010.[15] By communicating the existence of these programs and highlighting awards and employee success stories, management not only gains meaningful environmental and social impact ideas but also improves employee engagement. Promoting these types of initiatives via the organization's recruitment portal makes potential recruits aware of the organization's sustainability mission and helps to attract Millennial talent. As part of the communication strategy, define and select metrics to determine impact and measure success. Report frequently on key metrics using dashboards or leader boards so that employees understand the impact that they have both personally and collectively toward achieving sustainability goals. Employee feedback confirms receipt and comprehension of the message and provides data for plan modifications to improve the effectiveness of the sustainability program.

Table 11.5 is an example of a project communication plan to be used in conjunction with the change management plan. Focusing on the purpose of the action, the consistency of the message, the medium for delivery, as well as the metrics to measure impact facilitates the communication process.

Common communication vehicles include websites, intranet, emails, press releases, advertisements, brochures, internal bulletins, and social media forums. These methods are supplemented by in-person communications such as shareholder meetings, town hall meetings, staff meetings, and sales calls. Using a variety of techniques ensures that diverse stakeholders receive the message by including both push and pull techniques. Social media is a useful tool not only to communicate the message but also to track feedback from stakeholders. While stakeholder group type helps direct communication vehicle choices, stakeholders within those groups will have a variety of preferences and requirements for effective communication.

Table 11.5 Sustainability Communication Plan

Message	Date	Vehicle	Stakeholders/Audience	Author	Metric	Purpose	Approver

Another useful technique to promote adoption of sustainability is developing a sustainability commitment agreement that outlines specific goals and actions that support the sustainability initiative. Management and employees are encouraged to sign the commitment and to make a pledge for specific actions, such as reducing waste by placing paper in a recycling bin or reducing energy usage by turning off lights, shutting down computers at night, or changing their computer settings to go into energy-saving mode more quickly. Once employees make these types of commitments, their level of interest in sustainability and willingness to take action grows. As an organization matures in its sustainability journey, more formalized goals can be set for management and employees.

11.5 Engaging Stakeholders in the Change Process

Clear and frequent communications about the sustainability change process and the impact it is having in terms of reaching the organization's sustainability vision keeps the project on track. Both internal and external stakeholders want to understand the impact of sustainability programs and projects on the organization. They need to be given meaningful updates and data to track organizational progress against goals as well as to assess the impact of their contributions. Employees need to understand how their roles will be impacted, why the change is beneficial, and how they can become involved and make a difference. While incentives to promote sustainable behavior are impactful, empowering employees to identify opportunities to incorporate sustainability into their function embeds sustainability into core processes. Significant impacts include suggestions for materials conservation, energy savings, product reformulations, distribution redesign, and new sustainable products and processes. External stakeholders also want to know that their feedback has been heard and incorporated into the organization's sustainability plan, impacting strategy, operations, and policies.

"IKEA Sustainability Strategy for 2020: People & Planet Positive" is an example of using sustainability to transform the business through innovation and new opportunities to further embed sustainability in their own organization by driving sustainability into the daily lives of themselves and others. Their change drivers to transform their business are

1. Inspire and enable customers to live a more sustainable life through developing new products and solutions to help customers' reduce energy and water usage and to reduce waste while remaining within their lowest price possible framework.
2. Reach for resource and energy independence through securing long-term access to sustainably sourced raw materials, limiting resource usage, and reusing materials. Generate more renewable energy than energy consumed and ensure energy efficiency across the value chain.
3. Drive creating a better life for people and communities impacted by Ikea's business by extending their code of conduct to all participants in the value chain, support human rights especially the interests of children, and be a good community member.[16]

Table 11.6 shows commitments to action, absolute targets, and metrics to measure impact that IKEA's management has selected to transform their organization in line with their sustainability vision by 2020. These commitments and metrics are the foundation of the communication message to internal and external stakeholders. They also provide strategic guidance for decision making, resource allocation, and portfolio component selection.

IKEA is mature in its sustainability transformation. Their change management plan engages both internal and external stakeholders throughout its value chain, and it includes best practices and lessons learned from their years of experience on their sustainability journey. For example, IWAY, IKEA's code of conduct for suppliers, includes a dedicated provision to prevent child labor, which is a key initiative documented in the Children's Rights and Business Principles. Since IWAY launched in 2000, the

Table 11.6 IKEA People & Planet Transformation Commitments and Targets[17]

Commitments	Select Targets/Metrics
Make sustainability part of everyone's daily work.	FY15—95% of workers say sustainability is part of daily work, included in all personal development and incentive programs.
Empower & encourage employees to participate in volunteer activities during the workday to support sustainability programs.	FY15—95% of employees and suppliers and 70% of consumers see IKEA as an environmentally and socially responsible corporate citizen.
Utilize technology, innovation, & scale to offer new sustainable products and solutions.	FY20—4 times increase in sales from sustainable products and solutions for customers.
Invest in areas that support our core vision.	FY15—All stores are part of "IKEA Goes Renewable." New stores will be located, designed, equipped, & operated as the most sustainable IKEA facility at that point in time.
Increase collaborations & partnerships with other companies, suppliers, NGOs, and UN agencies.	FY15—Supplier compliance with IKEA code of conduct, IWAY.
Develop & enhance long-term partnerships with local communities to enhance their economic, social, & environmental development.	Support human rights using UN Guiding Principles on Business and Human Rights as a foundation.
Promote governmental policies that support positive change in the environment and society.	Use the Children's Rights and Business Principles to strengthen work in protecting the rights of children.

company reports 165,000 improvements within their supply chain that have positively impacted the environment and society.[18] The interrelationship of the core value of protecting children's rights, their supplier code of conduct, and their NGO and governmental outreach programs demonstrate the iterative nature of sustainable strategy adoption and refinement as an organization moves along the sustainability continuum.

While the complexity and reach of IKEA's program might seem intimidating, program and project managers charged with guiding sustainability programs should begin with a targeted approach and then build on that base. Successful completion of the initial sustainability project turns into multiple projects and then transforms into programs. As management makes more of a commitment to sustainability, a sustainability champion and then an office of sustainability is formed. Change is a slow process, and the organizations on the forefront of sustainability have refined their programs over decades. The good news for those just getting started is that the trail has been blazed and there are many good examples of successful programs and best practices. Remember that transforming into a sustainable organization is a journey, and it takes time. Follow the change management process and begin your organizational transformation.

The following checklist is a tool to help project management professionals facilitate employee engagement in the change management process.

Checklist for Engaging Internal Stakeholders in Sustainability

1. Enlist the C-suite to support the organization's sustainable vision through words and deeds.
2. Develop a clear sustainability mission statement and communication plan that articulates its alignment with core business goals.
3. Enlist senior management to launch the program, communicate sustainability goals and targets, and participate in incentive and award programs.
4. Align internal and external vision and messaging.

5. Clearly communicate employees' roles and responsibilities and how work roles will be impacted.
6. Meet employees where they are, by incorporating the message as part of the employees' work routine. Use existing structures such as the intranet, staff meetings, and signs in break rooms to let them know about the program.
7. Vary communication platforms, including push and pull formats, in-person meetings, prominent signage, email, newsletters, intranet portals, and private social media. Giving and getting information are both crucial.
8. Provide meaningful incentives for employees to encourage behavior change and alignment with sustainability mission and goals.
9. Provide opportunities for employees interested in sustainability to participate in leadership roles for special projects, green teams, or program steering committees.
10. Create a program seeking employees' ideas and contributions.
11. Create recognition, awards, and other incentive programs for employee participation and contributions.
12. Define clear metrics to measure progress and communicate results to employees through easily accessible dashboards.
13. Gather employee feedback through surveys and focus groups.
14. Incorporate the feedback into future sustainability plan and programs.
15. Be creative and have fun!

Focusing on the environmental, social, and economic benefits of the strategy and how they relate to servicing clients, improving processes, and improving communities helps to promote adoption. Creating a culture of sustainability requires both management and employees to incorporate the organizational vision of sustainability into functions, tasks, and routines. In order to reap the full benefit of sustainable strategy, it needs to become part of projects, processes, and policy. Effecting change within an organization is a challenging process, but adopting a sustainable strategy will deliver long-term benefits to all stakeholders.

11.6 Human Capital Advantage

HCM professionals are skilled in communication, employee relations, and change management. As a project management professional, a best practice is to leverage these skills by including an HCM professional on the project team. In order to engage employees in the process, consider the long-term impact that moving toward a sustainable organization has in terms of managing one of the organization's most valuable assets—its people. Involving HCM professionals goes beyond effective change management of a specific project. Including HR in the process begins their own sustainability transformation impacting recruitment, development, and management of the organization's people. As part of the change management process, consider how the HR function can support the sustainability transformation and further embed sustainability into the culture of the organization. The following are some survey questions for HR professionals on the impact of adopting sustainability on their functionality in order to build a more sustainable workforce.

HCM Survey to Build a Sustainable Workforce

1. How can HR proactively support the organization's sustainability strategy over the next three years?
2. Do employees understand the organization's sustainability strategy?
3. Is employee behavior in line with current sustainability policies? If not, what behavioral changes are needed?

4. What skills and knowledge are needed for leadership to support sustainability initiatives?
5. What skills and knowledge are needed for the workforce to support existing and new sustainability programs?
6. Does HR have a plan to address workforce planning and the skills gap created by the sustainability vision?
7. How will workforce planning (recruitment or downsizing) affect local communities? What are the risks and opportunities to our corporate social responsibility programs?
8. What are the best methods for engaging current employees on issues? Will these methods work for sustainability projects and programs?
9. What channels of communication are most effective with internal stakeholders?
10. What are the best ways to measure the extent to which sustainability is embedded into our culture?
11. Do HR staff members have the skills and competencies to support adoption of a sustainable culture, including changes in process and performance management? If not, what skills are required?
12. Is our organization's sustainability message conveyed to the prospective talent pool? If so, how?
13. Is training in our sustainability policy part of the onboarding process?
14. What changes need to be made to the incentive and compensation structures to reward sustainable behaviors?
15. What inputs have been received from external stakeholders that affect human capital management policies and procedures?

Responses to these questions provide the framework to develop a change management plan to focus on the organization's people as part of embedding sustainability into the culture of the organization. This approach helps to identify skills gaps in both leadership and the workforce so that HCM professionals can develop a plan to address these deficiencies through recruitment, training, education, and development projects. Identifying the skill set and requirements for a sustainable workforce provides the foundation for a project for HCM professionals to assess their own capabilities to drive the necessary changes. Working collaboratively with HR leadership helps to formulate standards and policies that will promote employee engagement in the change management process.

11.7 Conclusion

Creating a sustainable culture and unlocking the benefits of sustainable strategy requires organizational change. Understanding an organization's readiness for change and the many steps that need to be taken to effect change begins the planning process. As we learned from the Panera case, one of the greatest challenges to sustainability project management professionals is implementing a project in an organization that is not structured to support sustainability transformation. Creating an organizational structure such as a matrix or networked structure that removes silos, aligns resources, and drives the decision-making process forward creates a more adaptive organizational culture that aligns better with sustainable strategy.

Effecting change within an organization is a complex process. Utilizing project management tools such as change management and communication plans facilitates clearly defining the target stakeholders group, message, delivery mechanism, and approval process. Incorporating lessons learned and adopting best practices facilitates project management professionals in effecting organizational changes to drive adoption of a culture of sustainability. Engaging internal and external stakeholders in the process improves the likelihood of success, as demonstrated by the Levi and IKEA case studies. As an organization matures in its sustainability process, sustainability programs and projects morph into ongoing policies and procedures. Engaging with HR professionals to match the desired skills base for a sustainable workforce with the organization's sustainability vision provides a framework for HR to change their own functional area to better support building a talent pool that aligns with the organization's sustainability vision.

While sustainability concepts that focus on resource preservation and environmental impact such as the circular economy, product life cycle, and resource efficiency are drivers in the sustainability conversation, creating a culture of sustainability means changing the way people in the organization think and act every day. In order to effect change in core beliefs and values within an organization, the change management plan must engage employees and impact their actions and behaviors. To create a culture of sustainability, an organization must create a sustainable workforce.

Notes

[1] Arnulf Grubler, "Time for a Change: On the Patterns of Diffusion of Innovation," *Daedalus* 125, no. 3 (Summer 1996):19.
[2] Paul Rogers and Jenny Davis-Peccoud, "Networked Organizations: Making the Matrix Work," December 7, 2011, http://www.bain.com/publications/articles/decision-insights-12-networked-organizations-making-the-Matrix-work.aspx.
[3] Ibid.
[4] Alicia Kelso, "Panera, Chipotle Set Sustainability Standards by Starting Small, Thinking Big," *Www.qsrweb.com*, October 20, 2014, http://www.qsrweb.com/articles/panera-chipotle-set-sustainability-standards-by-starting-small-thinking-big.
[5] Ibid.
[6] "Fighting Hunger, One Bowl of Soup at a Time," *Panera Bread, in the Community*, accessed October 22, 2014, https://www.panerabread.com/en-us/articles/fighting-hunger-a-bowl-of-soup-at-a-time-video.html.
[7] Ibid.
[8] Tom Baker, Pedro Echeverria, and Kristina Kohl, "Survey Results: Project Managers and Sustainability," June 14, 2015, http://www.projectmanagement.com/white-papers/303957/The-Voice-of-Project-Managers-on-Sustainability-Projects-and-Requirements.
[9] "Our Approach to Sustainability," *Www.novonordisk.com*, accessed September 19, 2014, http://www.novonordisk.com/sustainability/how-we-manage/how-we-manage.asp.
[10] Chip Heath and Dan Heath, *Switch* (New York: Broadway Books, 2010), http://heathbrothers.com/books/switch.
[11] Mark Mc Graw, "Managing the Message," *Human Resource Executive* 28, no. 12 (December 2014).
[12] Levi's, "Water<Less, Doing Your Part," accessed May 19, 2015, http://store.levi.com/waterless/doingyourpart.html.
[13] Ibid.
[14] Heather Clancy, "Intel Rewards Employees for Thinking Sustainable. Do You?," *SmartPlanet*, accessed September 18, 2014, http://www.smartplanet.com/blog/business-brains/intel-rewards-employees-for-thinking-sustainable-do-you.
[15] Suzanne Fallender, "Getting Employees to See Earth Day as Every Day," *CSR@Intel*, accessed September 18, 2014, http://blogs.intel.com/csr/2013/04/getting-employees-to-see-earth-day-as-every-day.
[16] IKEA, "The IKEA Group Sustainability Strategy 2020: People & Planet Positive," n.d., http://www.ikea.com/ms/en_GB/pdf/people_planet_positive/People_planet_positive.pdf.
[17] Ibid.
[18] Ibid.

Chapter 12

Sustainable Strategy as a Lever to Attract, Engage, and Retain the Workforce

Sustainable strategy can be used in many ways as a lever to engage your workforce, program, or project team. The very act of creating a strategic sustainability vision and sharing this mission statement with employees begins the engagement process. Highlighting the sustainability vision establishes sustainability as a core value of the organization and begins the process of making it part of the organizational culture. Sustainability vision becomes a bridge unifying departments that are diverse in terms of geography, function, customers, and products. Devising and implementing a strategy of sustainability unifies the organization and provides a foundation on which to engage your current and prospective workforce.

Aligning sustainable strategy planning and implementation to leverage the core tenets of employee engagement, including clearly defining the mission and outlining employees' roles for successful execution of the strategy, facilitates bridging the sustainability performance gap. Begin by defining organizational goals, employees' roles, desired actions, and behaviors so that employees understand what the organization is trying to achieve and their role in the process. Evaluate the skills and knowledge base that employees will need in order to be successful in implementing the organization's mission of sustainability, and create training programs to address skill gaps. Identify the people who share the organization's values and who meet the technical and interpersonal skills required to build a sustainable workforce.

For project management professionals, employee engagement has a direct impact on the success of projects and programs. For program and project managers involved with sustainability, the impact of sustainable strategy as a lever for employee engagement is a benefit to promote adoption throughout the organization. In order to cascade sustainable strategy through the levels of the organization, program and project managers need to consider the organization's culture as they plan and implement all projects. As sustainability becomes part of organizational culture, it becomes integrated into all organizational programs, projects, and processes, not just sustainability projects. Portfolio component planning and implementation, such as scope, budget, and time including critical path and work breakdown structures, need to be modified in consideration of creating a sustainable workforce and workplace. In order to create these changes, the portfolio management process needs to incorporate the organization's commitment to sustainability in the program and project selection process. Proclaiming a mission of sustainability and then selecting programs and projects that don't include these priorities creates strategic misalignment and disengagement of the workforce and other stakeholders.

12.1 The Value of Engagement

Disengagement is a drag on organizational performance, costing them $500 billion per year. As indicated earlier in Figure 10.5, Impact of Employee Engagement, management teams that are successful in engaging their workforce outperform other organizations. World-class organizations have significantly higher ratios of engaged versus disengaged employees. According to Gallup research, approximately 30% of employees are actively engaged, 20% are disengaged, and 50% are merely present.[1]

Why is engagement so important in driving business success? Engagement improves workforce satisfaction, which drives up productivity and customer satisfaction, both of which improve profitability. Engaged workers translates into lower stress, fewer errors, lower absenteeism, and lower unwanted turnover. One key finding is that managers have a significant role in creating disengagement through poor management practices. Adopting a sustainable approach that details clear vision, action, and desired outcomes coupled with training and development helps to reduce the disengagement caused by poor management. Engagement levels are crucial to providing excellent customer service and bottom-line results. A Towers Watson study found that organizations with high employee engagement in terms of connection, enablement, and enthusiasm had operating margins that were 2.7 times those of low-engagement companies.[2] Engagement drives improved profitability, and sustainable strategy impacts employee engagement.

When management looks at what makes an organization successful, they look at three components:

1. Strategy
2. Assets
3. People

Given the competitive arena in which most organizations are operating, an organization's strategy provides direction and priorities but not necessarily exclusive competitive advantage. Strategies are readily accessible and easily broken down for duplication. Assets can provide competitive advantage if an organization controls certain resources or technology. However, even assets are subject to change in value. Consider the impact that lower oil prices have had on governments, corporations, and even communities. Also, technology can be duplicated and patents expire. Many CEOs have come to recognize that it is their people who give organizations their competitive advantage. In service organizations, the workforce is the primary asset. The key to a successful organization is their people's ability to execute the organizational vision. So, getting the people strategy right makes a significant difference in the success of an organization. Correspondingly, the importance of people cascades into program and project management in terms of knowledge, skills, and the overall ability of the team to execute project plans effectively.

Sustainable strategy drives engagement by aligning employees and company values, creating employee pride in both their organization and themselves through its social and environmental impacts. Organizational pride is demonstrated through, employees acting as ambassadors promoting the organization, both as a great place to work and as a valuable partner to the local community. Engagement is further supported by providing knowledge and tools to employees, enabling them to successfully incorporate sustainable strategy into their daily functions. Creating incentives that align with sustainable strategy and programs that promote employee feedback help generate engagement and enthusiasm within the workforce. Clearly communicating the message to employees about organizational goals and their role in achieving these goals addresses the most significant obstacles to employee engagement. It gets your people connected to the business goals while tapping into their own interests in the environment and community. Delivering a consistent message across the organization requires leadership to align organizational mission with actions and resource allocation. Through leadership acting as role models, the message is further cascaded throughout the organization, impacting, policies, project portfolio selections, and ultimately work streams. Engagement in sustainable strategy relies on feedback and continuous improvement to move the organization forward on its sustainability journey.

Table 12.1 outlines attributes of employee engagement and offers a translation into an employee's perspective of sustainable strategy. In addition, levers to drive sustainability are included to give

Table 12.1 Sustainable Strategy Supports Employee Engagement

Attributes of Engagement	Employee Perspective on Sustainable Strategy	Sustainable Strategy Levers
Understanding	I understand my organization's sustainability mission and my role relative to that mission.	• Create a culture of sustainability by aligning message and action throughout the organization. • Clearly explain the "why" for sustainable strategy.
Pride	I am proud to work for my organization because they are concerned about environmental and social impacts of our operation.	• Develop sustainable strategy with broad-based stakeholder engagement. • Give employees a voice in policies and programs.
Alignment	The values exhibited by my employer on the environment, community, and human rights align with my own personal values.	• Encourage leadership to model behavior in line with your sustainability mission. • Create shared value by engaging stakeholders in the process. • Encourage development of local green teams and affinity groups to support individual sustainability interests that align with organizational interests.
Outlook	The policies and long-term vision of my organization will meet my personal development needs in the long run.	• Develop policies and programs and allocate resources in alignment with sustainability mission. • Create an organizational structure to support a culture of sustainability. • Provide opportunities for personal growth and development.
Results	I have the skills, tools, resources, and authority to generate the target results.	• Align HR policies with sustainable strategy and embed throughout the employee life cycle. • Provide the organizational structure, tools, and resources to support the workforce.
Effort	I am willing to expend extra effort to achieve high-quality results.	• Create a matrix or networked structure designed for agility, collaboration, and high-quality results. • Align incentive programs with results.
Ambassador	I would recommend my organization as an employer to friends and family.	• Encourage employee participation in community volunteer programs. • Measure employee referrals for job openings.

tangible actions and programs to facilitate the alignment between a sustainable approach and employee engagement.

Understanding the components of engagement and how employee actions and behaviors demonstrate their engagement level is a starting point. Designing programs and projects to create and implement sustainability levers is the next step. Partnering with HR professionals on these projects facilitates assessing employees' level of sustainability engagement as well as providing expertise, tools, and techniques. HR provides subject-matter expertise on organizational culture, internal communications, compensation and incentive plans, policies and procedures, training and development, as well as experience with a cross-functional approach.

Including HR leadership in developing and implementing a program to engage internal stakeholders from the C-suite to the shop floor begins the process of creating a culture of sustainability that is designed to promote sustainability as a core values and to engage employees on these issues. Gathering information on requirements and perceptions from internal stakeholders is a good way to initiate this type of project. The following questionnaire is designed to be given to key internal stakeholders. The purpose is to focus on organizational human capital and leadership practices and how they support or need to be adapted to support a culture of sustainability.

Internal Stakeholder Questionnaire: Embedding Sustainability into the Culture

1. What core elements of the organization's sustainability strategy will affect people strategies and requirements over the next five years?
2. How can leadership proactively support the organization's sustainability strategy?
3. Does the CEO endorse, enforce, and exhibit behavior that aligns with the organization's sustainability vision?
4. Is C-suite leadership aligned around sustainability goals and targets? Are their actions and communications consistent with organizational strategy?
5. Does the organization's middle management understand the sustainability mission, including the risks and opportunities for their business function and people?
6. Do employees understand sustainability and what it means for them in their daily activities?
7. What programs, projects, or policies can facilitate organizational alignment?
8. What is our change management process for promoting sustainability as a core business value?
9. How are internal communications on sustainability handled? Are different communication channels and push/pull formats used to communicate with a diverse employee population?
10. How does workforce planning support skills development to provide the people to fill current and future roles to support sustainable strategy?
11. How are we engaging employees? Do they have opportunities to participate in corporate programs on environmental and social issues that align with their personal values?
12. What education and training are needed for employees to adopt behaviors in alignment with sustainability best practices?
13. Do recruitment policies actively support diversity and inclusion?
14. Are marketing campaigns adequately communicated internally alongside external promotion? Is there consistency between internal and external messaging?
15. How does the compensation and incentive structure reinforce sustainability goals through all levels within the organization?
16. What metrics are being used to measure the effectiveness of our sustainability program and the extent to which sustainability is embedded into the organization?
17. What information has been received from external stakeholders that may affect programs, projects, processes and policies designed to promote sustainability?

Responses to these questions provide an assessment for the project team of the organization's leadership practices, human capital practices, communications, policies, and behaviors, and how effectively each of these tools is being used to promote a culture of sustainability. With this information, project managers are able to make recommendations in order to manage sustainability portfolio components more effectively, including leveraging sustainability programs to attract, engage, and retain people as part of the process to create a more sustainable organization. (In Chapter 13, we will take a deeper dive into tools and techniques to embed sustainability into the culture of an organization at all stages of the sustainability journey.)

12.2 Building a Sustainable Workforce

The impact of an organization's sustainability vision on its employees begins even before a potential candidate becomes an employee. Highly successful organizations engage potential candidates on the topic of sustainability as part of their recruitment process, through applicant tracking portals, website information, and marketing messages. Creating a brand image that includes sustainability messaging is impactful for attracting top talent in order to create a sustainable workforce that values contributing to an organization that includes sustainability as a core organizational pillar.

One of the main drivers for CEO support of sustainability is the positive impact on brand trust and the image of their organization. While brand and image have always been recognized as crucial to customers, they are also very important to potential employees. In a Society for Human Resource Management (SHRM) survey, approximately 90% of respondents cited organizational vision and performance on sustainability issues as "very important" or "important" in creating a positive employer brand that attracts top talent.[3] Candidates are seeking information on an organization's sustainability profile to better gauge their alignment with their own personal values on environmental and social issues. They are reviewing sustainability reports, competitive rankings, and rating agency reports to determine how organizations define their mission, shape their policies, and report on their outcomes. A survey of job candidates found that those workers currently entering the marketplace, primarily Millennials, are highly influenced in their job selection choice by an organization's sustainability image.

The Millennial-generation workforce seeks alignment between their personal and an organization's values. According to a 2012 Net Impact survey, 72% of college students are looking for employment where they can make a difference in terms of the environment, community, or organization.[4] Incorporating an organization's sustainability story into its recruitment message helps attract and retain top talent. Millennnials want their daily work to reflect their personal values and to provide opportunity for meaningful impact. Today's college-age students are seeking a positive work environment and culture (91%), work–life balance (88%), an organization with similar values (74%), and the potential to contribute to society (65%).[5] These issues are so important to them that compensation structures based solely on financial incentives are not sufficient to attract and motivate Millennials. An Intelligence Group study found that 64% of Millennials would rather make $40,000 per annum performing a job they love than than make $100,000 per year in a position that doesn't motivate them.[6] Attracting top Millennial talent requires an authentic message and an organizational culture that focuses on building natural and social as well as financial capital. Organizations with policies and programs that support alignment between personal and corporate value are better positioned to attract top talent.

> **Who Are the Millennials and What Do They Want?**
>
> Adult Millennials, born after 1980 and now ranging in age from 18 to 33, make up 27% of the adult population. By 2025, they are estimated to make up 75% of the workforce. They are described as unattached to organized political parties and religion, digital natives linked by social media, with low level of social trust yet optimistic about the future.[7] Approximately one-third of Millennials aged 26–33 hold a four-year college degree, ranking this group as

> the best educated.[8] They are America's most racially diverse generation. Millennials differ from their predecessor generations in that they face greater economic hardships. This change in outlook may explain their interest in nonfinancial work motivations. "In that world, just as Millennials create communities built around shared interest not geographical proximity, causes create compatibility between otherwise disparate groups."[9] This shift in culture will be reflected in organizational priorities and decisions making.
>
> In a National Benchmark survey, 63% of Millennials responded that they would like their employer to contribute to social or ethical causes that align with their own interests.[10] This figure compares with around 50% for Generation X or Baby Boomers. In a National Society of High School Scholars survey of the best and brightest students (ages 15–27), their most desired employers, carving out the high-tech sector, were institutions whose missions include changing the world for the better.[11] As the Millennial generation grows to take over the workforce in the next 10 years, management needs to gain a better understanding of their values and drivers in order to engage them and integrate them into leadership. According to the Brookings Institute, Millennials are most focused on corporate social responsibility when making purchasing decisions. Millennial respondents reported increased trust (91%) and loyalty (89%) to companies that support solutions to social issues.[12] As the Millennial generation comes of age, they are demanding that organizations that seek their business demonstrate support for social and environmental causes in alignment with their interests. Understanding that the largest percentage of the 2025 workforce is going to make purchasing decisions based on sustainability criteria should be driving business leaders to adapt their product and service offerings to align with their customers' values. As the Millennials become the majority of the workforce, organizational beliefs and attitudes will shift to be more reflective of their values.

12.3 Creating a Sustainable Workplace

In order to embed sustainability into an organization, both the culture and the definition of work need to change. The change process needs to consider:

1. What is the organization's sustainable strategy?
2. How is the strategy translated into organizational goals?
3. How are these goals communicated, measured, and tracked?
4. How can employees be managed more effectively to reach these goals?
5. What actions and behaviors are required from employees?
6. How will processes and policies need to be changed?

The work structure needs to support managers who promote organizational agility and cross-functional solutions designed to support a sustainable culture. A sustainable work environment focuses on collaboration and teamwork. However, as concerns about greenhouse gas (GHG) emissions, traffic congestion, and work–life balance grow, more employees are working remotely and are physically present less frequently on the worksite. While a sustainable work environment is more inclusive, physically it may have greater separation, requiring new techniques and tools to manage a diverse and disparate workforce. Performance measures move from seeing people working hard or long hours to being based on work contribution. This change in focus results in better alignment between performance and strategic goals, with clearly defined performance metrics. Work outputs are increasingly defined and

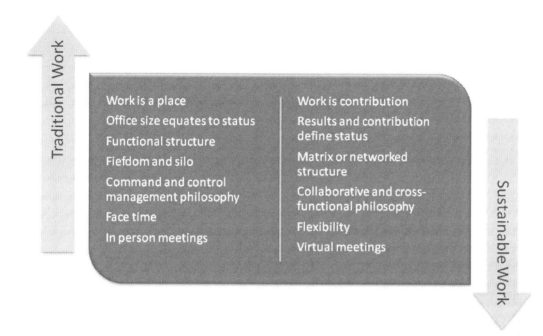

Figure 12.1 A Framework for Sustainable Work

structured to make good use of employees' time and skills. The culture moves in the direction of a supportive work community with a management culture of transparency on process and goals.

Consider a framework that promotes a sustainable workplace focusing on an employee's contribution relative to the organization's sustainability goals as the definition of success. Traditional work structure is compared with sustainable work structure in Figure 12.1. Traditional work is characterized by work as a location and performance based on face time. In a sustainable work environment, status is no longer based on location and office size, but rather on performance relative to clearly defined targets and goals that include environmental, social, and financial criteria. Metrics are identified, clearly communicated, and utilized to measure performance. A sustainable workforce promotes organizational agility and flexibility driving competitive advantage. Sustainable work is exemplified by collaboration and a cross-functional approach to developing products and solutions that help clients meet their own sustainability challenges. In order to create a sustainable work environment, new tools and techniques need to be provided for both managers and employees to support the needs of a sustainable workforce.

12.4 Employee and Team Engagement

As the structure of work changes, so will the project management techniques required for managing stakeholders and team members. In order to engage employees, program and project managers need to draw on the global sustainability vision and execute their plan in a way that promotes local action. Just like the workforce in general, team members' performance improves when they identify personally with the project's goals and objectives. Establishing a team that promotes environmental and social goals as well as financial performance, along with aligning incentives, creates meaningful engagement for team members. Research from Tuzzolino and Armandi suggests that sustainable organizations meet five developmental needs: physiological, safety, affiliative, esteem, and self-actualization.[13] "A meta-analysis on person–organizational fit found that values alignment is positively related to numerous positive outcomes such as job satisfaction and organizational commitment and negatively related to intention to quit."[14] This research further supports that sustainable strategy enhances employee engagement.

Engagement rises as employees better understand their role in the process, the impact they can have on the project, and how their project's sustainability goals align with the overall corporate mission. Meaningfulness at work is an important part of the engagement process. Does the employee find meaning in the organization for which she works? Some purpose-driven organizations such as NGOs and B Corps have missions of higher-purpose baked into to their organizational pillars. These statements clearly align with an employee's personal values, making the work inherently meaningful. Or does she find meaning in the actual work function that she does?[15] In other words, has a general-purpose organization embedded the concepts of sustainability into daily job functions, creating meaning through job design? As management develops a sustainable culture, these distinctions are helpful in formulating programs to engage employees who have job roles and functions with differing levels of sustainability contribution. These concepts relate to project team formation as well.

In order to leverage sustainable strategy as a tool for employee engagement, start the process as soon as the employee agrees to come on board. Once the employee accepts a position, incorporate sustainability messaging into materials and training into the onboarding process. Let new employees know about not only your sustainability policies but also your sustainability stories. If possible, have a sustainability team leader come and share some stories and projects with new recruits. Get new employees involved as quickly as possible by providing sustainability mentors so that they understand how sustainability initiatives impact both the organization and their job function. In addition, let them know about the organization's commitment to the community and about volunteer opportunities. Encourage them to participate by selecting voluntarism options that are of interest to them. To get new recruits to understand an organization's sustainability programs, have them join a sustainability project team. If you are having a campaign to reduce paper usage, get the new recruits involved right away. Explain how your group has decided to reduce their paper usage and help them identify ways that they can be involved in the campaign. Give the new recruits an opportunity to contribute ideas as well. Getting your new recruits involved in a current campaign with a mentor makes them feel part of the team and helps them understand the importance of sustainability to your organization.

A survey of HR leaders reports that 85% believe sustainable strategy is an important tool for improving employee retention.[16] While Millennials are the most-referenced category, other groups of employees are also seeking more meaning from work than just financial compensation. Work that aligns with one's own beliefs, values, virtues, and morals and that provides meaningful engagement motivates all employees.[17] In most organization, the compensation structure is aligned around a job or career orientation rather than a calling orientation, such as working for an organization with a higher purpose. At issue is that work design and performance management assumes all employees care about the same thing, such as pay and career path. Adding sustainability into the workforce design provides employees with a value option of caring for the environment or community. It provides employees with the opportunity to align work interests with self-interests in making the world a better place and bringing out the best in people. By creating networks of like-minded people such as green teams in organizations, sustainability grows and flourishes within the organizational culture. It becomes embedded in the organization.[18]

In creating programs to engage employees within your organization, the employees' ability and willingness to accept the messaging needs to be considered. The range is from the disengaged to the totally transformed employee who is fully engaged and who actively participates and contributes to the success of the organization's sustainability vision.

As indicated in Figure 12.2, employees have differing levels of experience with sustainability. The continuum ranges from disengaged to aware, engaged, and transformed. Fully transformed employees

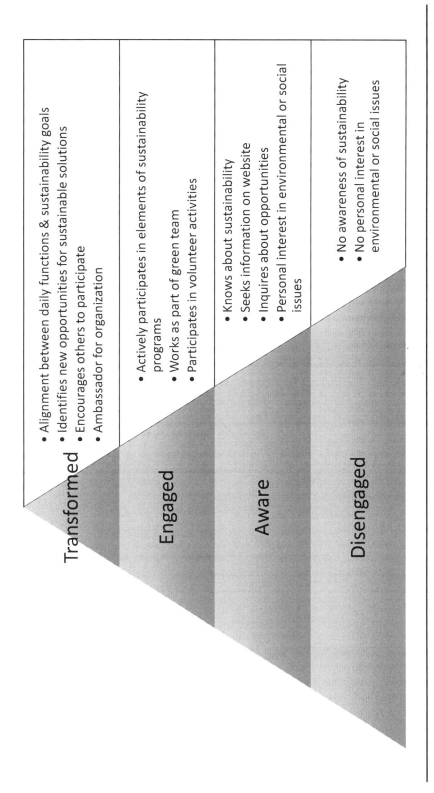

Figure 12.2 Sustainability Engagement Hierarchy

Table 12.2 Levels of Employee Engagement in Sustainability[19]

Category	Engagement Level	Identifiers	Measures/Metrics
Transformed	• Daily functions and activities align with sustainable strategy • Identifies new opportunities • Encourages others to participate • Ambassador	• Explains organization's sustainability mission • Personal value alignment • Ambassador for family, friends, community • Level of participation in volunteer programs supporting the environment or community • Identifies new opportunities to incorporate sustainable practices into job	• Employee surveys • Customer surveys • Stakeholder surveys • Industry rankings • Program participation (%) • Number of customers who rank organization higher than competitors on sustainability issues
Engaged	• Actively participates in elements of sustainability programs • Part of green team or volunteer efforts • Seeks new opportunities	• Understands sustainability mission • Number of employee green teams/affinity groups • Number of volunteer hours • Makes suggestions through incentives programs • Pride in organization's sustainability mission • Commitment to organization because of sustainability mission	• Employee surveys • Track employee social media • Track volunteer hours • Track number of green teams/affinity groups • Stakeholder feedback
Aware	• Knows about sustainability mission • Identifies sustainability statement	• Seeks information about organization's sustainability efforts on company website • Seeks information about opportunities to participate in green teams/affinity groups • Personal interest in the environmental and/or social issues	• Track website activity • Track percentage of employees who are aware of sustainability efforts • Track employee applications for green teams/affinity groups • Employee survey sustainability mission identification
Disengaged	• No awareness of sustainability vision	• No personal knowledge of ESG issues • No information on organization's sustainability program	• Employee surveys • Absentee rate • Unwanted turnover

embrace sustainability in all aspects of their lives, buying environmentally friendly products, using environmentally friendly transportation, and actively volunteering in their communities. At the other end of the spectrum are disengaged employees. They have no experience with sustainability. This concept is similar to that of consumers who are "dark green"—buying only environmentally friendly and socially responsible products—versus consumers who are "brown" and do not consider these issues in their purchasing decisions. Most employees' level of experience with sustainability falls into the aware and engaged levels, which are similar to shades of green for consumers. Of course, the employee population is not static, so you have people joining and leaving the organization as well as people moving between the levels as their knowledge and experience with the benefits of a sustainable approach grow. Table 12.2 considers employees at various levels of engagement and lays out some identifiers of the different levels and methods to measure employees' level of engagement with sustainability and specifically the organization's sustainability vision and programs.

Engaging people is about delivering a clear message, garnering commitment, and creating a sense of community. In order to leverage the engagement impact of sustainability, consider how employees break down into groups of commonality with similar characteristics. Some factors to think about include organizational level, business function, location, culture, access to technology, and level of engagement. Depending on the group, the approach will vary. If you are looking to build awareness, a project that focuses on information and communication is effective. Some organizations have found that very visible displays in high-traffic areas get employees attention. Setting up tables in high-traffic spots such as the lobby or cafeteria promotes awareness. Engaging employees in their natural environment and making it a fun experience creates interest. Never underestimate the value of free giveaways. Targeted emails and dedicated pages on internal websites are effective with those who are engaged and are seeking more information. Others find that games, contests, and campaigns create incentives to drive desired behavior. Design the program to meet your employee group where they are and base it on a point of connection. The case study at the end of this chapter highlights the importance of varying techniques to engage distinctive workforce groups.

Enlisting champions to share stories with peers is a very effective change management tool. Sharing personal experiences and the impact that a sustainable approach has on the organization, community, or environment is a compelling engagement technique. Sharing stories helps employees to understand how personal values can align with organizational values. This approach can be scaled across the organization if the champion is a senior leader such as the CEO via videos or large-format meetings. Asking the CEO to speak of her personal sustainability values and how they align with the corporate mission has a significant impact with both internal and external stakeholders. When organizational leaders ask co-workers to consider their own environmental and social impact, it makes a difference. If every employee selects an area in which to improve his own personal impact, collectively the workforce has a significant impact. For maximum engagement, create a mechanism to track and share employee efforts, such as a dashboard that allows employees to track their own progress and share it with others. As personal interest in sustainable practices grows, so does employee interest in corporate sustainability, moving the employees' sustainability engagement farther up the pyramid.

In order to reach the transformative stage, HR professionals and project managers must embed sustainability into the employee life cycle. Going beyond the sustainability message, all aspects of work must be realigned to support the organization's sustainable strategy. Reengineering job functions to promote achievement of individual, group, and divisional sustainability goals is a starting point. However, incentive structures must also be changed to align compensation with desired behaviors. These changes should take place at all levels of the organization so that managers' goals and incentives are changed along with their people's goals and incentives. The process begins with potential candidates and follows an employee all the way through to the exit interview. The project plan needs to provide support for the process through training, development, and leadership development throughout an employee's life cycle.

12.5 Best Practices to Embed Sustainability and Maximize Engagement

Table 12.3 lists several organizations that are leaders in engaging employees through sustainable strategy. These organizations represent a variety of industries, but they are all leveraging sustainable strategy to engage employees and improve their triple bottom line results—people, planet, and profit.

Each of the organizations listed in Table 12.3 has a slightly different program, but they all highlight the importance of sustainability with meaningful employee programs designed to drive employee engagement through sustainable strategy. Best practices include clearly communicating sustainability goals and desired actions while allowing for employee input and community impact. Establishing programs that encourage dialogue with employees about sustainability provides opportunities for employee engagement. Creating opportunities for employees to take ownership through developing their own ideas and contributions promotes success. Incorporating recognition programs that highlight employees who are leaders in sustainability programs and reinforcing desired behaviors with compensation incentives furthers employee acceptance of sustainable strategy.

In the Hyatt organization, for example, employees are encouraged to identify, develop, and implement their ideas for community involvement. Mark Hoplamazian, President and CEO of Hyatt Hotels Corporation, shares the organization's community commitment to creating better schools and better education for children in Chicago. This type of initiative is driven by both organizational and employee interest. It is good for families that work for Hyatt and good for Hyatt in that it has a high-quality workforce to draw on. While organizations promote a global sustainability vision, employees become fully engaged when they feel that their actions have an impact in their daily lives.

Table 12.3 Best Practices for Leveraging Sustainable Strategy

Organization	Best Practice	Impact
Intel	• Calculates a portion of every employee's annual bonus based on achievement of sustainability goals. • Recognition through "Intel Environmental Excellence Awards."[20]	• These initiatives challenge all employees to improve their department's sustainability performance through improved process and products.
Hyatt	• "Hyatt Thrive"—Over 85,000 associates in more than 450 countries act as ambassadors in their local communities, selecting and driving environmental, economic development, education, and health & wellness programs.[21]	• Employees' own their sustainability projects and are engaged in making a significant impact in their communities. • In the U.S., average tenure is 10 years. • The largest source of applicants comes from employee referrals.[22]
Clif Bar	• Benefits packages include personal sustainability incentives for purchasing a fuel-efficient vehicle and making eco-friendly home improvements. • Weekly staff meetings to share tips for living greener.[23]	• Aligns personal and corporate sustainability values to further embed sustainability into the culture of the organization.
Interface	• Encourages sustainability champions through a network of ambassadors that is an elite club with limited membership. • Senior executive ambassadors act as meaningful role models. • Valuable rewards.[24]	• Ambassadors choose to compete for the role. They gain high status within the company while promoting the message and importance of sustainability. • Some sales representatives have both "Sustainability Ambassador" and "Sales Account Manager" on their business cards.[25]

Companies that are leading the way in sustainability engagement are effectively embedding sustainability into their organization. They are making progress toward large environmental and social goals and they are delivering sustainable products and services to the marketplace. Unilever is a leader in this area with its sustainability framework, the "Sustainable Living Plan." Unilever's leadership refers to their strategy for sustainable growth as their "Compass Strategy," as it clearly lays out their vision for the future for continuing significant growth while reducing their environmental footprint and enhancing their social impact. "We have found that doing business sustainably is possible and that brands that build sustainability into their offer are more appealing to consumers. We realize that we can make a bigger difference to some of the world's major social, environmental, and economic issues if we leverage our scale, influence and resources to drive transformational change."[26] As part of this process, they are empowering women, including building a gender-balanced organization by enhancing training and skills as well as opportunities across their value chain.[27]

Employees need their fingerprints on projects and programs in order to make them meaningful to them. In our practice, we have spoken with employees who had the opportunity to volunteer through their corporation, but who didn't feel compelled to participate. A significant barrier was that they didn't feel personally connected to the recipient charities or the type of work performed. These employees were looking for options that spoke to their own personal values. They were seeking options to select their own charities and direct donations of their time and resources to areas that aligned with their core values.

Volunteer programs that align corporate social initiatives with core corporate mission as well as speak to employees through purpose or opportunities for skills advancement have significant impact on engagement. As illustrated in Figure 12.3, voluntarism drives employee engagement. The greater the strategic alignment of the voluntarism program, the greater will be the impact on employee engagement.

Volunteer program participation is one of the most impactful metrics for evaluating a firm's level of corporate social responsibility. Companies that support volunteer efforts through paid time off, flexible work schedules, and use of company resources generate 45% more volunteer hours per year. Approximately 80% of managers view corporate volunteer programs as important.[28] These programs help attract qualified applicants, build knowledge and skills, improve employee morale, and encourage retention. However, volunteer programs are not created equally. In order to drive sustainability into a

Figure 12.3 Voluntarism Drives Employee Engagement

corporate culture, employees need not only to participate in the programs but also to be invested in the process. The most sustained and impactful programs are organized and implemented by employees. When employees participate in corporate volunteer programs, they actually become more committed to their employers.[29] A well thought-out and employee-centric volunteer program is a powerful tool to support sustainability vision and develop a sustainable workforce.

12.6 Project and Portfolio Perspective

As a project manager, it is vital to understand motivational drivers for your team. All employees are positively engaged through sustainability; however, the motivation for Millennials is most significant and growing. As Millennials become a larger portion of the workforce, the composition of project teams will begin to change. While financial motivators such as completion bonuses and overtime pay may have worked in the past, research indicates that Millennials need to be engaged in more purpose-driven projects. They are looking to have a meaningful and valuable experience and to be involved in projects that promote environmental stewardship and social responsibility. As project managers create teams and assign work roles, these motivational needs of employees need to be considered and planned into the work structure.

Table 12.4 shows an example of one way to measure the impact of current sustainability programs and projects on employees' perceptions of organizational sustainability and the impacts of their

Table 12.4 Project Proposal: Measuring the Impact of a Sustainable Workforce

Title	Benchmark employee awareness and engagement in organizational sustainable strategy and the impact of a sustainable workforce on customers' perception of organizational sustainability.
Executive Summary	Measure employees and customers' understanding of sustainable strategy in order to gauge the impact of current programs, and identify best practices and areas for improvement. Identify employees' level of engagement with sustainable strategy and the impact it has on customer messaging and engagement.
Sponsor(s)	CSO, VP HR, VP Sales
Project Manager	HR Project Manger Level 3
Benefits	1. Measure employee understanding of the organization's sustainability vision. 2. Measure customers' perception of the organization's sustainability message. 3. Measure employees' engagement level and the impact of sustainability programs on engagement. 4. Establish a baseline for sustainability engagement for employees. 5. Identify and gain insight into customers' perception of sustainability program from employee behaviors and actions. 6. Identify best practices and effective programs.
Stakeholders	Sustainability steering committee Leadership Business Function Leaders Employees Customers
Project Team	HR professional with employee survey responsibility Marketing representative with customer focus group experience Business analyst Focus group facilitator Project coordinator Sales team representative Vendor survey developer

(Continued on the following page)

Table 12.4 Project Proposal: Measuring the Impact of a Sustainable Workforce (*Continued*)

Milestones & Key Deliverables	Identify survey tool—3/31/16 Identify customer focus group service—3/31/16 Steering committee approval of survey and focus group—3/31/16 Development of employee survey—4/30/16 Development of focus group format—4/30/16 Annual employee survey—6/30/16 Customer focus groups—7/31/16 Analysis of data—8/31/16 Final recommendations—9/30/16 Revisions to programs—12/31/16 Survey and focus group revision—3/31/17
Budget	$75,000/annum Costs allocated to HR and Sales budgets (50%)
Resources	Project team Survey software—external vendor Survey provider Focus group service Focus group facilitator
Project Risks	Gathering feedback not in alignment with management expectations Lack of employee participation Lack of customer participation
Success Criteria	1. Establish an employee engagement baseline for sustainability 2. Percent of employees who can identify sustainability statement 3. Percent of employees who believe that sustainable strategy positively impacts organizational outcomes 4. Establish a customer baseline for organizational sustainability profile. 5. Percent of customers who can identify the organization's sustainability statement 6. Percent of customers who indicate that the organization's sustainability profile is important in making product/service choices 7. Identify relationship between employees' sustainability engagement level and customers' perception. 8. Employee participation of 50% in survey 9. Identify best practices and effective programs. 10. Identify areas for improvement.
Assumptions & Constraints	1. Clear identification of survey focus, purpose, and impact. 2. No survey has ever been done to measure the engagement of employees in our sustainable strategy. 3. Focus groups will represent a cross section of customers. 4. Employee survey response rate average 30–40% 5. All department heads will encourage participation. 6. Leaders and managers will participate. 7. Senior management support
Location	U.S. facilities
Organizational Templates & Standards	Utilize organizational project management templates Weekly status reports Steering committee presentation format and summary
Monitor & Control	Feedback loop Surveys Metric tracking Lessons learned

behaviors and actions on customers' sustainability perceptions. The project includes gathering feedback and measuring both employees' and customers' baseline understanding of the organization's sustainability mission and goals as well as engagement levels. After establishing an initial baseline, annual benchmarking will be used to measure the impact of sustainability programs and projects on engagement. The project is designed to gain insight into the level of engagement by management and employees relative to the organization's sustainability mission (disengaged, aware, engaged, or transformed) and the impact that employee sustainability engagement has on customer perceptions of the organization's sustainability vision. The objective is to identify drivers to facilitate closing the gap between sustainable strategy and organizational impact.

Information gathered from this project informs leadership about the perception of their organization's sustainability programs from two key stakeholder groups—employees and customers. With this feedback, leadership is able to select sustainability portfolio components to create better alignment between sustainability vision and key stakeholder requirements and perceptions. Creating sustainable change within an organization impacts people and process in terms of daily actions and behaviors. Employees' sustainability behaviors go beyond internal actions and impact key stakeholders' perceptions such as customers. Creating a feedback loop is crucial to the ongoing success of the transformation so that lessons learned and best practices are incorporated into the change process. As project management professionals embrace sustainability, understanding the broad impact that sustainability vision and action has on managing people changes how work is done. The process of how an organization performs its core functions sends a message to stakeholder groups as to the level of sustainability embedded into organizational culture.

12.7 Case Study: How the City of Cambridge, Massachusetts, Engaged Employees on Commuting Habits and Effected Change[30]

This case study outlines an effective change management strategy to alter employees' commuting habits across a broad and diverse organization. It also demonstrates how sustainable strategy acts as a lever to improve employee engagement.

What began as a compliance project evolved into a value-add project for both the community and the City of Cambridge (City). Engaging the workforce on commuting habits is not only good for the environment, it is also good for the health and well-being of employees. In a recent study from the U.K., it was found that commuters who switched from driving to walking, cycling, or public transit lost more than 2 pounds in two years.[31] While the majority of U.S. commuters continue to drive alone, the U.S. Census Bureau reports that walking, biking, and use of public transit are growing at a faster rate than single-occupancy vehicle usage.[32] Jennifer Lawrence, Sustainability Planner, City of Cambridge, shares their story of how embracing environmental sustainability engaged their municipal workforce and impacted commuting choices.

In 1995, GHG emissions generated by daily commutes came on the radar for many Massachusetts organizations, when the Commonwealth of Massachusetts issued a compliance standard requiring employers with more than 1000 employees to participate in a bi-annual survey of employees' commuting habits. The goal was to establish a baseline and then to monitor commuting habits while encouraging employees to consider alternative forms of transportation to the single-occupancy vehicle. Once they became aware of their own employees' commuting habits, the City of Cambridge thought that they could provide resources and information in order to assist with transportation alternatives to car usage within their city. In 2009, the Community Development Department (CDD) launched the "CitySmart" program to effect change in the commuting habits of City residents. Their website provides information and resources to help residents get around the metropolitan area without driving. Whether walking, biking, using public transit, or even driving, the CDD has options for safe, convenient, and affordable transportation for residents, commuters, and visitors.

Based on the success of this project, in 2012 the team focused their attention on their own municipal employees' commuting habits. This new project focused on revamping their internal employee transportation demand management program and remarketing it as "CitySmart for City Employees." On their radar were issues such as reducing "drive alone" commutes, educating employees on employer-offered sustainable commuting options, and reducing the impact of the municipal fleet on emissions and traffic congestion in the City. The program focused on making City employees more aware of sustainable commuting options such as carpooling, transit, biking, and walking. Changing the commuting behavior of 3300 employees is a major change management project, but the team was up to the challenge. They devised ideas for stakeholder engagement and promoted alternatives to the single-occupancy vehicle commute.

Techniques to Engage Employees

They began gaining traction with employees through in-person information sessions by attending staff meetings. Initial responses were favorable, so they developed new engagement techniques and added programs. Over the next three years, the impact of the project on modifying employees' commuting habits grew significantly. (A list of metrics measuring success is included at the end of the case.)

The project budget is self-sustaining to allow for ongoing development. The project is funded through the Community Development Department budget with payments related to programs (such as subsidized Hubway membership) going back into the CDD budget in order to fund continued programming.

The group offers the following types of education and outreach programs to build stakeholder alliances and grow employee participation. While all of these programs are effective in driving change, Jennifer has found that tailoring programs to meet the specific needs of the diverse employee population creates the most impact.

Successful Techniques to Engage Employees

a. Monthly email communication to employees who have opted in, giving tips for getting around town, contests, and events
b. Contests and competitions between internal employees and departments as well as with external organizations in the region
c. Educational and informational workshops
d. Free bicycle tune-ups
e. Bicycle education series:
　i. Urban Cycling Basics
　ii. Rules of the Road
　iii. Bicycle Maintenance
　iv. On-Bike Basic Training
　v. Women-Powered Cycling
f. Staff meeting information sessions
g. Individual commuter consulting sessions
h. "On-site Outreach"—tables in city building lobbies to share information with employees as they arrive for work
i. Annual transportation fairs for City employees

Contests have been an effective way to reach employees. A popular contest hosted by the Green Streets Initiative is the "Walk/Ride Day Corporate Challenge," a competition among Boston and Cambridge area employers to determine which organization wins the title of "Green Commuting

Champion." The challenge runs monthly from April through October with a goal of encouraging staff to choose a sustainable commuting option on the last Friday of the month. The City of Cambridge encourages participation in the program, and the project team uses email, posters, and contests to make employees aware of the program and to encourage them to register online. The friendly regional competition gets lots of attention and participation. The City also participates in other competitions, such as "Car Free Day" and "Bike Week." During "Bike Week" in 2014, there were 119 events organized by community groups, companies, and local governments and 12,000 participants. It is clear that this event has garnered widespread stakeholder support. City employees have fun with the competitions and enjoy winning prizes such as water bottles, biking gear, and T-shirts. Rewarding employees' behavior with recognition and even small prizes such as water bottles and T-shirts has a meaningful impact.

In order to engage the diverse City employee population, the project team varies their selection of communication channels to best match the employee audience. They found that younger employees respond well to email communications, contests, and events available through online registration. However, other techniques are needed in departments such as the Department of Public Works, where not all employees have access to the email system. Staff meetings and in-person visits work best for reaching this group of employees. An effective engagement technique for the Public Works employees was an "On-site Outreach" that focused on biking as a commuting option. The Public Works employees who bike to work regularly were empowered by the support from their organization and pleased to get some tips and advice on biking. Others were excited to learn about the wide variety of employee programs and options available to them. The event provided positive reinforcement for sustainable commuting as well as for the City of Cambridge as an employer of choice.

Another campaign focused on reaching school-based employees. While the school district has an email system, it isn't linked with the City of Cambridge system, so reaching school employees with monthly email blasts isn't possible. By attending on-site staff meetings, Jennifer was able to meet with school staff members and to explain the "CitySmart for City Employees" programs and employee benefits available to them, such as transit passes and Hubway bike-share memberships. Given the parking challenges in Cambridge, the staff was thrilled to learn that they could request that the City create designated carpool parking spots near the school for their use. Once school staff members became aware of the various programs, the staff and teachers found the "CitySmart for City Employees" programs to be a valuable employee benefit. One of the key lessons learned is that communication plans need to consider the stakeholder audience and then tailor a communication platform to engage with them. One approach doesn't fit for all.

Senior management support for the program is robust and visible. The City Manager sends monthly emails that include upcoming events and programmatic details of the "CitySmart for City Employees" offerings. In addition, the CDD team has formed strategic alliances with other organizational functional leaders such as with the Departments of Traffic & Parking, Public Works, and Community Development in order to better engage with employees. They have also formed strategic alliances with external stakeholders such as area businesses, surrounding communities, and governmental agencies to create additional opportunities and impact in their community.

Commuting Alternatives

The "CitySmart for City Employees" programs focus on providing viable commuting alternatives for employees in order to meet their lifestyles, changing schedules, and need for flexibility. The following programs reflect a broad programmatic offering to meet a wide variety of commuting requirements.

Commuting Alternatives and Programs with Impact
   a. Hubway bike-share subsidized memberships
   b. Free employee shuttle

c. Walking guides
   d. Carpooling parking spots
   e. Transit pass subsidies
   f. Regional and local competitions and events
   g. Tools for getting around Cambridge

A popular offering is Hubway, a bike-sharing program with 140 stations throughout the cities of Boston, Brookline, Cambridge, and Somerville. Hubway works with a variety of stakeholders such as businesses, institutions, and agencies to offer and continue to expand their network. Within the City of Cambridge, 33 bike-sharing stations are currently available. It is easy and convenient to grab a bike near your office, class, or home and ride to a meeting, job, shopping trip, or a gathering of friends and family. In Cambridge, Hubway operates in all four seasons, as bike paths are cleared of snow for winter access.

The bi-annual commuting survey results indicate that the "CitySmart for City Employees" program has really changed employees' commuting behavior. Even employees who thought that they would never change their commuting behavior have found that Hubway provides a viable alternative to driving alone. Employee stories depict their changing perspectives about commuting options. One example is that, rather than driving, employees can walk their children to school; then grab a bike at a nearby Hubway station and head for the office. In the past, these same employees drove because of the lack of transit alternatives to match their multifaceted commutes. Other employees have indicated that they take a train for the first leg of their trip, then grab a bike for the last leg, as it is quicker and more flexible than other commuting options.

As part of this educational process, the team has found that automobile drivers in Cambridge are learning to share the road with sustainable forms of transport. One of the "CitySmart" core messaging themes is about respecting one anothers' usage of shared roadways. The end result is that both the employees and other citizens of the City of Cambridge have great options for getting where they need to go without adding to GHG emissions, congestion, and traffic.

Project Results

By multiple measures, the "CitySmart for City Employees" program has been a success. The team is able to share many employee stories, which demonstrate the meaningful impact of the program on employees' personal and work lives. The number of people biking in Cambridge has tripled over the past decade partially as a result of this program. The following metrics demonstrate the beneficial impact and the progress made through this program.

Metrics that Matter (2011–2014)

1. Employee shuttle ridership rose 300%.
2. Email recipient list grew from 40 to 400.
3. "Bike Week" participants rose by 100%.
4. Workshop participants rose from 3 to 214.
5. "Walk/Ride Day" participation rose from 90 to 311.
6. Hubway membership rose from 58 to 148.

In addition, Jennifer shared some of her insights as the program manager for best practices and lessons learned over course of the program. A list of resource websites follows for program and project managers who wish to offer similar programs in their communities or workplaces.

Best Practices/Lessons Learned

1. Ongoing and demonstrated senior leadership support is crucial.

2. Relationships matter. Take the time to build them.
3. Strategic alliances with stakeholders can make or break a project.
4. Surveys and metrics give credibility to your mission.
5. Engaging employees requires utilizing a variety of channels and communication platforms.
6. Small changes generate large impacts.
7. Have fun!

Conclusion

The "CitySmart for City Employees" program has been a resounding success in changing commuting behaviors, reducing carbon emissions, lessening traffic congestion, and engaging employees. While driving change in any project is complicated and time-consuming, the CDD team has demonstrated that effecting behavior change on a large scale is possible through creative programs and effective communication.

Resource Links for the City of Cambridge Programs and Events

The City of Cambridge-CitySmart Program
http://www.cambridgema.gov/CDD/Transportation/CitySmart.aspx Walk/Ride Day Corporate Challenge
    http://gogreenstreets.org/2014-walkride-day-corporate-challenge
Bay State Bike Week
    http://massbike.org/blog/tag/bay-state-bike-week
City of Cambridge Hubway Program
    http://www.thehubway.com/partners/cambridge

12.8 Conclusion

Creating a culture of sustainability provides organizations with a valuable lever to create a sustainable workforce, attracting and engaging employees and improving organizational performance as measured by the triple bottom line—people, planet, and profit. Human capital professionals are valuable partners in developing a culture of sustainability. As the Millennial generation matures and grows to become the majority of the U.S. workforce, the importance of creating a sustainable workplace will increase. Those organizations that focus on building a sustainable workforce will attract the most talented and skilled workers, providing them with a competitive advantage. Engaging in best practices to promote a culture of sustainability and offering opportunities for impact through work and volunteer opportunities creates business value. Undertaking a project to better understand your organization's perceived culture of sustainability identifies successful programs and areas for improvement. Closing the gap between sustainability vision and action centers around the ability to transform an organization's core beliefs to align with sustainability values and practices. The real measure of sustainability comes from what your key stakeholders perceive as being demonstrated through behaviors and actions of both your leaders and your workforce.

Notes

[1] Gallup, Inc., "State of the American Workplace: Employee Engagement Insights for U.S. Business Leaders," Gallup, Inc., 2013, http://www.gallup.com/strategicconsulting/163007/state-american-workplace.aspx.

[2] Towers Watson, "Global Workforce Study: Engagement at Risk: Driving Strong Performance in a Volatile Global Environment," accessed June 2, 2015, https://www.towerswatson.com/en/Insights/IC-Types/Survey-Research-Results/2012/07/2012-Towers-Watson-Global-Workforce-Study, 2012.

[3] SHRM, BSR, and Aurosoorya, "Advancing Sustainability: HR's Role," Society for Human Resource Management, 2011.

[4] Cliff Zukin and Mark Szeltner, "Net Impact: Talent Report: What Workers Want in 2012," Rutgers, May 2012, https://netimpact.org/sites/default/files/documents/what-workers-want-2012.pdf.

[5] Ibid.

[6] Morley Winograd and Michael Hais, "How Millennials Could Upend Wall Street and Corporate America," Brookings Institute, May 2014, http://www.brookings.edu/~/media/research/files/papers/2014/05/millennials%20wall%20st/brookings_winogradv5.pdf.

[7] Pew Research Center, "Millennials in Adulthood," *Pew Research Center's Social & Demographic Trends Project*, March 7, 2014, http://www.pewsocialtrends.org/2014/03/07/millennials-in-adulthood.

[8] Ibid.

[9] Winograd and Hais, "How Millennials Could Upend Wall Street and Corporate America."

[10] Ibid.

[11] Ibid.

[12] Ibid.

[13] Ante Glavas, "Employee Engagement and Sustainability," *Journal of Corporate Citizenship*, no. 46 (Summer 2012):13–29.

[14] Ibid.

[15] Ibid.

[16] SHRM, BSR, and Aurosoorya, "Advancing Sustainability: HR's Role."

[17] Glavas, "Employee Engagement and Sustainability."

[18] Ibid.

[19] TD Bank, Environmental Defense Fund Climate Corps, and Borwnflyn, "Environmental Employee Engagement Roadmap," n.d., accessed June 4, 2015, https://netimpact.org/sites/default/files/documents/eeeroadmapfinal.pdf.

[20] "Intel Sustainability Initiatives and Policies," *Intel*, accessed June 8, 2015, http://www.intel.com/content/www/us/en/corporate-responsibility/sustainability-initiatives-and-policies.html?wapkw=sustainability.

[21] "Hyatt Thrive," accessed June 8, 2015, http://www.hyatt.com/gallery/thrive/interactiveMap.html.

[22] "Associate Experience | Hyatt Workplace," accessed June 8, 2015, http://www.hyattworkplace.com/our-workplace/associate-tenure.

[23] "Employee Engagement: Five Companies That Get It," *Triple Pundit: People, Planet, Profit*, accessed June 8, 2015, http://www.triplepundit.com/2012/02/employee-engagement-five-companies.

[24] Interface, "Embedding Sustainability: One Mind at a Time," March 2010, http://www.interfacecutthefluff.com/wp-content/uploads/2010/05/EmbeddingSustainability.pdf.

[25] Ibid.

[26] "Unilever, Our Approach," accessed October 6, 2014, http://www.unileverusa.com/sustainable-living-2014/our-approach/index.aspx.

[27] Ibid.

[28] Adam M. Grant, "Giving Time, Time After Time: Work Design and Sustained Employee Participation in Corporate Volunteering," *Academy of Management Review* 37, no. 4 (October 2012):589.

[29] Ibid., 590.

[30] Jennifer Lawrence, How the City of Cambridge Massachusetts Engaged Employees on Commuting Habits and Effected Change, Interview, September 23, 2014.

[31] Rachel Bachman, "Your Commute Could Help You Lose Weight," *Wall Street Journal*, August 11, 2015, sec. Life, http://www.wsj.com/articles/your-commute-could-help-you-lose-weight-1439315593.

[32] Ibid.

Chapter 13

Tools and Techniques to Embed Sustainability

We have discussed why sustainability is important for long-term organizational success. Now, let's take a deeper dive into how to implement projects designed to promote organizational sustainability. This chapter focuses on providing resources, tools, and frameworks for program and project managers as well as highlights cases and best practices to engage employees to create a culture of sustainability. The chapter begins with educational resources, early-adoption projects, and first-stage employee engagement techniques, then moves to more complex and interconnected projects. As an organization moves forward on the sustainability continuum, management realizes that in order to have significant impact on global sustainability challenges, they need to pursue collaborative solutions with both internal and external stakeholders.

13.1 Resources

There are a variety of resources providing information, tools, and best practices to leadership teams, sponsors, and project management professionals. In order to select the most useful resources to meet the requirements of your organization, consider the organization's experience with sustainability. Initially, sustainability champions are trying to get project buy-in from leadership; resource needs focus on building the business case for sustainability, regulatory and compliance requirements, and success stories from industry leaders to educate and inform management. After the initial phase of successful sustainability projects, management buy-in begins to solidify as the business value of sustainability becomes more apparent. At this point an organization enters the integration stage, where sustainability challenges become increasingly complex, interconnected, and driven by stakeholder requirements. Resource needs are more specific, technical, and geared toward stakeholder management, cross-functional projects, and substantial organizational change. As an organization moves into the transformational stage, in which sustainability and business strategy are in full alignment, resource needs expand to include recognized reporting standards, third-party verification requirements, and partnerships. Programs focus on engaging with key stakeholders as well as industry and academic partners to promote sustainable solutions. Table 13.1 outlines resources that are useful during the different phases of the sustainability journey.

Table 13.1 Resources for Information to Develop Sustainability Programs Based on Journey Stages

Exposure	Integration	Transformation
Articles on Benefits of Sustainability	Customer Reporting Requirements, Industry Standards, First-Level Sustainability Reporting	Sustainability Standards—GRI, CDP Third-Party Verification Requirements
Competitors' Information on Sustainability	Sustainability Conferences, Industry Research	Strategic Partnerships with NGOs, Academia, Government, & Others
Regulatory & Compliance Requirements	Federal, State, & Local Government Websites	SASB Reporting Protocols
Industry Events Ted Talks	Internal & External Stakeholder Engagement Techniques	UN Agencies & Industry Agreements & Standards
Energy Star Programs	NGO Research	Internal Training & Development Programs
General Case Studies	Project-Based Case Studies: DOE, EPA	Industry Leader Case Studies—Best Practices
Consultants	EPA Sustainability Analytics	Stakeholder Advisory Group

While all of these resources have benefits to project management professionals, selecting resources that reflect the organization's sustainability level of experience facilitates effective communication of the vision and benefits of a sustainable strategy.

Moving from left to right in Table 13.1, the addition of new resources is meant to be an additive process. While all of these resources are useful, ensure that your audience has the proper foundation to accept and understand the information before forging ahead with detailed protocols such as Global Reporting Initiative (GRI) reporting standards. It is easy to derail a sustainability program by trying to introduce too much information to key internal stakeholders without building the right structure and foundation for them to be open to adopting sustainable strategy.

In order to facilitate building a resource base for program and project managers, the following information resource list has been provided. While the list isn't meant to be exhaustive, it offers a broad representation of resources that include frameworks, best practices, and case studies. All of these resources are useful tools for program and project managers tasked with managing projects to embed sustainability into their organizations.

Information Resources

1. Energy
   - U.S. Department of Energy (DOE): Energy analysis—tools, maps, data and publications on energy, http://www1.eere.energy.gov/analysis/index.html
   - Energy Star: http://www.energystar.gov
2. Sustainability Analytics
   - U.S. Environmental Protection Agency (EPA): Assessment tools and approaches to support sustainable decision making around the interdependencies among people, planet, and profit, http://www.epa.gov/sustainability/analytics
3. Sustainability Case Studies
   - EPA: Examples from a wide variety of industries and projects, http://www.epa.gov/sustainability/analytics/illustrative-approach.htm
   - DOE: Documented energy savings for large manufacturing, http://www.energy.gov/eere/amo/case-studies-system

4. Laws and Regulations
   - EPA: Regulatory information by industry segment and topic, http://www2.epa.gov/laws-regulations
5. Sustainability Articles
   - Environmental Leader: Variety of topics and industries, http://www.environmentalleader.com
   - GreenBiz: Tools and techniques, http://www.greenbiz.com
   - Ethical Corporation: Global focus, http://www.ethicalcorp.com
   - Triple Pundit: http://www.triplepundit.com
6. Non-Governmental Organizations (NGOs)
   - Environmental Defense Fund (EDF): Protecting natural systems, http://www.edf.org
   - World Wildlife Organization (WWF): Promoting sustainable choices and people's impact on forestry, marine, freshwater, wildlife, food, and climate, https://www.worldwildlife.org
   - Human Rights Watch: Defends the rights of people worldwide, http://www.hrw.org
7. Sustainability Standards and Reporting
   - Sustainability Accounting Standards Board (SASB): Industry materiality standards, http://www.sasb.org
   - Global Reporting Initiative (GRI): Reporting protocols for sustainability issues, https://www.globalreporting.org
8. United Nations Initiatives
   - UN Sustainable Development: Global collaborative initiatives, https://sustainabledevelopment.un.org
   - Principles for Responsible Investment (PRI): ESG-focused investors, http://www.unpri.org
   - UN Environmental Programme (UNEP): Global environmental conditions and trends, http://www.unep.org
9. Industry Consortiums
   - The Sustainability Consortium (TSC): Diverse stakeholders building science-based tools to address sustainability issues, http://www.sustainabilityconsortium.org
   - Sustainable Apparel Coalition: Trade organizations for global apparel and footwear market, http://www.apparelcoalition.org
   - U.S. Green Building Council: LEED certification, http://www.usgbc.org
10. Conferences
    - Sustainable Brands: Global conferences, http://events.sustainablebrands.com
    - Delaware Valley Green Building Council: Regional events, http://dvgbc.org
    - Wharton Initiative for Global Environment Leadership: Academic and industry partnerships, http://igel.wharton.upenn.edu

13.2 Benchmarking Organizational Sustainability

Begin by measuring a baseline of your organization's culture of sustainability. In order to gain an accurate baseline, best practice recommends an employee survey. As discussed in Chapter 12, employees come in different shades of green based on their own personal experience. In order to manage a sustainability transformation effectively, it is important to understand where employees fall in terms of their knowledge of and experience with sustainability issues. As an organization moves forward in its sustainability journey, the survey should measure organizational alignment between business and sustainability missions. The survey also gauges employee perceptions of the effectiveness of existing programs, their engagement, and their perceived impact of sustainability on their roles. The following questions are broad in scope, but they give a framework for gathering employees' perceptions about sustainability within the organization, with a deeper dive into particular engagement programs.

Checklist to Measure Organizational Sustainability—Baseline, Program Effectiveness, and Program Development

1. Are employees personally knowledgeable about environmental, social, and community issues?
2. Are employees aware of the organization's sustainability mission?
3. Can they identify the organizational mission statement?
4. How do employees describe the culture of the organization?
5. Do employees see the organization as a sustainable organization?
6. Can employees explain the organization's sustainability mission and its role in achieving business targets?
7. Do employees understand their role in sustainability and how it impacts their job function?
8. Are employees interested in sustainability issues in their personal lives?
9. Do employees have the opportunity to contribute to environmental, social, and governance goals in their roles?
10. Are job functions structured to align with the organization's sustainability mission?
11. Does the compensation structure promote alignment with the sustainability mission?
12. Do employees have the tools and resources to contribute to the sustainability goals?
13. Are managers supporting or impeding employees' efforts to incorporate sustainability into daily functions?
14. Does the leadership team act as role models to promote a sustainable approach across the organization?
15. Do employees have an opportunity to select and participate in volunteer programs that align with their interests and the sustainability mission?
16. Are employees taking personal responsibility for environmental stewardship and social responsibility?
17. Are employees aware of work/life balance programs? Are these programs being offered consistently across the organization?
18. Are employees aware of diversity and inclusion programs such as flexible work arrangements and affinity groups?
19. Are employees offered opportunities to participate in green teams, community outreach, and other sustainability events?
20. Do employees feel connected to the organization's sustainability programs?
21. Are employees rewarded for meeting sustainability goals?
22. Are incentives and award programs effective in generating sustainability ideas that save resources, reduce costs, reduce risks, or create new opportunities?
23. Do employees promote our organization to family and friends as a sustainable organization, a positive corporate citizen, and a great place to work?
24. Are development plans in place at the manager and employee level to develop skills needed to build a sustainable workforce?

Responses to these questions provide a baseline for the effectiveness of existing sustainability programs as well as identify opportunities for new programs and projects. Using a survey gives an assessment as of a point in time. Using these surveys over an extended timeframe allows for tracking progress against desired goals for employee awareness, understanding, and participation in sustainability programs, and projects. Incorporating this information into the portfolio component assessment process facilitates better alignment between the component selection process and creation of a more sustainable workforce.

While surveys are useful to take the temperature of an organization in terms of establishing a baseline and measuring the impact of programs and projects on creating a sustainable culture, deeper dives into specific areas are best handled through employee roundtables or focus groups. Depending on the program or project, the group chosen for a more in-depth review of issues varies. Focus groups can be structured to include both internal and external stakeholders. To gather a broad internal perspective, select group participants from cross-functional areas. For a more in-depth perspective on a particular topic, the project manager may want to consider addressing specific issues with affinity groups that are impacted by the issue. For example, a focus group to assess the organization's diversity and inclusion program may include affinity groups such as Women Executives, African American Engineers, and Lesbian, Gay, Bisexual, Transgender, and Allies (LGBTA) to gain a better perspective on the effectiveness of current and proposed programs from this segment of the employee population. Information gathered from these sessions helps to inform program and project selection and implementation.

13.3 Opportunities for Employee Engagement

Well-structured green teams are effective vehicles to engage employees on sustainability issues. While the term was originally adopted because of the environmental focus, they can focus on both environmental and social issues. Green teams that are structured to promote organizational strategic objectives but allow for local ideas, input, and involvement are the most effective. This approach ties into organizational goals encouraging leadership support and resources while promoting personal interest and initiative.

Managing a green team is a balancing act between top-down direction and bottom-up inspiration. The grass-roots nature of green teams empowers employees to think creatively and to take ownership of projects while working in a collaborative format to achieve significant sustainability goals. The Chapter Model has been an effective management approach to facilitate employee engagement while providing corporate oversight. The Chapter Model designates a corporate resource to provide guidance and coordination for the local green teams. This approach allows for sharing of best practices and lessons learned across the organization's green teams. It also leverages relationships with internal resources that become allies in achieving the goals of the green team's projects. Forming partnerships with facilities, procurement, and human resources facilitates sharing success stories, recruiting new team members, changing behaviors, and garnering management support. When management engages with employees through the green team structure, they change employees' perception about their relationship with management and the impact that they can have on their organization. Employees perceive sustainability as an important organizational strategy, and they feel empowered by management to address both large and small sustainability challenges.

At eBay, their employee green teams started small but grew rapidly because of the impact of their projects. Their success stemmed from making the business case for green team projects. Some impressive results include the development of a solar array at their headquarters in San Jose, California, achieving a 99% waste diversion rate at the San Jose facility, and creating an alternative commuting program in Bern.[1] Management was so inspired by the impact that their employees were having on sustainability issues that they invited eBay's buying and selling community to join the green teams to further expand the global reach of programs. More than 300,000 members have joined and are engaged in buying, selling, innovating, and thinking in ways that promote environmental and social responsibility. Green teams are supported through an online community that promotes new ideas, provides inspiration, offers deals to help members to achieve their own sustainability pledges, as well as a forum to connect and share ideas. Community benefits include receiving free reusable boxes, hearing about new partnership opportunities, and having access to contests and promotions that support sustainability.[2] When structured and managed in the right way, green teams can have significant impact on sustainability goals and promote a culture of sustainability with internal and external stakeholders.

Table 13.2 illustrates a project that utilizes green teams to reduce office power usage by changing the behavior of co-workers through creating awareness of issues and offering solution options. The benefits of using a green team approach are that it is fellow employees making the recommendations, some impactful changes are low- or no-cost, and the benefits are tangible and realized quickly. This approach empowers green team members who are involved in this pilot project and potentially creates new opportunities for them with organization-wide project adoption. The visibility to management and the other areas of the organization based on a successful pilot project promotes green team member engagement and empowerment.

Table 13.2 Green Team Project Plan

Title	Pilot Project NJ Green Teams to Support GHG Reduction Target
Executive Summary	Pilot Green Team development at the NJ facility in order to test impact of education and behavior change on reducing energy usage. Based on the pilot results the program will be scaled to facilities across the US in support of our target to reduce organizational GHG emissions by 10%. The first phase of the project is benchmarking energy usage in the facility. The second phase is forming departmental Green Teams to make recommendations to reduce energy usage and to engage employees in the process. The pilot project timeline is 6 months.
Sponsor(s)	CSO and NJ Facilities Manager
Project Manager	Sustainability Project Manager
Benefits	1. Engage employees in organizational sustainability goals. 2. Educate workforce on behaviors and habits that result in reduced energy usage both in the office and at home. 3. Empower employees by demonstrating how their actions impact the organization's GHG emissions. 4. Reduce energy consumption and GHG emissions. 5. Identify further opportunities for capital improvements, alternative energy sources, and equipment changes to further reduce GHG footprint. 6. Engage external stakeholders on environmental concerns.
Stakeholders	NJ Employees National Plant Managers VP of Operations Procurement CSO Communities NGOs, Environmental Groups Utilities Project Team
Project Team	Project Manager Sustainability Subject Matter Expert NJ Plant Manger Energy Procurement Resource IT Resource NJ Business Functional Leader Green Team Leaders Green Team Members

(Continued on the following page)

Tools and Techniques to Embed Sustainability 279

Table 13.2 Green Team Project Plan (*Continued*)

Milestones & Key Deliverables	Benchmark facility energy usage Project kick-off Gather stakeholder requirements Survey employees Employee focus groups Unplug weekend/energy audit Equipment utilization assessment Removal of nonutilized equipment Green team formation Training and development of green team Green team project implementation Green team recommendations Establish departmental goals Gather feedback Post-project employee survey Share lessons learned with stakeholders Develop green team model to scale Scale program for US-based facilities
Budget	$100,000 Sustainability Office budget Survey and focus group Travel, materials, incentives
Resources	Internal sustainability SME Energy Star Building benchmarking Team members Facility crew IT Department Survey vendor Focus group vendor Utilities
Project Risks	Insufficient impact to reach 10% reduction target Employees' lack of participation Behavior change temporary Green Team recommendations not feasible Stakeholders not engaged Results not scalable
Success Criteria	1. Measure an energy usage baseline for NJ facility 2. Identify energy efficiency opportunities 3. Identify non-essential equipment drawing power 4. Establish Green Teams for all departments 5. 100% of Green Teams make recommendations 6. Post-project employee survey indicates awareness of Green Teams and GHG reduction targets 7. Development of departmental energy saving goals 8. Benchmark a 10% reduction in energy usage on an annualized basis 9. Develop a model for Green Team formation to scale across organization 10. Identify best practices and effective programs 11. Identify areas for improvement 12. Ongoing stakeholder engagement
Assumptions & Constraints	1. Senior management support and participation 2. Employee participation and interest 3. Awareness of organization's sustainability agenda 4. Ability to scale across organization
Location	New Jersey Facility

The project manager used this approach to create awareness internally and externally about the organization's commitment to reducing its energy usage in support of lowering its organization's GHG emissions. Engaging with both internal and external stakeholders clarified the scope of requirements. Empowering the employees through departmental green teams allowed them to focus their attention on specific departmental actions. Creating a format to share best practices with other green teams facilitated the change process. While some recommendations focused on alternative energy sources, facility upgrades, and equipment modification, the green team also recommended the following inexpensive behavior changes that are scalable across all segments of the organization.

Green Team Recommendations for Reducing Energy Usage

1. Measure facility energy baseline using Energy Star Portfolio Manager (http://www.energystar.gov/buildings/facility-owners-and-managers/existing-buildings/use-portfolio-manager)
2. Unplug Weekend—Conduct a plug load audit. Unplug all nonessential equipment power over the weekend. On Monday, plug in only what you need to use.
    a. Working with facilities management, donate or recycle equipment identified as no longer useful, such as old mail sorters, postage machines, fax machines, etc.
3. Plug all essential equipment into properly rated power strips. Turn off the power strip nightly.
4. Engage employees on ways to reduce energy usage. Utilize brown-bag lunches, contests, and campaigns to motivate employees and encourage participation.
5. Make Green Team updates part of staff meetings.
6. Ask employees to sign an energy-reduction commitment.
7. Encourage employee suggestions through an online suggestion box. Provide incentives for those ideas selected for implementation.
8. Best Practice Recommendations
    a. Change internal computer setting to go into sleep mode more quickly.
    b. Eliminate personal space heaters.
    c. Agree to raise or lower the thermostat 3 degrees.
    d. Reduce the number of printers in a department.
    e. Turn off all computers at night.
    f. Turn off lights when you leave the room.
    g. Reduce business travel through teleconferencing.
    h. Relocate desks to accommodate personal preferences for warmer or cooler environments.
9. Building Improvement Recommendations
    a. Install motion-sensitive lighting.
    b. Utilize a building management system.
    c. Install programmable thermostats.
    d. Consider alternative energy sources such as solar panels over employee parking lot.
    e. Install electric vehicle charging station.
10. Share lessons learned and best practices across the organization
    a. Identify and engage a cross-functional management team.
    b. Gather requirements from all stakeholders.
    c. Measure baseline, identify SMART goals, and select metrics to monitor progress.
    d. Incorporate education and training into the project.
    e. Allocate sufficient resources.
    f. Reinforce program through goals and incentives.
    g. Allow Green Teams to have input, make recommendations, and be part of implementation.
    h. Recognize Green Teams for their contributions.

The project focused on engaging employees and other internal stakeholders on energy usage. While many energy projects focus on equipment, in order to create long-term change it is important to remember that people—not facilities—are the users of energy. Educating employees on their role in energy usage and giving them opportunities to make the process their own not only reduces the organization's GHG footprint, it also provides an opportunity for employee empowerment and engagement. From an external stakeholder perspective, cost savings from lower energy usage improves operational performance for investors. Less air pollution and a reduced draw on the electrical grid improves local air quality and helps to preserve the energy infrastructure for communities. The key to success is an effective communication plan combined with tools and techniques that make the process both fun and impactful.

13.4 Continuous Improvement Plan

The Energy Star resource goes beyond energy benchmarking and provides useful resources for project managers looking for tools and techniques to make continuous improvement in energy performance by engaging key stakeholders such as management, employees, vendors, and customers in order to improve facility energy usage, reduce GHG emissions, and saving money. The website provides checklists and interactive tools to get programs started and employees interested. Energy Star's many years of experience reflect that energy programs that are consistently implemented over time generate more effective results than one-off projects. In addition, their research and experience reinforces that senior management support is crucial for effecting sustainability change.[3] Creating organizational change requires resource commitment, and that can rarely happen without senior management support. Budgets for energy projects need to consider both the initial project as well as the ongoing maintenance of the new equipment and the change management programs.

Successfully creating an energy management plan depends on the program manager's ability to promote a culture of energy efficiency within the organization. Energy Star resources include a guide to build an energy management team, including recommended structure, kick-off event, and project rollouts, with case studies and checklists from partner organizations. In addition, a tool kit is provided to help manage behavior change of stakeholders such as employees, tenants, and customers. The process for creating a continuous-improvement energy management plan is highlighted in Figure 13.1. It begins with engaging internal and external stakeholders to gather requirements in order to make an informed and relevant organizational commitment. This commitment takes the form of a project plan, which is shared across the organization along with SMART goals (see Chapter 6) and defined metrics. To begin the process, measure the facilities energy baseline so that all stakeholders understand the starting point on the journey. Next, develop an energy management plan to address the particular needs of the facility. Considerations include the building type, use, purpose, condition, and age. Select the energy savings or efficiency project. Ensure engaging key stakeholders through a project kick-off and ongoing project updates. Empower employees to make recommendations and encourage participation through incentives and feedback opportunities. Measure progress using the defined metrics, and assess results against the SMART goals. Gather data and feedback; then, review and revise the plan based on lessons learned. Recognize success, address shortfalls, and scale the energy management project for use across other areas of the organization.

The following best practices are based on the tenets of change management to promote successful behavior change and policy adoption. This process is less about the energy efficiency equipment and technology selection and more about the process of educating, informing, and engaging managers and employees. The objective is to develop a project plan that has senior management support, encourages cross-functional engagement, and delivers in terms of long-term changes to the facilities energy usage. Consider who should be involved in the project and what their roles and responsibilities will be for the project. We know that C-suite support is crucial to the success of the project, so include either a C-suite sponsor or a champion. The final team configuration will depend on the organization, energy management objectives, and the scope of the project. Table 13.3 highlights project roles. As a project manager,

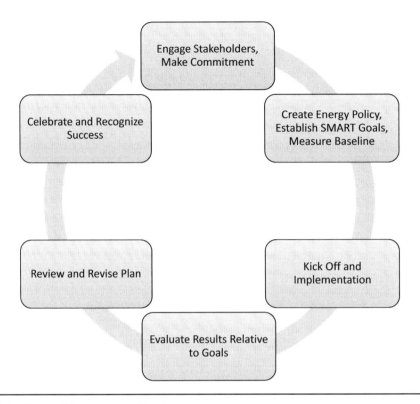

Figure 13.1 Energy Management Plan[4]

Table 13.3 Energy Management Project Roles

Senior Management	C-Suite Sponsor/Champion
Key Stakeholders	• Facilities & Operations Management • Supply Management • Financial Management • Human Capital Management • Employees • Customers • Community • Project Team • Other
Project Team Members	• Project Manager • Engineering • Operations & Maintenance • Facility Manger • Corporate Real Estate • Procurement Specialist • Environmental Health & Safety • Utilities • Contractors/Vendors • HR • Finance • IT • Marketing • New Product Development

Table 13.4 Best Practices for an Energy Management Plan[5]

Category	Best Practice
Energy Program Manager	• Senior level supported by leadership • Ability to work with a broad range of staff from maintenance to C-suite
Senior Management	• Energy project sponsor reports to senior leadership team • Senior management champion • Outline benefits, plan, & proposed team
Facility Engagement	• Engage facility mangers in two-way communication • Create on-site green teams • Identify on-site energy specialist to champion initiative
Partners	• Identify consultants, vendors, customers, NGOs, joint venture opportunities
Structure	• Identify champions or leaders by division/function • Utilize existing structure and networks
Resources	• Ensure that the team has adequate time to participate in project • Ensure that the energy program manager has decision-making authority. • Ensure that budget has been allocated for project development & maintenance
Plan Creation	• Refine goals, creates plan, & prepare for kick-off • Identify energy projects • Budget for energy project creation & continuation • Identify roles, responsibilities, & work streams
Measure Baseline	• Conduct energy audit & identify opportunities

it is important to include team members who bring a broad set of skills and resources to promote successful adoption and continuous improvement.

Table 13.4 provides recommendations for best practices from organizations that have already implemented an energy management plan. These suggestions range from the type of program manager to select to the best ways to engage and ensure senior management support. While these recommendations are based on an energy management project, many of the best practices hold true for other types of sustainability programs.

Ensure that key people are included in the energy policy development to promote organizational buy-in. Having the CEO issue the official energy policy adds credibility and attention to the energy policy. Craft the policy in accord with the organizational culture, ensuring that the initial phases of the policy are within the skills and abilities of current management and employees. Drill down to the daily activities of workers and communicate desired actions and behaviors in a clear and understandable manner.

With energy objectives and goals identified, and a roadmap laid out, the next step is implementation. Table 13.5 suggests tools, techniques, and training that improves the likelihood of success. Education and training are key components of a successful sustainability project implementation.

This plan focuses on engaging stakeholders in a dialogue rather than following a top-down mandate. Project managers must look both internally and externally for partners to leverage expertise and create opportunities. The plan uses education and training as a way to share best practices and to influence behaviors. It also identifies training gaps to build organizational capacity. Green teams are empowered to encourage fellow employee commitment and participation. The focus is on changing both management's and employees' behavior to work toward a shared solution for ongoing energy management improvement. While technology, equipment, and building improvements are part of the process, significant focus is placed on changing behaviors and attitudes about energy consumption and empowering employees to identify savings and recommend suggestions. Sustainability projects impact

Table 13.5 Project Kick-off and Implementation[6]

Category	Requirements
Implement Program	• Team rolls out project in accord with the plan • Communicate with key stakeholders • Raise awareness • Build capacity • Motivate • Track & monitor
Education	• Team members attend energy-related events, summits, fairs, etc. • Utilize green teams to promote awareness • Create awareness of energy initiative via intranet, posters, surveys, and contests • Cross-functional networking & sharing of best practices
Training	• Build capacity • Identify training requirements & hold training sessions • Leadership development • Team leader professional development • Adopt best practices
Monitoring & Tracking	• Gather data • Identify & track metrics • Select energy tracking system
Communicate	• Track, report, & publish results to all stakeholders • Promote awareness through internal & external communication networks
Subject-Matter Experts	• Engage outside resources for subject-matter expertise • Identify projects for outsourcing
Green Teams	• Identify & engage leaders • Encourage employee participation • Promote acceptance of energy-savings pledge • Embed sustainability into culture
Celebrate Success	• Involve leadership in recognition program, e.g., Chairman's Club • Create & develop reward/incentive programs • Identify external awards & opportunities for recognition
Review Plan	• Gather feedback • Identify lessons learned • Adopt best practices • Modify plan for continuous improvement

stakeholders across an organization's product or service life cycle. Even something that appears to be an internally focused initiative, such as an energy management project, has impacts on both internal and external stakeholders. A key to closing the sustainability performance gap is effective engagement of all stakeholders.

Interactive energy management tools are useful to engage employees and to help them identify areas in the office or plant in which they can have an impact on energy usage. These ready-to-use tools are a great resource for sustainability officers and project managers seeking to educate their organizations on the impact of employees' actions on energy usage at work. Another resource is an employee education toolkit complete with everything a project manager needs to host an educational "lunch and learn" for employees on the organizational, environmental, and social value created by energy efficiency. The program outlines how employees can conserve energy at work. The kit includes an interactive 30- to 60-minute training session, an email invitation format, informational flyers, posters, and tip cards. Basically, it includes everything that is needed to invite employees and to engage them on energy usage and conservation. While Energy Star's resources, checklists, and best practices focus on energy conservation and emission reduction projects, the tools and techniques are useful to engage stakeholders of a wide variety of sustainability projects.

13.5 Impact of Maturing on the Sustainability Continuum

Fully transforming into a sustainable organization requires management to move beyond the basics such as energy management in order to effect meaningful change in their operations. As an organization matures in its sustainability journey, leadership realizes that decisions that are made both within their own organization and along their value chain, impacting the environmental and social performance of their organization's products and services. Supplier decisions impact the products' or services' environmental and social footprint. While suppliers don't sell directly to end users, their decisions impact the final consumer. Downstream on the value chain, consumers make decisions to purchase and use a product or service based on the complete life-cycle impact. They also impact the product life-cycle footprint with their own decisions about use and disposal of products. All of the decisions and actions of the full value chain, from design to materials sourcing, production, transportation, labor policies, marketing messages, and actual product usage, impact the overall sustainability of a product or service. Decisions along the value chain impact stakeholder groups both individually and collectively.

In the life-cycle assessment of stakeholder impact depicted in Figure 13.2, each role in the value chain process has its own set of stakeholders. A supplier has its vendors, employees, communities, and customers to consider as their management devises its sustainability programs and policies. However, the decisions that are made by this supplier's management will impact not only its own sustainability profile but also that of organizations and customers farther along the value chain. Each of the parties on the value chain is connected and impacted by the others' decisions and actions. With sustainability maturity, management seeks to integrate sustainability not only into its core operations, programs, and projects but also into the organization's value chain.

Management analyzes their products' and services' life cycles across their value chains to better understand the environmental and social impacts of their decisions and actions as well as those of their suppliers and customers. As management seeks to address significant sustainability challenges such as global labor rights, clean water, and an adequate and safe food supply, the solutions to these issues require leveraging diverse talents and resources and necessitates greater collaboration both across and between organizations.

A significant global challenge is increasing levels of carbon dioxide in the atmosphere, rising to 400 parts per million (ppm). While this number represents a historical high, more concerning is that the rate of increase has been growing at an annual rate of 2.73 ppm CO_2 per year.[7] According to scientists, this high level of carbon in the atmosphere is causing climate change resulting in superstorms, flooding, and droughts.[8] These environmental changes are impacting business operations, public health, and food and water security. Leading organizations such as government agencies, cities, and private companies are looking for collaborative solutions to this complex issue. To this end, the U.S. Department of Energy (DOE) initiated its Clean Cities Program, a voluntary government and industry partnership. The program provides resources and support for cities and local governments to work with private enterprise in order reduce greenhouse gas emissions in their regions. Their approach focuses on alternative fuels, idle reduction, and fuel efficiency. One of the groups, New Jersey Clean Cities, has undertaken initiatives including stakeholder outreach and education, annual report of petroleum displacement, quarterly alternative fuel price reports, and maintenance of data for the DOE "Alternative Fuel Station Locator." Specific program contributions include DOE Clean Cities support, EPA diesel emission reduction program such as the marine vessel engine replacement program, compressed natural gas (CNG) fleet and infrastructure program, and the regional Plug-in Electric Vehicle (PEV) network development program.[9] Taking a collaborative approach to address globally accelerating GHG

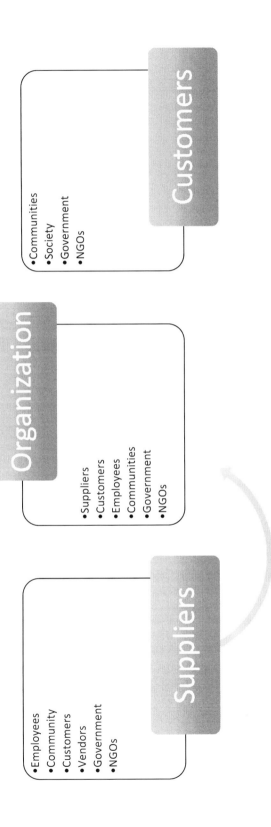

Figure 13.2 Value Chain Stakeholders

emissions creates global impact through local action. It leverages collective resources for greater impact on local and global air quality issues.

> **EV Everywhere Grand Challenge**
>
> In order to further support the goals of the Clean Cities Program, the U.S. Department of Energy developed a new program to promote Plug-in Electric Vehicles usage called the "EV Everywhere Grand Challenge." The Challenge's goal is to make PEVs as affordable and convenient as gasoline-powered vehicles by 2022. The program provides information about electric vehicles, charging requirements, charging resources, and steps that the DOE and local communities are undertaking to support PEV adoption and deployment.
>
> In evaluating the obstacles to widespread adoption of PEVs, the DOE identified a weak link in the availability of workplace charging for vehicles. Initial charging scenarios focused on home and public charging; however, analyzing a day in the life of a car shows that cars are parked at workplaces much longer than at public locations. Chuck Feinberg, Chairman and President of the New Jersey Clean Cities Coalition, shared the following research on PEVs. "With more than 260,000 sold, PEVs are being adopted roughly three times as fast as hybrid vehicles during their first three years on the market." The market is changing rapidly as automakers embrace PEV technology. In order to promote PEV usage, New Jersey offers a sales tax exemption for zero-emission vehicles. New Jersey is a member of the Transportation and Climate Initiative. In addition, the state has placed PEVs on their procurement list as they seek to increase the PEV percentage composition of their own vehicle fleet. Using a Clean Cities Grant, the New Jersey agency developed a plan for recharging station design and site guidance, model codes, permits, and ordinances, as well as stakeholder engagement through information and outreach.[10]
>
> Increasing, both private and commercial PEV usage has generated a positive impact on air quality, especially in heavily populated states such as New Jersey. The Vehicle Technologies Office provides resources for employers, building owners, employees, and others who want to promote, support, or install workplace charging.

Local air quality related to greenhouse gas emissions remains a concern across the globe. Figure 13.3 identifies common sources of GHG emissions by the Scope 1, 2, 3 classifications. Scope 1 emissions are under the direct control of an organization and come from production activities, fleet emissions, and on-site landfills and waste water treatment. Scope 2 emissions come from an organization purchasing heat, power, and electricity. Scope 3 emissions are not directly under an organization's control, stemming from sources such as business travel, employee commuting, and upstream and downstream value chain activities. Leading sustainable organizations are looking beyond their own Scope 1 and 2 emissions and are seeking to reduce Scope 3 emissions as they impact total global emissions. From a stakeholder perspective, Scope 3 emissions impact the final product's sustainability story. (Refer to Figure 13.2.) An organization can have great internal sustainability policies, but if its suppliers don't follow the same rigorous protocols, the end product or service is not sustainable. In the case of GHG emissions, these Scope 3 emissions impact the overall carbon footprint of the product. Similarly, employees commuting to work may not be classified as direct organizational GHG emissions; but their drives to work significantly impact global GHG emissions, local communities' air quality, traffic, and congestion. In order to address their stakeholders' concerns, sustainable organizations are offering programs to reduce employee-commuting emissions, such as offering flexible work arrangements, encouraging ride sharing, promoting mass transit usage, and encouraging adoption of PEVs by their employees.

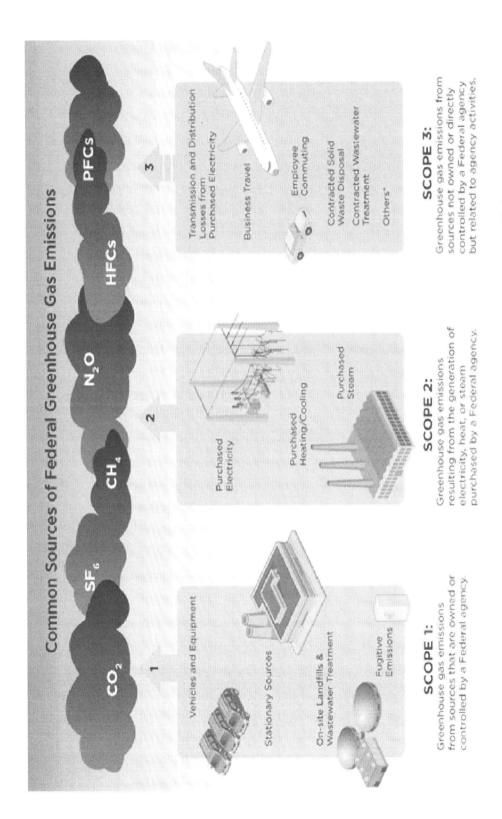

Figure 13.3 Greenhouse Gas Emission Drivers by Scope Classification[11]

An example of a successful Scope 3 carbon reduction programs is an HR-driven sustainability program to promote flexible work arrangements. The program's benefits include reducing traffic, congestion, and carbon emissions generated by employees' daily commuting. At Cisco operations in the UK and Ireland, the average employee telecommutes 2 days per week, and those using Cisco Virtual Office technology work from home 3 days per week. The impact is a reduction of 35 million miles of commuting per year. Not only does this reduce Cisco's Scope 3 emissions by 17,000 tonnes/year, it also reduces infrastructure usage and vehicle congestion in local communities. In terms of productivity, Cisco reports that the flexible work program generates productivity savings of $333 million per year.[12] This successful program has a positive impact on the environment, promotes a more diverse workforce, improves work/life balance, and adds value to the bottom line. Making the program a triple bottom line win!

Other organizations are addressing stakeholder concerns on air quality by promoting the use of alternative transportation, including encouraging employee use of PEVs. As with any project, engage with key stakeholders to ensure success. Both support and obstacles come from surprising places. While charging stations support an organization's overall target of lowering Scope 1 and 3 GHG emissions, some organizations have internal conflicts with building energy managers tasked with minimizing building energy usage. In fact, these facilities energy managers often have performance goals tied to building energy usage, which generates Scope 2 emissions. Not surprisingly, they are concerned that PEV charging stations might cause building power usage to increase. Based on lessons learned by the DOE, most organizations that install charging stations have not experienced a significant increase in electrical usage. As a selling point for facilities managers, the inclusion of workplace charging provides points for building LEED certification.

Despite the benefits of adopting workplace charging, navigating an organization's sustainability infrastructure and policies can be a challenge for the project manager. The project described in the following section outlines a successful approach to reduce employee commuting GHG emissions by promoting PEV adoption supported by workplace charging.

13.6 Promoting Plug-in Vehicle Adoption by Employees by Developing Workplace Charging Stations

13.6.1 Project Manager Tools

Based on research on vehicle usage, the DOE has identified work locations as an ideal scenario for charging vehicles. Vehicles spend sufficient time at workplaces to complete the charging cycle. In order to encourage workplace charging, the DOE has launched a program with details available on their website (http://energy.gov/eere/vehicles/ev-everywhere-workplace-charging-challenge). The site provides information about PEVs, charging station basics, locations of charging stations across the country, as well as tools and resources to assist sustainability and facility managers with charging installation and management decisions. It also provides content on how employers can engage employees on PEV adoption.

13.6.2 Impact of a Workplace Charging Station

From an employee perspective, a workplace charging station reduces employee anxiety over PEV range-of-driving concerns. It makes PEVs a viable alternative for those who don't have a residential charging station or those with long commutes. Some employers add benefits such as designated parking, free parking, and free electricity to encourage PEV adoption. From an employer perspective, charging stations offer a relatively low-cost benefit for employees. In addition, workplace charging stations provide a visible and tangible commitment to the environment, helping to attract and retain employees who are engaged through personal alignment with an organization's sustainability message. Workplace charging supports an organization's sustainability message by aligning projects and programs with its mission and goals.[13]

From a vehicle fleet management perspective, PEVs offer lower operating costs and contribute toward GHG reduction targets. However, commercial fleet users have reported challenges such as driving-range limitations for delivery, inability to reload and redeploy a vehicle because of battery charging requirements, and concerns about the financial health and long-term viability of PEV battery suppliers.

DOE Recommended Best Practices for Workplace Charging Projects[14,15]

- Gain senior management support
  - Educate on the environmental and social impact
  - Use case histories to alleviate safety concerns
  - Explain energy usage and impact on demand charges
  - Utilize existing templates for liability concerns
  - Standard permitting process
- Gain key internal management support, such as facilities manager
  - Ensure that the facility manger understands that convenience and access is a key aspect
  - Highlight positive impact on employee engagement and retention
  - Discuss benefit of fleet conversion with logistics
  - Address similar concerns as senior management
- Conduct an employee survey on PEV use and planned use
  - Provide employee information on PEV facts
  - Identify how your employees will use workplace charging stations
  - Workforce transportation habits
  - Encourage leadership participation
  - Parking facility
    - Suburban office complex
    - Urban parking garage
    - University campus
  - Parking protocols
    - Designated spots
    - Permit
    - Frequency of use
- Select a convenient location
  - Proximity to electrical
  - Multiple-car use of charging station
  - Utilization of charging station if parking location is remote
  - Incentive for EV if charging and parking is premier location
- Evaluate the site electrical system structure and required changes
  - Survey the facilities manager about the location of wiring and circuit breakers
  - Existing location for charging
  - Creating a location for charging
    - Concrete or asphalt requirements

- Average infrastructure cost per charging station $500–$15,000
- Select an appropriate charging structure
  - Level 1: Direct plug in
    - Best serves full-time employees who commute 25 miles or under one way
    - Hardware cost, $300–$500
  - Level 2: Charging station
    - Best serves part-time employees or those who travel to and from the office location during the day
    - Hardware cost, $500–$6000
  - According to the DOT, Level 1 charges a 32-mile trip in 8 hours while Level 2 charges the same trip in 2 hours
  - UL rated
  - ADA compliance
  - Use of a EVSE network access system for billing
- Install system
  - Finalize site selection
  - Estimate electrical load
  - Contact EVSE suppliers
  - Contact utility to ensure adequate current electrical supply and options including special EV programs
  - Obtain permits
  - Engage contractor
- Develop internal procedures for maximum usage
  - Payment policy
  - Level of access
  - Systems optimization
    - Email communication among users
    - Time schedule to avoid demand charges
  - Share best practices
- Monitor and evaluate
  - Metrics
  - Incentives

13.6.3 Determine Organizational Readiness

While charging stations are a great way to promote PEV usage, building a station for an organization whose employees do not own PEVs or who commute primarily by mass transit doesn't add value to the community, the organization, or employees. In order to assess your organizational readiness for a charging station, use the following questions to create an employee PEV interest survey.

PEV Usage and Workplace Charging Survey Questions[16]

1. Do you own a PEV?
2. If yes, how often do you use it to commute to work?
3. How far is your one-way commute?
4. What is the amount of time typically required to recharge your PEV to 90%?

5. What is your timeframe for leasing or purchasing a new vehicle?
6. Are you considering purchasing or leasing a PEV in the future?
7. If no, what are the impediments to purchasing or leasing a PEV?
8. Would workplace charging impact your decision to purchase a PEV?
9. Are you willing to pay for workplace charging?

Project managers can find a sample employee survey for workplace charging planning from the DOE at http://energy.gov/eere/vehicles/downloads/sample-employee-survey-workplace-charging-planning. In addition, this site offers resources targeted toward employees to help them make more informed decisions about purchasing or leasing a PEV. The tools available on this site facilitate project managers planning their PEV adoption project in order to maximize engagement and promote workplace charging. Partnering with HR to survey and communicate with employees and to promote the PEV program to new recruits further embeds an organization's commitment to sustainability.

One of the most significant shared best practices from those with effective workplace charging programs suggests that project managers promote the benefits of PEV usage to employees in order to increase the success rate. The following are some of the most effective ways to promote the program.

Tips to Promote PEV Usage in the Workplace[17]

1. Use internal communication networks to inform employees about the program, new developments, and to provide updates. Inform new employees of the program during the onboarding process.
2. Provide informational resource links such as eGallon and vehicle cost calculators to help employees compare the cost of ownership and emissions for a wide variety of vehicles.
3. Encourage PEV drivers to share their experience and stories with other employees via presentations or informal discussions.
4. Conduct workplace charging station demonstrations to educate employees on the process and benefits.
5. Provide optimally located parking spots for PEV-driving employees.
6. Use posters in common areas to promote awareness of the program.
7. Work with business travel coordinator to promote rentals of PEVs for business travel.
8. Coordinate with fleet manager to promote fleet conversion to PEV.
9. Invite local PEV dealers to participate in events to promote PEV usage, such as a Ride and Drive Day.

While this list contains specific recommendations for promoting PEV adoption, these promotion and engagement ideas are relevant for other sustainability engagement programs as well.

13.6.4 Workplace Charging Station Success Stories

The following are a few stories from Workplace Charging Challenge partners and other employers on the impact that their charging programs are having on their employees, communities, and sustainability goals.

Evernote, a technology company located in Northern California, launched its workplace charging program as part of an employee engagement initiative. The company has installed 10 Level 2 charging stations. It also provides employees with a monthly $250 allowance to lease or buy any vehicle that qualifies for a California carpool lane sticker. The concept behind the program is that speeding employees' commutes will improve productivity. Through a specially negotiated deal with Nissan for its PEV called LEAF, the allowance fully covers the employee cost of a lease for a LEAF. They further supplement the program by covering the cost of charging through the use of a Blink card given to employees when they purchase their PEV. In order to manage charging station usage among its employees, they have created a charging calendar based on the distance the employees commute. Changes in scheduling due to absences or vacations are handled through a group email. About 20% of Evernote's Redwood City employees are using this offering. The program has been working well, and the company plans to add 10 more Level 2 charging stations in the next few years.[18]

Raytheon's workplace charging program serves more than 20,000 employees in seven locations across the United States. Their 30 Level 2 charging stations are part of Raytheon's sustainability program, designed to achieve 15 long-term sustainability goals by 2015, such as reducing GHG emissions, energy usage, water usage, and waste, as well as the greening of Raytheon's supply chain and product offerings. Raytheon employees have shared their stories and lessons learned on the installation and use of charging stations at industry conferences and forums, both to highlight the firm's commitment to workplace charging and to share best practices with other firms interested in workplace charging.[19]

MetLife provides alternative commuting options for employees as part of its environmental sustainability programs. As part of this program, MetLife has installed 32 PEV charging stations at 14 U.S. corporate offices. Comments from employees include that the Met Life charging station program was a significant part of their decision to purchase PEVs. Other comments include one employee who felt that Met Life's commitment to provide workplace charging demonstrates the organization's forward thinking and interest in meeting the needs of its employees. From these employees' perspective, this program contributes to making Met Life an employer of choice because of its environmental and social commitments.[20]

13.7 Creating a Sustainable Culture

Engaging employees on sustainability issues is a crucial piece of a program to successfully introduce long-term sustainability into an organization. Throughout the book, we have discussed how engaging with human capital professionals can facilitate embedding sustainability into an organization's culture. One of the major challenges in this engagement process is providing a framework that incorporates levers to drive and metrics to define and measure employee engagement on sustainability. Leveraging a relationship with HCM facilitates developing an organizational infrastructure that supports building a sustainable workforce. Table 13.6 highlights the impactful role that HCM professionals have in the process and suggests human capital goals to promote alignment between HR functionality and sustainable strategy, metrics to measure impact, and the corresponding business benefit.

This framework facilitates engaging HR in order to further embed sustainability into the organization and to reap the benefits from an aligned and integrated workforce management approach. Incorporating these goals and metrics into a balanced scorecard to promote sustainability is shown in Figure 13.4.

Sustainability goals should be part of all business leadership's balanced scorecards so that compensation is tied to performance on sustainability. The scorecard in Figure 13.4 focuses on HR leadership. Selection of specific metrics for inclusion in an organization's scorecards is based on the industry, sustainability goals and targets, and designated areas of impact. The balanced scorecard approach evaluates each of the performance components—finance, customers, education and development, and internal processes—equally. This approach recognizes that long-term organizational success relies on more than financial metrics.

As we have seen, embedding sustainability into an organization is accomplished through programs and projects that harness the power of an organization's people to effect change. Incorporating sustainability requirements into managerial performance measures creates a leadership culture that prioritizes sustainability goals. With this focus, the organizational environment becomes much more conducive

Table 13.6 HR Scorecard for Sustainability

HR Role	HCM Goal	Metrics	Business Value
Embed Organizational Sustainable Strategy, Values, Ethics	Employees understand and follow corporate mission, values, & ethics.	• Percent of employees trained in ethics, policies, values • Percent of employees on annual survey acknowledging mission, values, ethics	• Protection of brand & reputation, license to operate • Mitigate risk of fines & penalties • Promote employee engagement
Workforce Planning/ Recruitment	Recruit the best talent and provide a diverse workforce to meet new challenges & opportunities.	• Turnover ratio • Percent of employees who self-identify as belonging to a diversity category • Percent of women & minorities in leadership roles • Number of executives who self-identify as belonging to a diversity category	• Lower unwanted turnover • Develop talent & leadership to support strategic growth • Diverse workforce opens new markets, promotes innovation, & improves customer satisfaction
Total Rewards	Compensation tied to sustainability goals and metrics. Gender pay equality.	• Number of employees with sustainability goals in performance plans • Sustainability metrics tied to corporate & divisional goals • Ratio of male to female executives • Compensation by gender	• Performance incentives align with sustainable strategy • Build a senior management pipeline that includes diversity • Improves engagement & trust
Health & Wellness	Employees are in good health & are able to consistently contribute to the corporation.	• Frequency of absence • Percent of employees enrolled in health & wellness programs • Employee survey happiness & engagement responses.	• Reduce business insurance cost • Lower absenteeism • Improve productivity • Improve employee engagement
Engagement	Employees understand corporate mission on sustainability. Employees serve as ambassadors. Employees incorporate sustainability into daily activities.	• Percent of employees trained in sustainability • Percent of employees who believe volunteer opportunities support core corporate mission & align or build skill development • Percent of employees who report on employee survey that they understand how sustainability is incorporated into their role	• Embedding sustainability into people & process • Improve employee engagement • Promote positive community image • Improve community through volunteer program • Engage community stakeholders
Performance	Define sustainability performance goals in job description. Utilize life-cycle analysis to identify areas for improvement. Comply with industry and governmental standards & regulations.	• GHG emission reduction targets • Sourcing & product material requirements • Design specifications • Supplier scorecard • Logistic & transportation targets • Human rights standards	• Promote savings on energy, resources, & processes • Reduce risk through life-cycle analysis • Avoidance of fines & penalties • Develop new markets, products, services • Promote continuous improvement

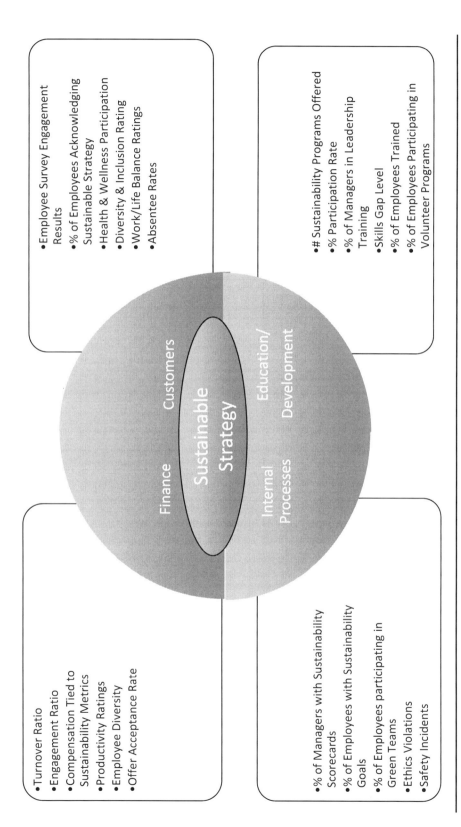

Figure 13.4 Sustainability Balanced Scorecard for HR

to program and project managers tasked with managing programs and projects to promote a more sustainable organization.

13.8 Diversity as a Cornerstone of Sustainability

Creating an inclusive environment is the foundation for creating effective internal and external stakeholder engagement. Creating sustainable value begins with the culture of an organization and requires managers to consider their organization's policies and practices to support a diverse workforce and an inclusive environment. Creating an environment that allows employees to bring their full selves to work fosters creative thinking, results-oriented action, and engagement. Diversity in the workplace includes race, ethnicity, national origin, religion, age, social status, and education, to name a few elements. Understanding the role that diversity plays in sustainability strategy begins to open the door to sustainability-centered opportunities.

The global population is becoming more diverse. This population dynamic presents opportunities for organizations, but leveraging these opportunities requires leadership to reevaluate the makeup of its organizational composition, including its board of directors and its leadership team. Embracing diversity and developing opportunities for inclusion at all levels within the organization creates business value. Employees and managers with diverse backgrounds provide unique perspectives for innovation and problem resolution, thereby improving productivity and service response levels. From a client service viewpoint, employees who have similar backgrounds and experiences as your clients can often communicate more effectively, providing new opportunities and opening new markets. From

Table 13.7 Diversity and Inclusion (D&I) Program Guide

Area of Impact	Actions	Benefits
Employee Diversity	• Recruit from more diverse sources—job boards, schools, and networking groups. • Support diverse populations through employee resource groups. • Create policies & processes that promote inclusion. • Offer flexible working arrangements.	• Recruiting the best and the brightest • Enhanced revenues from employees with a broader skill set to provide new solutions & products • Greater capacity to understand client needs & to address new client requirements
Management Diversity	• Create leadership development & mentoring programs for women & minority candidates. • Develop training programs for leadership on the benefits of D&I.	• Diverse leadership team's ability to access new markets • Diverse leadership team serves as a role model for diverse customer and employee base
Board Diversity	• Develop a diverse slate of candidates for board openings. • Provide board training on the benefits of D&I. • Engage with stakeholders on the topic of D&I.	• Improved stakeholder engagement • Expanded variety of skills & experiences to contribute to organizational strategic vision
Supplier Diversity	• Promote supplier diversity through minority supplier registries. • Engage with diversity network groups hosting supplier fairs. • Establish D&I standards & training for supply chain.	• Improved product & service offerings • Improved customer service and support • New ideas & solutions for clients • Improved supply chain diversity

an employee engagement and retention perspective, creating an environment that is welcoming to all employees allows the organization to recruit and retain the most talented individuals.

Management that accepts that employees have varying lifestyles, beliefs, and backgrounds and that one size does not fit all is taking the right approach. In order to create a more diverse workforce, management needs to create a culture of inclusion in which all people feel welcome and valued.

Table 13.7 offers a program guide to creating a more inclusive workplace environment, including areas of impact, action steps, and business benefits for an organization that plans and implements diversity and inclusion programs.

The goal of D&I programs is to encourage individual differences and empower employees to recommend change in order to make them feel that they have the right fit with the organization. From a project management perspective, this concept touches every sustainability portfolio component across the organization. How are project teams formed? Are the same people tapped regularly for new projects? As a project manager, how are you encouraging diversity and inclusion on the project team? Is sufficient time allotted to collect requirements from a diverse group of stakeholders? The diversity of project teams is a direct reflection of management's commitment to D&I as part of sustainable strategy. It is about walking the walk, not just talking the talk.

13.9 Programs to Support D&I

While there are a variety of programs to promote D&I in the workplace, one that focuses on promoting diversity by creating a more inclusive environment is a flexible work program. At its core, this program recognizes that individuals have different work requirements and that a traditional workday, week, or even on- site location may not be the best solution for employees with diverse needs. For successful adoption, flexible work programs require broad organizational acceptance, ranging from the C-suite to the most junior employee. The program benefits include cost savings, innovation, better customer service, and employee empowerment and engagement. Figure 13.5 highlights the business value creation benefits of flexible work programs. Developing an environment that promotes flexibility for collaboration to develop new ideas, products, and solutions for clients creates business value. Offering employees alternatives to address their individual lifestyle requirements engages top talent and retains top performers. Program results led to improved customer satisfaction and greater scheduling flexibility to manage peak workflows. Turnover ratios improved and productivity measures rose. In addition, the office footprint to support the workforce became smaller, as less office space is required to support a flexible work structure.

Best Practices for Launching a Flexible Work Program

1. Enlist C-suite sponsorship and find a C-suite champion.
2. Make a clear business case for the program, including metrics to monitor progress and measure success.
3. Pilot the program before scaling for the full organization.
4. Identify flexible work options such as telecommuting, compressed schedules, and extended shifts that work for your organization.
5. Select a technology-based tool to facilitate offering a consistent program across business sectors and to track metrics.
6. Train employees on new skills and provide employees with tools and resources to better understand how the flexible work program will operate as well as the tools and resources that they will need to perform duties such as connecting to an off-site, secure Internet connection.
7. Train managers in the new skills they will need to promote the programs and to manage remote workers.

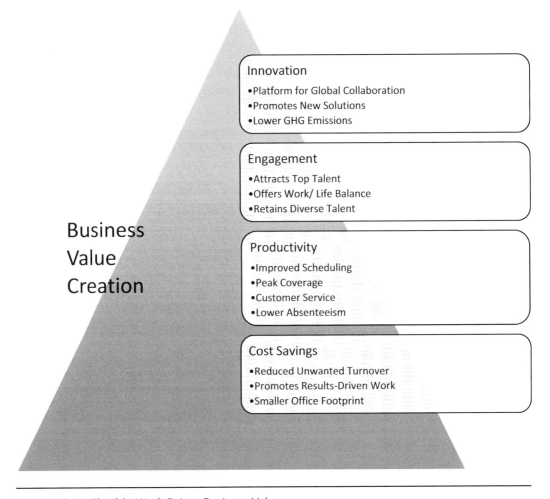

Figure 13.5 Flexible Work Drives Business Value

8. Create a process to monitor results across the organization to ensure consistent delivery to all parts of the business.
9. Kick off the project in a highly visible way, such as setting up information tables in the cafeteria, and providing "swag" to get people talking and asking about the program.
10. Create a communication plan to inform managers and employees about the program. Create a dedicated intranet page, affinity groups, and online chat forums to allow sharing of information and best practices.
11. Modify office layout to support flexible working arrangements.
12. Create a feedback mechanism via an online portal to allow questions to HR and to make program suggestions or modifications.
13. Tie management compensation to D&I goals and flexible work metrics.

Flexible work programs support a "sustainable work" approach in which work is the contribution and employees are valued according to their contribution to organizational goals (see Figure 12.1). Work performance is measured by employees' results, not the amount of time in the office. The work structure is collaborative and cross-functional, with reporting relationships more in line with a matrix structure. With five generations in the workforce, each will have different requirement and needs. Providing a flexible working environment promotes an inclusive environment that helps to address the various needs of a diverse workforce.

13.10 Conclusion

This chapter has focused on resources, project plans, and actual case studies in order to provide project management professionals with practical examples of embedding sustainability within an organization. A list of resources has been provided to give sustainability champions tools from which to develop a business case for recommending sustainable strategy to organizational leadership. Initially, the focus was on establishing a baseline to understand your organization's current level of sustainability. If the organization has already started on its sustainability journey, the baseline assessment is expanded to determine current program effectiveness and opportunities for future development. Then, we took a deeper dive into employee engagement and the value of green teams to empower employees. The key to green team success is alignment with core business strategy and effective management. The Green Team Project Plan highlights the impact that these employee groups have on organizations. The Energy Star case demonstrates that energy efficiency projects are more effective when they are viewed as a change management project with broad stakeholder engagement, clear communication, as well as technical expertise. This concept has been taken a step further by considering the impact of sustainable strategy on the value chain stakeholders and the impact that each of these stakeholders has on the selection of portfolio components in support of sustainability. The DOE Clean Cities program provides perspective that to solve large global challenges such as air quality, organizations must look beyond their own operations and engage with other strategic partners to solve these challenges. The EV Everywhere Challenge case gives detailed recommendations to project management professionals about working collaboratively to address the challenge of reducing Scope 3 GHG emissions generated by employee commuting.

Lastly, tools have been provided to facilitate embedding sustainability into the human capital management function of an organization to foster a culture of sustainability. Another detailed case example was provided for project management professionals tasked with engaging a more diverse employee population through offering a flexible work program. While each of these areas are very specific in terms of the programs and projects undertaken, several common themes run through them. In order to effect change, employees must be engaged. A cross-functional approach has a much broader impact than a siloed approach. In order to solve significant sustainability challenges and to create opportunities, organizational leaders must look beyond their own internal operations and forge partnerships with value chain stakeholders as well as community and government stakeholders. In order to engage stakeholders, the project manager must have an effective communication and change management plan. Organizational leadership must support a culture of sustainability, including developing specific goals and metrics for business functions that are tied to compensation. Human capital professional are invaluable allies in creating a framework to support a sustainable culture.

Notes

[1] "Engaging People," *Green Shopping, Eco Friendly Products*, accessed August 31, 2015, http://green.ebay.com/greenteam/ebay/blog/Engaging-People/25.

[2] Ibid.

[3] Energy Star, "Teaming Up to Save Energy: Protect Our Environment Through Energy Efficiency," accessed June 15, 2015, http://www.energystar.gov/sites/default/files/buildings/tools/Teaming_Up_To_Save_Energy_508_0.pdf.

[4] Ibid.

[5] Ibid.

[6] Energy Star, "Guidelines for Energy Management," 2013, http://www.energystar.gov/sites/default/files/buildings/tools/Guidelines%20for%20Energy%20Management%206_2013.pdf.

[7] "Global Climate Change: NASA Scientists React to 400 ppm Carbon Milestone," *Climate Change: Vital Signs of the Planet*, accessed June 23, 2015, http://climate.nasa.gov/400ppmquotes.

[8] Ibid.

[9] Chuck Feinberg, "NJ Sustainable Business Initiative & Clean Air Council," Creating a Sustainable Infrastructure for Electric Vehicles in the Workplace, Trenton, NJ, November 14, 2014.

[10] Ibid.

[11] Climate Change Division U.S. EPA, "Greenhouse Gas Emissions: Greenhouse Gases Overview," Overviews & Factsheets, accessed May 21, 2014, http://www.epa.gov/climatechange/ghgemissions/gases.html.

[12] Paul Swift and Andie Stephens, "Homeworking Is Where the Savings Are," Carbon Trust, May 2014, http://www.carbontrust.com/media/507270/ctc830-homeworking.pdf.

[13] CALSTART, "Best Practices for Workplace Charging," September 2013, http://evworkplace.org/wp-content/uploads/2013/10/Best-Practices-for-Workplace-Charging-CALSTART.pdf.

[14] Ibid.

[15] U.S. Department of Energy, "EV Everywhere Workplace Charging Challenge: Resources," accessed November 17, 2014, http://energy.gov/eere/vehicles/ev-everywhere-workplace-charging-challenge-resources#case.

[16] CALSTART, "Best Practices for Workplace Charging."

[17] U.S. Department of Energy, "PEV Outreach Resources for Your Employees," November 2014, http://energy.gov/sites/prod/files/2014/11/f19/Toolkit_EmployerGuidance_Final_11-14-14.pdf.

[18] CALSTART, "Best Practices for Workplace Charging."

[19] U.S. Department of Energy, "U.S. Department of Energy's EV Everywhere: Workplace Charging Challenge," 2014, http://energy.gov/sites/prod/files/2014/11/f19/progress_report_final.pdf.

[20] Office of Energy Efficiency & Renewable Energy, "Workplace Charging Success: MetLife," accessed June 16, 2015, http://energy.gov/eere/vehicles/articles/workplace-charging-success-metlife.

Chapter 14
Selecting Goals and Metrics that Matter

While program and project managers agree that metrics are important in measuring project success, identifying and selecting meaningful sustainability criteria can be challenging. Metrics are important because they provide clarity about key goals, shed light on progress, and lead to process and behavior change. The saying, "That which gets measured, gets changed," provides the foundation for the value of metrics in sustainability. Of course, the act of knowing doesn't necessarily translate into meaningful sustainability changes. The process is complex and requires senior management buy-in and commitment to organizational sustainability and support for programs and projects to drive the transformation.

14.1 Alignment of Strategy, Goals, and Metrics

Often, sustainability-related projects begin with "feel good" aspirations rather than tangible business-related metrics. While the intentions are admirable, projects with vague goals don't flourish because no one understands what constitutes a successful outcome, the organizational benefits, and their roles in the process. In order to create a successful sustainability program, clear and meaningful organizational goals must be established. A process must be agreed on to gather data and establish a baseline, and then identify and measure key performance indicators (KPIs). Developing sustainability programs and projects and reporting on progress with industry-accepted materiality standards provides for meaningful engagement with both internal and external stakeholders. Goals and metrics provide project managers with a vital communication tool to convey project direction, demonstrate progress, and align team members' actions and behaviors.

According to Green Research, goals signal organizational intentions and help to drive employee behavior. To be effective, goals should be specific, quantitative, time-bound, and forward-looking[1] (see Figure 14.1). Goals that are not clear or are undefined create a myriad of problems. Issues can range from undesired outcomes to lack of project funding, as sponsors tend to evaporate when project outcomes are unclear. In addition, nonspecific goals cause executive dismay because the executives are being asked to take responsibility for something they don't know how to measure progress or success. Employees take a similar view of unspecific goals. If they don't understand their role, employees become

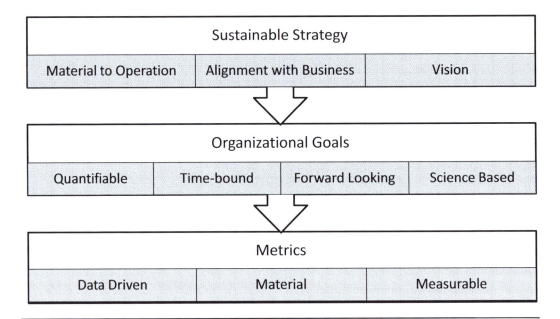

Figure 14.1 Alignment of Sustainable Strategy, Organizational Goals, and Metrics

disenfranchised with program. Clear strategic initiatives lead to clear goals and meaningful measurement targets.

The solution is to make your goals SMART (Specific, Measurable, Achievable, Realistic, and Time-bound). The concept of SMART goals was discussed in Chapter 6. Organizations that have clear strategies are able to develop SMART goals with clear action plans and meaningful metrics to support initiatives.

Organizations have a variety of sustainability requirements, and drivers for developing sustainability goals vary. Research conducted on the factors that drive organizations to set sustainability goals reported that the following were the top drivers:

- Significant environmental impacts (74%)
- Alignment with strategy (67%)
- Customer-driven (59%)
- Position as industry leader (52%)
- Other stakeholder requirements (30%)
- Employee motivations (30%)[2]

Organizations have different motivations to establish sustainability goals and different requirements for tracking and reporting on their progress. As a result, one organization's reporting focuses on gathering data and preparing scorecards to meet the needs of its customer stakeholders group, while another is driven by environmental regulatory compliance. Others seek to meet sustainability goals such as science-based GHG reduction targets or the U.N. Sustainable Development Goals (https://sustainabledevelopment.un.org/).

From a long-term perspective, aligning sustainability goals with core strategy provides a meaningful link for managers and employees, allowing for behavior change and adoption of a sustainable culture. Identification of a sustainability goal helps to clarify strategic direction and to communicate desired actions.

According to the Governance & Accountability Institute, 72% of companies included in the S&P 500 Index published a sustainability or corporate responsibility report in 2013. This figure compares to fewer than 20% of the same group reporting on sustainability in 2011.[3] As the number of organizations reporting on sustainability continues to rise, the differentiator between these organizations' sustainability programs becomes quality of programs and the reporting content. Those organizations reporting on material issues that address stakeholder requirements will lead in developing a sustainable culture.

14.2 Reporting Standards and Frameworks

Reporting paradigms such as the Global Reporting Initiative (GRI) and the CDP provide standards against which organizations can be measured and compared. (Refer to Figure 10.1 to review GRI guidelines and Chapter 2 for additional CDP guidelines.) The GRI reporting structure is divided into environmental, social, and economic categories, with specific guidelines for each category. Reporting using accepted sustainability standards provides an organization with a threshold of credibility. Defined standards and benchmarks facilitate tracking progress over time. Using common standards allows for clearer conversations about sustainability performance and lets stakeholders conduct more meaningful comparative analysis between organizations.

> **Global Reporting Initiative (GRI)**
>
> GRI is a not-for-profit whose mission is to encourage sustainability reporting on environmental, social, and economic impacts by all companies and organizations. It has developed the Sustainability Reporting Framework, which includes reporting guidelines, industry sector guidelines, and additional resources to promote greater organizational accountability and transparency in reporting. Thousands of global organizations of varying size use the GRI Framework to report on their sustainability activities and results. G4 is the latest version of the GRI Reporting Guidelines.[4]

Other resources come from the International Standards Organization (ISO), including ISO 14000,-Environmental Management; ISO 50001, Energy Management; ISO 20121, Sustainable Events and Guidance Standards; and ISO 26000, Social Responsibility. ISO 14000 is a set of guidelines to help organizations manage their environmental responsibilities, including life-cycle analysis, audits, communications, and product labeling. According to a survey of the over 5000 users of ISO 14000, the ISO environmental management protocols provide the following benefits:

- Meeting compliance and regulatory requirements and the organization's environmental performance (75%)
- Garnering management commitment and employee engagement (60%)
- Improving public image and meeting stakeholder requirements (50%)
- Providing competitive advantage (75%) and financial advantage (63%)[5]

Clear guidelines with measurable results provide significant business value. These survey results reinforce the fundamental project management concept that well-planned projects that are clearly defined in terms of goals, actions, and metrics are more impactful and successful.

ISO 26000 offers guidance on social responsibility for organizations, including recognizing the importance of social responsibility and engaging stakeholders. Core subjects focus on organizational governance, human rights, labor practices, the environment, fair practices, consumer rights, and community engagement.

Some industries have created their own reporting tools in order to facilitate meaningful industry competitive comparisons. An example is the Sustainable Apparel Coalition's "The Higg Index," which is a self-assessment tool for brands, retailers, and manufacturers to measure their environmental and social impacts at every stage of their sustainability journey (http://apparelcoalition.org/the-higg-index). These reporting standards reflect leading industry sustainability reporting protocols. Even if management does not wish to comply fully with these guidelines, they provide detailed categories and suggested standards that build a foundation for credible sustainability reporting.

The Sustainability Tracking Assessment & Rating System (STARS) is a self-reporting framework for academic institutions at the college and university level to measure and disclose their sustainability performance transparently. The purpose of this framework is to provide meaningful comparisons and incentives for continual improvement, as well as to facilitate sharing of lessons learned and best practices. The framework is based on performance indicators and criteria in four categories: Academics, Engagement, Operations, and Planning & Administration. Rankings are also divided into four categories: Bronze, Silver, Gold, and Platinum. This international framework facilitates transparent and meaningful sustainability program comparisons for students, faculty, and communities. The University of Vermont case at the end of this chapter takes a deeper dive into the STARS methodology and organizational impacts. Another resource for detailed industry-based metrics is the Sustainability Accounting Standards Board (SASB), which was discussed in detail in Chapter 4.

Additional reporting guidance comes from rating indices standards such as RobecoSAM Dow Jones Sustainability, FTSE4Good, and the CR's 100 Best Corporate Citizens. Each of these tools provides guidance on key areas for consideration, information on industry leaders, and guidelines to establish best practices. As a program or project manager, these resources provide guidance and direction for sustainability projects, especially about selecting measures for success.

If your organization is undertaking a project to improve sustainability reporting to comply with generally recognized standards with third-party verification, the GRI Framework is the most widely accepted standard. Project managers new to the rigors of sustainability reporting may wish to bring in a subject-matter expert experienced in complying with this level of reporting.

Table 14.1 is a list of Corporate Responsibility (CR) magazine's top corporate citizens. Reviewing the sustainability reports from these sustainability leaders provides project managers with an excellent framework for identifying industry-specific materiality issues and metrics chosen to track progress.

Table 14.1 CR's 100 Best Corporate Citizens 2015[6]

Top 10	
Rank	Company
1	Microsoft Corporation
2	Hasbro
3	Johnson & Johnson
4	Xerox
5	Sigma-Aldrich Corp.
6	Bristol-Myers Squibb Co.
7	Intel
8	Campbell Soup Co.
9	Ecolab, Inc.
10	Lockheed Martin Corp.

Table 14.2 2014 Microsoft Sustainability Goals and Metric Excerpts[7]

Category	2014 Goal Selection	Metric
Ethics/Governance	Engage in cross-industry forums to facilitate development of best practices.	Engage with organizations through associations such as World Economic Forum Partnering Against Corruption Initiative (PACI)
People	Increase diversity with a focus on senior-level positions.	Ratio of senior executive women and minorities rose from 22% to 27%. Ratio of women and minorities on board of directors rose from 33% to 40%.
Community	Empowering 100 million more youth through the Microsoft YouthSpark program.	124 million additional youth engaged through YouthSpark.
Human Rights	Privacy and data security transparency by reporting on U.S. National Security Orders Report.	Publish data every 6 months on the number of legal requests from the U.S. government related to national security laws.
Supply Chain	Improve working and living conditions in Tier 1 factories through Model Factory Program.	Tier 1 Supplier Scorecard rating of 95%. Six suppliers reached goals by end of FY14. Third-party verification audits.
Environment	Increase sourcing of renewable power.	Purchased 3 billion kWh of green energy, equal to 100% of global electrical use.

The process used for generating the list and ranking the companies is an involuntary audit of all companies on the Russell 1000 Index for seven categories: Environment, Climate Change, Employee Relations, Human Rights, Governance, Finance, and Philanthropy/Community Support. The data source is publically disclosed information from company websites, annual reports, sustainability reports, CDP disclosures, and other publically available sources. Each of these firms reports on progress against goals and states future commitments. As an example, take a deeper dive into the leader, Microsoft. In its 2014 Citizenship Report, Microsoft considered its material components of sustainability and reported on 2014 highlights and 2015 future commitments with specific and measurable targets and metrics. Excerpts from the report that demonstrate goal selection and metrics to measure success are shown in Table 14.2.

As the Microsoft report indicates, specific, measurable, actionable goals lead to meaningful changes in behavior. By understanding the objective, both management and employees were able to make significant contributions to the achievement of these goals. In addition, external stakeholders are able to assess their impact and monitor organizational performance against stated goals. As an organization moves forward on its sustainability journey, it is a process of peeling back layers and constantly asking the question, "How can we do it better?" The more specific the goals and more clearly defined the metrics, the greater success project managers will have in managing stakeholders, motivating project team members, and delivering successful outcomes.

Developing a framework and language for sustainability facilitates communication across all functions within an organization. Clear goals and metrics facilitate communication of intentions and actions to all stakeholders. Once terms and definitions are agreed on, communications become much clearer and more effective throughout the process. Refer to Chapter 11 for further details on sustainability communication plans.

The University of Vermont case at the end of this chapter highlights the value of a creating common language of sustainability. Also, the Dow Sustainable Chemical Index discussed in Chapter 4 reinforces the benefits of clear, measurable goals and metrics to promote communication. Many issues and problems can be avoided or resolved by ensuring that all team members and stakeholders are defining terms, projects, and outcomes in the same way.

ing Meaningful Metrics

rgets and goals, the next step is to ensure that the right things are being measured. ...ted need to align with organizational goals in order to reinforce commitments and to achieve desired outcomes. In a way, metrics help to tell an organization's sustainability story, allowing for objective analysis of results by all stakeholders.

In order to have effective metrics, systems and protocols must be established to ensure data accuracy and to gather and track data. Metrics need to have rigor in order to provide meaningfulness and credibility. Increasingly, stakeholders such as governments, customers, investors, and community members are requiring increased transparency from organizations. They expect to see their feedback reflected in the metrics that are tracked and measured in order to believe that their input is being valued.

Sustainability reporting requirements consider economic, environmental, and social impacts, supplemented with disclosures such as strategic outlook, organizational profile, materiality, stakeholder engagement, and governance policies. Within these areas, a plethora of sustainability key performance indicators are possible; the important first step is to select those that are material, or meaningful, to both your industry and your organization. Some considerations for selecting metrics include:

1. Environmental impact
2. Strategic alignment
3. Customer concerns
4. Industry leader standards
5. Stakeholder requirements
6. Employee motivation
7. Competitive pressure
8. Regulatory requirements
9. Anticipated future regulatory requirements
10. Organizational readiness for action
11. Timeframe

The sample metrics described in Tables 14.3 and 14.4 are from a report on industry-based sustainability KPIs that are considered the most relevant sustainability metrics for inclusion in annual reports. These examples provide areas of sustainability focus, definitions for how success will be measured, and specific time-bound metrics to measure progress and ultimately success.

According to research on common sustainability practices, annual goals and targets are most common, but a 5- to 10-year timeframe is frequently used when setting specific target improvement goals.[10] For example, becoming a zero emission organization may take 10 years or more, but setting annual targets helps focus management and employees on the incremental steps needed to ultimately succeed. Most organizations use a structure of formal commitments to achieve improvements in metrics. In

Table 14.3 Health and Safety Metrics[8]

Category	Definition	Metric
Employee Safety	Performance in terms of accident prevention	Lost-time incident frequency rate = number of lost-time injuries/1 million person hours
Occupational Health & Safety	The existence of a health & safety monitoring protocol & system	Trend line for injury or accidents rate/working hours
Working Conditions	Factory working conditions, wages, & working hours	Supplier scorecards verified by third-party audits

Table 14.4 Environmental Metrics[9]

Category	Definition	Metric
Innovation	Organization offers products & services that support sustainability.	Percent of products & services targeted toward promoting sustainability
Management of the Environment	Policies & procedures are in place to meet environmental requirements.	Number of environmental fines & penalties reported annually & as a trend line
Hazardous/Toxic Emissions or Waste	Hazardous or toxic materials as defined by available government databases disposed of in waste streams or emissions.	Disclosure of hazardous waste & emissions by kind, weight, and disposition to air, water, & landfill
Design for the Environment	New products are designed in a manner consistent with reducing energy as well as toxic materials & ingredients.	Percent of new products designed in accordance standards such as EPA Design for the Environment standards
Energy Usage	Effectiveness of energy management initiatives.	Track energy usage over 3–5 years
GHG Emissions	Effectiveness of programs to reduce Scope 1, 2, 3 emissions.	Measure & track GHG emissions over 3–5 years

other words, these formal commitments become part of performance scorecards for compensation. The most common goal categories are safety, waste recycling, energy usage reduction, and GHG emission reduction targets. Many organizations encourage business units to fund their own environmental or sustainability projects. This practice encourages business units to take ownership of their sustainability programs, embedding the process into the business unit culture. In order to receive funding, business unit sustainability projects may need to meet financial threshold requirements, such as providing a payback within 5 years or less.[11] As a project management professional, it is important to consider organizational requirements as part of developing a sustainability program or project.

There are a variety of resources for guiding sustainability program and project managers in the selection of metrics. Reviewing sustainability leaders in your industry and sustainability best practices provides guidance on metric selection. Selecting metrics based on their alignment with organizational goals is crucial. However, a key to success is to understand your organizational readiness to track, measure, and report on sustainability metrics. Selecting metrics that require data and reporting that the organization's information systems and people are not equipped to provide can quickly derail the project. Consider the organization's readiness for this type of project and the incremental changes that may be necessary to achieve readiness.

Important considerations for metric selection include their relevance in garnering and maintaining C-level support for both current and future sustainability programs. Also, metrics should clearly convey the organizational commitment to sustainability to both internal and external stakeholders. Clearly communicating sustainability goals and demonstrating progress toward those goals builds credibility with stakeholders. Gathering data and reporting on metrics provide the link to make this connection.

As goals and metrics are developed, consider how targets will be set at the corporate level and then cascaded down through the business units. Are there structures in place to gather and track data in order to measure progress on the selected metrics? Identifying people within the organization who have the data and convincing them to update and share the data can be challenging. Reporting on metrics without meaningful data brings into question the credibility of a sustainability program.

As program targets are being set, is managements' focus on internal operation, or does it include a life-cycle approach? Do program and projects require supplier or customer participation to reach stated sustainability goals? Is the organization ready in terms of resources and systems to accept value chain types of sustainability goals?

The process of developing sustainability goals and meaningful metrics becomes more complex and interconnected as an organization's sustainability program evolves. While aspirational sustainability goals can provide motivation for long-term change, most managers and employees need more concrete and measurable goals in order to effect meaningful change. In addition, tying metrics to performance measures requires them to be achievable, not just aspirational. Otherwise, managers and employees become disenfranchised with the sustainability program.

14.4 Meaningful Metrics Drive Engagement

Research in the field of workforce management reveals that workers want to be respected and to understand what it is that an employer wishes them to do in order to be a viewed as a valued employee. As basic as that premise sounds, research from Norton & Kaplan indicates that 95% of the workforce does not understand their organization's goals and strategy.[12] As a result, employees don't understand how their actions and behaviors create or detract from creating value for their organization. Communicating effectively with employees about strategic goals and the ways that their behaviors and actions can help to achieve the goals is crucial to successful sustainable strategy adoption. An example of a plan to influence employee behavior by selecting clear objectives aligned with sustainable strategy is given in Table 14.5.

From a portfolio perspective, leadership needs to identify sustainability goals and objectives as part of the long-term planning process. As an organization matures in its sustainability journey, goals and strategies need to be evaluated and reprioritized. The identification, evaluation, and prioritization of

Table 14.5 Aligning Sustainable Strategy and Business Objectives[13]

Strategic Area	Objectives	Business Value Impact
Financial	• Increase Revenue Mix of Sustainability Products • Reduce Risk Related to Climate Change & Reputation	• Increased Revenue • Improved Performance • New Product & Service Offerings • Lower Operational & Compliance Risk • Lower Cost of Capital
Stakeholders/Customers	• Offer New Sustainable Products, Services, & Solutions for Customers • Increase Engagement of Stakeholders on Material Issues • Open New Markets & Expand Opportunities	• Improved Stakeholder Relations • Higher Customer Satisfaction • New Product & Service Offerings • New Partnership Opportunities
Internal Processes	• Product Life-Cycle Assessment • Encourage New Ideas & Innovations to Save Money & to Improve Offerings • Improve Environmental Performance	• Reduced Operating Costs • Improved Human Rights Performance • Improved Community Relations
Organizational Capacity	• Form Strategic Sustainable Partnerships to Collaborate on Solutions • Training & Development of Workforce on Ethics & Governance Issues	• Creation of Sustainable Culture • Improved Ethics/Governance Compliance • Improved Community & Stakeholder Relations

Table 14.6 Tying Sustainable Objectives to Metrics, Targets, and Projects[14]

Objective	Metric	Target	Project
Increase Portion of Revenues from Sustainable Products	Percent of Units of Green Products Sold/Total Units Sold	10% Year 1 15% Year 2 20% Year 3	Innovation Incubator
Increase Engagement of Stakeholders on Material Issues	Stakeholder Engagement Survey Score	75% Year 1 85% Year 2 90% Year 3	Stakeholder Focus Group
Improve Community Air Quality Through Reduced Emissions	GHG Emissions kWh/$ of sales	30% Reduction 30% Reduction	Green Manufacturing Program
Increase Workforce Awareness of Code of Ethics	Percent of Workforce Attending Ethics Training	75% Year 1 90% Year 2 100% Year 3	Ethics Training Program

sustainability goals drive program and project selection. Ultimately, this process drives organizational change toward a more sustainable culture and operation. Creating and measuring the right indicators drives this change. The first step in the process is to benchmark key indicators and identify the drivers. Effective sustainable strategy requires a cross-functional approach that often includes measures that are interconnected across business functions and that relies on a matrix reporting relationship. In this non-linear workplace structure, balanced scorecards are often used to facilitate clear communication of both financial and nonfinancial metrics. This approach facilitates translating enterprise-level plans to the business units. The objective is to have all departments working toward a unified strategic sustainability vision. The balanced scorecard is a tool that facilitates cascading goals throughout the organization and aligning employee training and development with strategic goals. In addition, it provides clear alignment between compensation and incentives and organizational strategy. Using benchmarking, metrics, and scorecards facilitates communication with employees from the C-suite to the shop floor about the organization's goals and the actions and behaviors that are desired of them to achieve these goals.

The next step is to align business objectives with metrics (see Table 14.6). Once the metrics have been agreed on, performance targets can be created. Consider that the learning curve is steepest in Year 1, so more progress may be achieved in later years. Lastly, programs and projects to support these targets are identified, planned, and implemented.

Sustainability programs and projects are agile in nature, with lessons learned being incorporated into projects to improve outcomes. (An example of this agile nature is the continuous improvement energy management plan in Chapter 13.) Modifications are made as management learns from past programs and projects. A pathway to developing a sustainable organization is created by linking strategy, people, and objectives through projects designed to deliver on targets and organizational goals.

Through the process of translating the strategic vision into business objectives, leadership builds employee engagement and strategic alignment. Figure 14.2 is an example of a balanced scorecard of metrics to promote sustainable strategy. Many of the metric categories, such as EBITDA and customer satisfaction ratings, appear regularly in management scorecards. The difference is that the foundational assessment tools for a sustainable organization include questions and assessment criteria relating to sustainable strategy.

Other metrics such as scope-level emission reductions, energy intensity reductions, and water savings relate directly to environmental sustainability goals. Organizational capacity focuses on creating a culture and infrastructure to support the organization's sustainability program. The concept behind the balanced scorecard is that each of these categories is equally important in achieving organizational success, so each component is equally weighted for performance assessment.

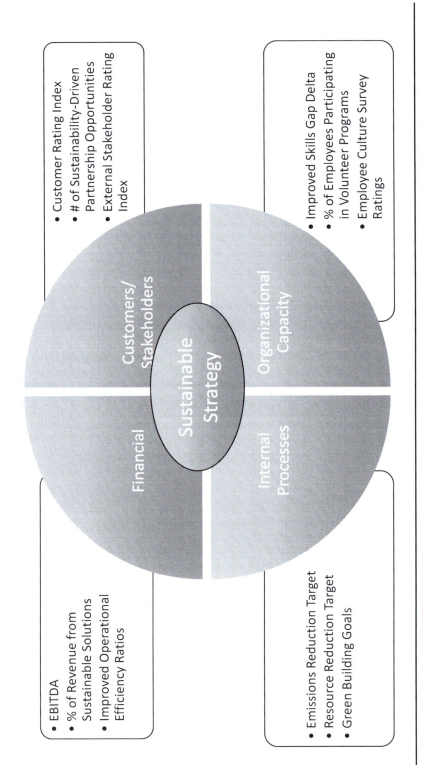

Figure 14.2 Sustainability Balanced Scorecard

14.5 Data Quality and Accuracy

Gathering accurate and relevant data is a vital step in reporting on an organization's sustainability progress through metrics. Technology-supported reporting frameworks promote data accuracy and timeliness. The quality of your program is judged by the quality of your data and adherence to reporting under an accepted framework. Table 14.7 shows an example data quality matrix.

The quality of data is vital in providing meaningful interactions with stakeholders. It must be presented clearly in a format that can be understood by the stakeholders. The information needs to be relevant in terms of aligning with the project scope and goals and be delivered on a timely basis so that it can be incorporated into stakeholders' decision-making processes. In addition, the source and process of data gathering must be disclosed to promote transparency of process. If certain projects or locations have been excluded from the information presented, full disclosure of this omission from the project scope should be provided.

Data accuracy also improves the portfolio planning process. Decisions that are based on incomplete or faulty data tend not to be effective decisions. If the data used to make these decisions is not accurate, the benefits of sustainable strategy such as risk mitigation, brand protection, partnership opportunities, and employee engagement become diminished. In addition, lack of data integrity may cause an entire program of project to be called into question. When building a sustainability program, take the time to create a data gathering and reporting structure that contributes to the credibility of the program. In order to facilitate the data management process, consider the role for technology in gathering data to improve accuracy and reduce costs.

Table 14.7 Data Quality Matrix

Criterion	Leaders	Average	Laggards
Quality	Independent Third-Party Verification of Internal Results	Nonverified Internal Data Based on Standard Reporting Framework	Estimate or Nonstandard Reporting Framework
Timeliness	Current Reporting Cycle	Previous Reporting Cycle	2 or More Years Old
Relevance	Data Generated from Specific Programs or Projects	Data from Similar Sites with Similar Products/Services	Industry Averages

14.6 Role for Technology

Initially, most organizations track sustainability data using spreadsheets. While spreadsheets maybe sufficient for small to midsize organizations, managing data in complex and globally disperse organizations on a spreadsheet is difficult. Challenges abound, such as chasing after people with responsibility for the necessary data, spreadsheet errors, multiple reporting periods, unclear definitions, and timeliness of information. As a result, the demand for global sustainability management software is rising. Table 14.8 lists a selection of sustainability software providers drawn from Gartner's Environmental

Table 14.8 Environmental Health and Safety Software[15]

SAP	IHS	CMO Compliance
SiteHawk	Gensuite	3 E Company
Enablon	Intelex Technologies	Enviance
Enviance	UL Workplace Health and Safety	Medgate

Health & Safety (EH&S) Magic Quadrant. Another source for EH&S software providers is published by Verdantix in its Green Quadrant EH&S Software report.

Enablon is considered by Gartner as one of the largest best-of-breed EH&S software vendors with a significant presence in North America and Europe. They also provide services to the broader sustainability and social responsibility markets. Enviance is very strong on air quality compliance, including the U.S. Clean Air Act and the EPA's mandatory GHG tracking and reporting requirements. Some providers, such as SAP, offer EH&S reporting as part of their broader ERP solutions. Some of the solutions are cloud or software-as-a-service (SaaS) solutions, while others are on-site software.

Understanding the focus of the software solution and ensuring that it will meet the organizational data gathering and reporting needs is an important first step. Environmental management software is designed to manage the process of complying with environmental regulations such as those from the EPA for waste and emissions. Health management software focuses on tracking and managing data related to employee health in the workplace, such as exposure to toxins and sound decibel level impacts on hearing. Safety management tracks accidents, monitors systems and processes, and audits compliance. Product safety software tracks the use, handling, and transportation of hazardous materials. Risk assessment software helps organizations assess and monitor risk related to these issues. Other classifications of software include enterprise risk management, governance and compliance, energy management software, and building management software.[16]

Table 14.9 highlights the leaders in the global sustainability management software market. Many of these providers are also mentioned as leaders in providing software to promote employee engagement by a Verdantix report.

Table 14.9 Top-Rated Global Sustainability Management Software[17]

CA Technologies	Enablon
CarbonSystems	IHS
Cloudapps	PE International
IBM	Schneider Electric
Verisae	Credit360

Ranking reports from groups such as Verdantix, which produces the Green Quadrant Sustainability Management Software reports, are useful in helping sustainability professionals identify which software offerings best meet their needs. This report relies on product benchmark data from suppliers, questionnaires, and interviews with customers across industry segments.

The primary driver behind sustainability software selection has been to enhance data quality. When selecting software, there are a number of considerations such as:

1. Regulatory content for your industry
2. Reporting frameworks and compliance standards
3. Workflow and notifications process
4. Documents management
5. Configuration versus customization
6. Reporting and analytics
7. Integration with enterprise resource planning (ERP) and other systems
8. Scalability and global compatibility

Selecting software to support and scale to an organization's sustainability program is very similar to other software selection processes. As with most projects, the business value of the project needs to be demonstrated to the funding steering committee. The primary business drivers for sustainability software are improved data quality, saving time, faster time to disclosure, assurance cost savings, and consultant cost savings.[18] Best practice suggests focusing the business case justification on improved data and controls, internal efficiencies, and long-term savings.

As part of the process, assess the requirements of internal and external stakeholders. Gain a sound understanding of both today's requirements and future plans to ensure that the system selected meets both current and future needs. Based on the needs assessment, technical requirements, and system requirements, develop selection criteria. Using these criteria, identify vendors that meet these needs and arrange for system demonstrations. While viewing the demonstrations, consider your organization's sustainability program and system requirements. Software providers often have industry-specific capabilities that may not support your requirements. Conduct due diligence on the software providers, including obtaining customer references, before making a vendor selection.

If big software solutions are too expensive for your organization, there are several sustainability software services that offer accessible tools to streamline data management and support growing sustainability requirements. Some examples are listed in Table 14.10.

Table 14.10 Sustainability Reporting Software[19]

Name	Service	Target	Delivery
OneReport	Reporting system to streamline the management of data query, collection & distribution	Customer, regulatory, & reporting surveys	Cloud
Scope5	Effectively & transparently track sustainability performance & reduce costs	Track & analyze sustainability data Generate common standard reports (GRI, CDP)	Cloud
Measurable	Collect data, generate reports, meet reporting requirements for carbon, water, & energy usage of building portfolios	Real estate developers, REITS	Cloud

Each software system has strengths and weaknesses. The selection process should be driven by organizational business requirements and the overall value proposition. Figure 14.3 identifies the major steps, including needs identification, vendor identification, and software selection and implementation to facilitate technical support of an organization's sustainability program.

As with any business decision, the selection of sustainability software comes down to making the business case for a sustainability software selection and implementation project. Some points to consider in making the decision are as follows.

Functionality. Does the software provide the capabilities needed by the organization, such as tracking sustainability data, metrics, environmental and safety compliance; managing documents; and providing audit capabilities, sustainability data management, and reporting? Are reporting formats able to meet stakeholder requirements?

Integration. How easily can this system be integrated into existing systems and processes? How will data be transferred and captured? Is the system user-friendly to promote user acceptance? Are internal subject-matter experts required? Does the plan include ongoing leadership support and sufficient budget for maintenance?

Training. What kind of training is provided by the vendor? How intuitive is the system? Are there ongoing training support, dedicated customer service, user communities, and wiki resources?

Change Management. Is vendor support provided to transition from spreadsheets or department-specific systems of recordkeeping to an enterprise-wide system? What are best practices to engage stakeholders?

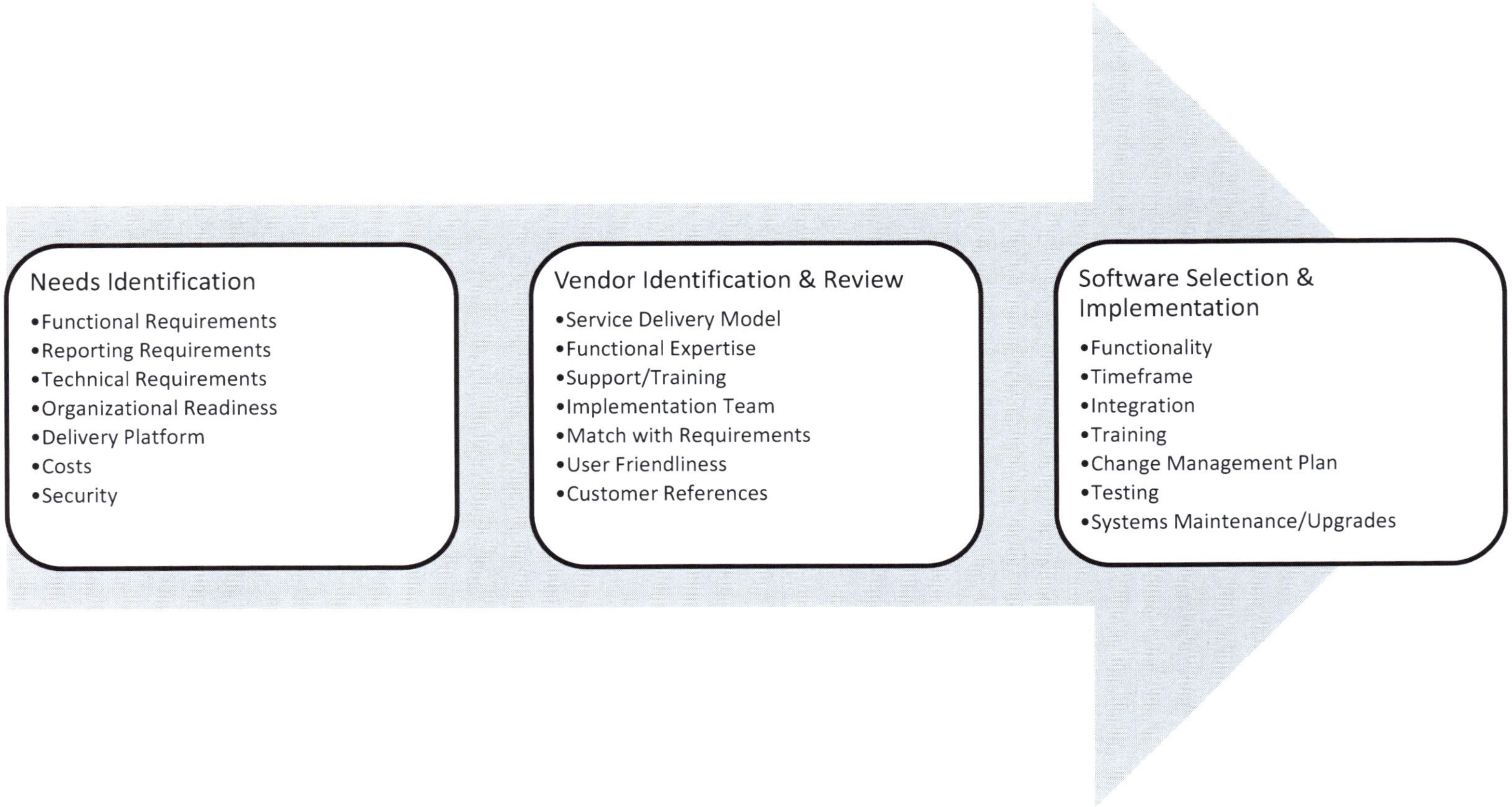

Needs Identification
- Functional Requirements
- Reporting Requirements
- Technical Requirements
- Organizational Readiness
- Delivery Platform
- Costs
- Security

Vendor Identification & Review
- Service Delivery Model
- Functional Expertise
- Support/Training
- Implementation Team
- Match with Requirements
- User Friendliness
- Customer References

Software Selection & Implementation
- Functionality
- Timeframe
- Integration
- Training
- Change Management Plan
- Testing
- Systems Maintenance/Upgrades

Figure 14.3 Sustainability Software Selection

Testing. How is testing and data verification handled to ensure accurate data transfer and quality standards?
System Maintenance. What are best practices to assign responsibility for systems maintenance, timely and accurate data collection, and data quality?

The selection process comes down to picking software that has the best fit for organizational needs, at the best price, with a good implementation process, and ongoing training and support. The accuracy of the sustainability report is only as good as the data quality and system used to create it.

14.7 Portfolio, Program, and Project Management Impact

Translating organizational strategy into business outcomes is a key role for portfolio, program, and project managers. From a portfolio perspective, translating strategic vision into criteria for program and project selection is the initial step in the process of communicating sustainability priorities throughout the organization. Programs and projects are then undertaken to effect organizational change, develop sustainable products and services, improve product life-cycle performance, and engage internal and external stakeholders in order to deliver the desired sustainable business outcomes. Selecting reporting frameworks and metrics is related to an organization's maturity in its sustainability journey. Figure 14.4 identifies characteristics of reporting, data management, and projects based on an organization's progress in its sustainability journey. For project management professionals, this figure helps to identify

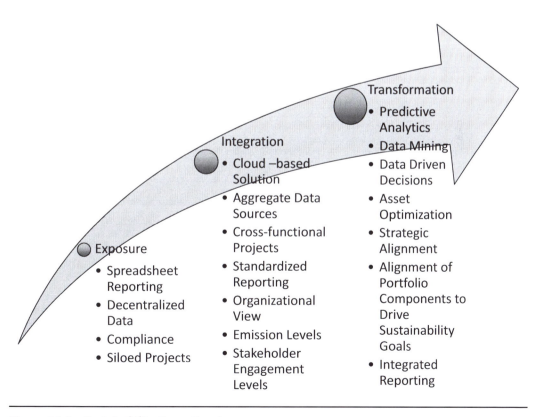

Figure 14.4 Sustainability Reporting Journey

where an organization is in its journey as well as data management and reporting considerations to move forward.

The linkage in this process is the clear establishment of goals and the translation of the goals into metrics with specific time-bound targets based on credible data that is reported using acceptable standards. Sustainability software facilitates the process of gathering data and tracking metrics, allowing project managers to better communicate sustainable strategy to key stakeholders. Sustainability reporting protocols add credibility to project performance results for both internal and external stakeholders. Project management professionals who identify clear project goals and then measure and report on project outcomes using defined metrics and reporting protocols drive adoption of sustainable strategy within the organization.

> **Integrated Reporting**
>
> Integrated Reporting (IR) is a standard for financial reporting that focuses on the long-terms value creation of companies. The objective is to provide stakeholders with relevant information about material issues that impact the organization. Through this international framework the International Integrated Reporting Council (IIRC), a global group of companies, regulators, investors, NGOs, accounting professionals, and others, seeks to promote clear and concise communication about an organization's strategy, governance, and performance relative to value creation over the short, medium, and long term.[20] Corporate leaders in sustainability have adopted this approach to better communicate the interconnectivities between natural, social, and financial capital.

14.8 UVM Case Study: STARS Project[21]

Planning and Implementing the First STARS Report for the University Vermont
Mieko A. Ozeki, Sustainability Officer, UVM

The University of Vermont (UVM) has a long history as a leader of sustainability in higher education. As sustainability began to move to the mainstream in the late 2000s, the University noticed students becoming increasingly interested in the institution's performance on its sustainability efforts. During this timeframe, college-ranking publications, such as *Princeton Review* and *Sierra Club Magazine*, were sending surveys to colleges and asking about the sustainability efforts of these institutions. Sustainability offices were flooded with survey requests; many had similar questions but required separate responses. Significant time and resources were expended to complete these survey responses. The problem with these surveys was that they were neither consistent nor transparent, because ranking results appeared to be based on varying criteria. Campus sustainability professionals were becoming frustrated and voiced their concerns through a joint letter to the ranking publishers. Some institutions took a stand and refused to continue to complete these surveys.

The Sustainability Tracking, Assessment & Rating System (STARS) concept began in 2006 with a goal of providing a transparent, self-reporting framework for higher-education institutions to consistently track and measure their sustainability performance. The performance indicators and criteria for the latest version fall into four categories: academic; engagement; operations; and planning and administration. The goals of the framework are to allow public access and meaningful comparisons between institutions, provide incentives for continual improvement, facilitate sharing of information and best practices, and build a stronger, more diverse campus. The STARS program has grown and now has over 668 registered institutions and is on the STARS 2.0 version of the program.

In 2011, the Sustainability Office at UVM began a pilot of STARS to determine its effectiveness in tracking performance, providing information to outside groups, offering uniform standards, and managing and engaging their broad-based community. A Campus Sustainability course was taught annually in the spring, focusing on campus operations performance indicators. Students conducted service-learning projects, whereby the Sustainability Office served as their client. They collected information from interviews with operations personnel and researched best practices from other colleges and universities. Their final projects included posters, maps, and videos, which were presented to senior administrators during finals period. During these final presentations, the students outlined their findings and recommendations for improvement. Through this process, senior administrators became familiar with sustainability concepts and much more interested in sustainability as their customers, students, were vocalizing the means for improving the University's performance. These initial sustainability courses provided an excellent framework for piloting the STAR data collection process and a structure through which to engage students and faculty on the program.

Based on the preliminary evaluation, the Sustainability Office estimated that UVM could receive a Silver rating from STARS. With the pilot results, the Sustainability Team approached the Provost to solicit project approval to adopt the STARS reporting framework. Their initial project approval request was unsuccessful because of a misunderstanding of the program proposal. It was a case of the best intentions being misunderstood because of a lack of common definitions and language about sustainability.

The next opportunity came with the transition of a new University president. Part of the STARS pilot project management plan had been to engage with stakeholders beyond the traditional realm of sustainability professionals to include faculty, staff, and students. In the past, most efforts focused on facilities and energy managers. During an introductory meeting with the new University president, the environmental faculty mentioned the STARS program as a means of enhancing the quality of sustainability education and as a solution to the current vagaries of the published ranking reports on "green" colleges and universities. In 2012, the President and Provost approved the formation of the Envisioning Environment Work Group to make recommendations in the areas of environmental research, education, and outreach. The following year, the group's recommendation was that UVM enroll in the STARS program. The President was aware of UVM's position as a leader in the field of sustainability in higher education and agreed to participate in STARS. The framework of STARS provided a means to accurately portray the depth and breadth of sustainable curriculum and practices to their current and potential students. It would be a cornerstone in branding UVM as an institution that is committed to sustainable practices.

In February of 2014, UVM issued its first STARS report and received a Gold rating. UVM was among a few pioneer institutions to establish an Environmental Program, report annual GHG inventories since 1990, create an Eco-Reps program, establish a Clean Energy Fund, and establish a Sustainability Faculty Fellowship program. Their programs reflect a broad-based community effort, which includes input and actions from students, faculty, and staff.

While the overall project has been a success, the project team faced numerous challenges, including the data gathering process, data accuracy and integrity, launch of a new version of STARS during the project rollout, and a flood that damaged their office and displaced their team. The success of the project required engaging diverse stakeholders, building an institutional foundation of sustainability, providing a sufficient timeframe for complex and cumbersome data collection, and adhering to a unified standard. Lessons learned include piloting the program to develop a process, gathering data to build your business case, ensuring sufficient budget, testing current information for accuracy, and managing expectations. STARS has given the Office of Sustainability a framework for ongoing conversation with other members of the university community, creating room for change and improvement.

The benefits of the STARS program include engagement of both internal and external stakeholders. The use of a common framework, unified definitions, and consistent metrics allow for meaningful and insightful dialogue across the University and throughout the academic community.

14.9 Conclusion

Metrics are useful to determine impact and document progress, but they are also useful to engage internal and external stakeholders. The metrics that are chosen by a project manager reflect the areas of focus and impact of the project. From a stakeholder perspective, the metrics that are selected by an organization to document their sustainability story indicate the organizational priorities. Once a sustainable strategy has been developed, progress and results need to be defined, measured, and reported in a manner that reinforces that strategy. Reports and metrics become tools to facilitate both internal and external conversations about progress toward sustainable strategic goals.

The process of gathering accurate and relevant data and reporting the results in a meaningful report should not be underestimated. The process is time-consuming and challenging, requiring sustainability champions to consider carefully organizational goals, programs, and projects as well as what measures will be used to determine success. Accurate data presented transparently promotes conversations with stakeholders on issues such as continuous improvement to realign resources, selection of new technologies, or realignment of a project to move an organization toward its sustainability vision. A variety of software solutions are available to facilitate the process of improving data quality and timeliness of reporting. Selecting a reporting framework provides a meaningful platform to share results with stakeholders over time. Using a common language to discuss sustainable strategy with employees, suppliers, investors, and customers improves stakeholder engagement and facilitates understanding and ultimately achievement of the sustainability mission.

Notes

[1] David Schatsky, Principal and Founder, Green Research, "Benchmarking Environmental Sustainability Goals," *Environmental Leader*, accessed June 25, 2015, http://www.environmentalleader.com/2011/07/25/benchmarking-environmental-sustainability-goals.

[2] Elaine Cohen, "How to Set Sustainability Goals: The Do's and Don'ts," *The Green Economy Post: Green Careers, Green Business, Sustainability*, accessed April 1, 2015, http://greeneconomypost.com/set-sustainability-goals-18531.htm.

[3] Governance & Accountability Institute, Inc., "Seventy-Two Percent of S&P Index Publish Sustainability Reports in 2013," June 2, 2014, http://www.ga-institute.com/nc/issue-master-system/news-details/article/seventy-two-percent-72-of-the-sp-index-published-corporate-sustainability-reports-in-2013-dram.html?tx_ttnews[backPid]=1&cHash=8e53ff176eb49dc3b7442844c65833ac.

[4] GRI, "What Is GRI?," n.d., https://www.globalreporting.org/information/about-gri/what-is-GRI/Pages/default.aspx.

[5] ISO, "ISO 14000-Environmental Management," June 29, 2015, http://www.iso.org/iso/home/standards/management-standards/iso14000.htm.

[6] *Corporate Responsibility Magazine,* "CR's 100 Best Corporate Citizens 2015," April 2015, http://www.thecro.com/files/100BestList2015.pdf.

[7] Microsoft, "Microsoft 2014 Citizenship Report," 2014, file:///Users/kristinakohl/Downloads/Microsoft_2014_Citizenship_Report.pdf.

[8] Axel Hesse, "SD-KPI Standard 2010-2014," accessed June 29, 2015, sd-kpi_standard_2010-2014.pdf.

[9] Ibid.

[10] NAEM Research, "Leading-Edge Metrics and Programs," 2014, http://c.ymcdn.com/sites/www.naem.org/resource/resmgr/Docs/survey-2014-gmtm-eb.pdf.

[11] Ibid.

[12] Robert S. Kaplan and David P. Norton, "The Office of Strategy Management," *Harvard Business Review*, accessed July 2, 2015, https://hbr.org/2005/10/the-office-of-strategy-management.

[13] Howard Rohm, "Link Sustainability to Corporate Strategy Using the Balanced Scorecard," n.d., http://www.balancedscorecard.org/portals/0/pdf/linkingsustainabilitytocorporatestrategyusingthebalancedscorecard.pdf.

[14] Ibid.
[15] Leif Eriksen, "Magic Quadrant for Environmental, Health, and Safety Management Systems," May 6, 2014, Gartner Group, http://www.gartner.com/technology/home.jsp.
[16] Ibid.
[17] "Schneider, PE, Enablon, CA Top Sustainability Software Market," *Environmental Leader*, accessed June 25, 2015, http://www.environmentalleader.com/2013/07/03/schneider-electric-pe-international-top-sustainability-software-market.
[18] Ibid.
[19] Michael Ansaldo, "3 Services That Simplify Sustainability Reporting," *GreenBiz*, February 10, 2015, http://www.greenbiz.com/article/3-apps-simplify-sustainability-reporting.
[20] Integrated Reporting, "What? The Tool for Better Reporting," February 4, 2016.
[21] Mieko A. Ozeki, Project Management of the First STARS Report for the University of Vermont, September 10, 2014.

Chapter 15
Celebrating Success

In order to build an effective sustainable strategy, the framework must include tools and techniques to reinforce desired behaviors for both internal and external stakeholders. Establishing a clear sustainability vision and communicating it to your employees, suppliers, customers, and other stakeholders sets the framework for creating a culture of sustainability. Employees need to understand the organization's sustainability goals and their role in achieving those goals. Recognizing and rewarding ideas, actions, and collaborations that support sustainable strategy reinforce the organizational priorities to all employees. Similarly, external stakeholders such as suppliers and customers relate to rewards and incentives because they feel that they are being given clear information and that the organizational process is transparent and credible. Celebrating sustainability milestones and achievements creates a positive and rewarding experience for both the organization and its stakeholders.

The process of moving from sustainability vision to changes in business operations and outcomes requires fundamental changes in how work is done within an organization. Translating senior management vision into desired employee actions and behaviors to move the needle on sustainability initiatives is a crucial step in the process. In order to effect this change, programs and projects must incorporate sustainability standards and criteria and also include rewards and incentives to promote desired outcomes.

Employee engagement is driven by understanding of the organizational vision and the employee's role in the delivery of that vision. In order for employees to be engaged, they must feel that their actions and behaviors related to sustainability are meaningful to the organization and to their own performance evaluation. Incorporating sustainability metrics as part of compensation is a crucial long-term change management step. As indicated in Figure 15.1, incentives drive sustainable outcomes. Recognition and rewards are integrally tied to changing behaviors, because these types of programs reinforce management's commitment to sustainability vision and organizational change.

The objective is to avoid having sustainability targets and goals become just another item on an employee's "to do" list. Demonstrating the impact that employee actions can have on sustainability goals and empowering them through vehicles such as green teams drives change. Creating meaningful incentives that provide both organizational recognition and financial rewards garners attention. In addition, celebrating successful sustainability programs and project teams provides benefits not only for the teams involved but also for others within the organization. Employees who are not directly involved in the project see the excitement and recognition for their co-workers and will want to become involved. Recognition, acknowledgement, and incentives drive long-term success of sustainability programs and projects.

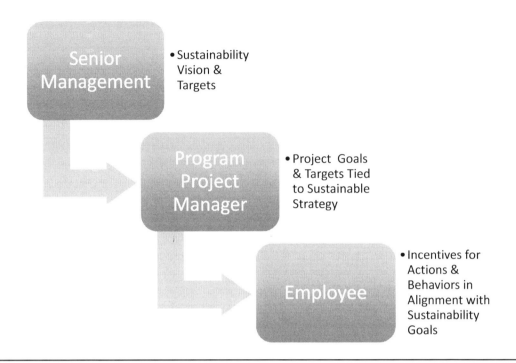

Figure 15.1 Incentives Drive Sustainable Outcomes

15.1 Impact of Incentives

From a project perspective, incorporating incentives into the change management process gets results by sending clear messaging about desired behaviors and outcomes. While incentives are important throughout the sustainability journey, they become much more integrated into core competencies and compensation structures as an organization moves along the sustainability continuum.

Early on in the sustainability adoption process, incentives such as contests, special recognition, and departmental competitions provide a carrot for promoting desired behaviors. Incentives will vary by organization and type of project. To generate awareness of a health and wellness program, small promotional items such as branded exercise bands or a contest to win a Fitbit® gets employees' attention. If an organization is seeking input from employees on ways to improve their organization's sustainability performance, they may create a multiyear program with annual winners awarded cash prizes and invitations to senior management recognition events.

Incentives are useful for garnering project attention and changing employee behavior. Celebrations bring focus and attention from across the organization. This attention opens the door for greater cross-functional interest and promotes sharing of success stories and best practices across departments, business units, and geographically separated groups. Ultimately, the best celebration is reaching the transformation stage and successfully improving an organization's environmental and social impact, thereby creating an organizational legacy of sustainability.

According to Jeff Rice, Walmart's Director of Sustainability, financial rewards and recognition send clear signals to stakeholders about the importance of sustainability to an organization.[1] Walmart uses this approach to engage both internal and external stakeholders. Walmart's My Sustainability Plan (MSP), launched in 2010, encourages associates to focus on personal goals such as healthier living, resource conservation, protection of the planet, and quality of life. This voluntary program was devised to create opportunities and incentives for Walmart associates to better understand the impact of sustainability through their own personal experience. As a result, they are better able to embrace sustainability in the workplace.

In order to create incentives, the platform uses "gamification"-gaming concepts to encourage behavior change and recognize success, rewarding employees for achieving goals and encouraging them to share their progress with other associates. Recognition awards for various goal categories are incorporated into the program to celebrate accomplishments and to share success stories. Rewards are unlocked when personal goals are achieved. The MSP informs and engages associates in order to promote broad-based acceptance of sustainability throughout Walmart's geographically and culturally diverse workforce. As a result, Walmart's associates are better able to understand the organization's sustainability goals and their role in impacting meaningful environmental, social, and economic value.

In 2013, sustainability goals became part of Walmart buyers' performance scorecards. To celebrate their success, Walmart awards a "Sustainable Buyer of the Year" in each of its business units. The award includes both recognition and financial incentives.[2] The program provides clear messaging on desired business goals and incentives to generate desired outcomes. These programs are examples of effective motivational tools to drive internal stakeholder behavior. Celebrating success reinforces the message to your workforce that they are engaging in the right behavior and that they are generating the right outcomes.

At Walmart, external stakeholders are offered incentives as well. For suppliers, Walmart uses both internal and external recognition programs. High-performing suppliers are offered benefits such as exclusive access to executives. Offering an opportunity for a supplier to make more than an "elevator pitch" to senior executives is a real incentive. Public recognition comes in the form of supplier acknowledgment and sustainability recognition on e-commerce sites that focus on sustainability leadership. In addition, suppliers have access to support materials, training, and guides to improve their own sustainability performance. Suppliers that are lagging in performance are offered additional support from the Walmart team known as "Family Meetings" to work on areas that need improvement.[3] Most important, Walmart has built a bridge between supplier scorecards and organizational action and behavior. Linking strong sustainability performance and increased business with Walmart is the most effective means of celebrating supplier sustainability success.

15.2 Gamification

Providing feedback and corresponding incentives to change behavior is an important aspect of change management. Acting in a sustainable manner may initially take more time, effort, or resources. Or, it may just be about learning alternatives to accomplish the same thing in a slightly different manner. According to Kevin Werbach, Wharton Associate Professor of Ethics and Legal Studies, Fortune 500 companies are using gamification as a tool to motivate employees in areas such as project management, health and wellness, and sustainability. The game encourages participants to engage in desired business processes by using gaming motivational devices such as badges, progress status, competitions, and leader boards.[4] A variety of institutions use gamification to motivate consumers, employees, and community members to modify their behavior to promote more sustainable actions. The benefit of this approach is that users are frequently and consistently rewarded for desired actions and behavior through the gaming process. It is a constant celebration of success!

The following examples shed light on the gamification process and impact. Recyclebank encourages recycling by rewarding participants for making smarter choices about trash disposal. Communities join the program to encourage residents to recycle in order to reduce the amount of trash going to landfill. The reduced volume of trash going to landfill lowers municipal tipping fees and reduces the resource stress of trash going into already overloaded landfills. Participants earn points based on their volume of recycling, which can then be redeemed for gift certificates at both national and local establishments.[5] It is fun, and many communities have had significant success in changing residents' behavior by using the program.

Nissan uses gamification in their Leaf electric vehicle (EV) line to provide drivers with real-time feedback on their driving behavior, including how it is impacting their vehicle's fuel efficiency. The display is behind the steering wheel, and it uses symbols shaped like trees to demonstrate driving behavior that promotes saving battery charge.[6] While this tool doesn't really promote competitive interaction, it does provide usable information to drivers to make immediate changes in driving behavior that will reward them with greater EV driving range.

Clorox offers a consumer-facing app that provides customers with a tool to check product ingredients while in the store so that they can be better informed about the sustainability impact of their purchase. Starbucks offers a "Buck for Stars" program to encourage coffee cup reuse. Each time a customer brings her coffee cup in for a refill, she receives a star and a discount on coffee ranging from 10 cents on the first refill to 40 cents for the fourth.[7] The program is an outgrowth of a crowdsourcing campaign aimed at getting customers to help Starbucks reduce the environmental impact of disposable coffee cups.

SAP offers "Twogo," a mobile cloud-based ride-sharing application to reduce the number of single-occupant cars on the road. While SAP employees use the app to save time, share costs, and reduce commuting emissions, the service is available to other organizations to provide a corporate ride-sharing program for their employees.

While these gaming incentives target different stakeholder groups, they all provide feedback and rewards for choices and changes in user behavior. Gamification is a great tool for encouraging employees to identify and submit ideas for resource savings, to collaborate across business functions, and to encourage innovative products and solutions. Gamification can also be an effective method to engage with supply chain vendors on key sustainability initiatives or to engage with customers about sustainable products and to create brand loyalty. Table 15.1 highlights top benefits for organization that embrace gamification to support sustainability.

Table 15.1 Benefits of Gamification for Sustainable Strategy

Increase Employee Engagement	Increase Stakeholder Engagement
Build Brand Loyalty	Increase Participation in Health & Wellness Programs
Provide Employee Training & Development on Sustainable Processes	Promote Sustainable Actions & Behaviors in Employees
Promote Collaboration	Encourage Innovation
Provide Immediate Feedback	Promote Competition for Continuous Improvement

15.3 Employee Resource and Affinity Groups

Offering employees a chance to create groups to pursue diverse interests on company time with company budget and senior executive support or mentorship sends a clear message about celebrating diversity and creating an atmosphere of inclusion within an organization.

The Hasbro Employee Network is made up of over 800 employee-driven groups created to give employees a voice, offer networking opportunities, provide leadership development opportunities, and track trends and issues that are important to their employees. Teams develop their own charter and are sponsored by an executive. They meet monthly to discuss issues that are relevant to the group. In addition, they identify opportunities and participate in Hasbro's volunteer programs. Hasbro's Equality

Awareness Resource Team (HEART) was created to promote an inclusive and respectful environment for LGBT employees, provide networking and community outreach, and drive business value. Their activities include forming partnerships with Youth Pride and the LGBT College Resource Group to collaborate on community outreach. One of their success stories is securing Hasbro as a sponsor for a Rhode Island Pride event. In addition, Hasbro's green teams focus on creating a culture of environmental responsibility by creating awareness through education and promoting colleagues' behavior changes. A major program was a month-long celebration of Earth Day, including tips for employees on recycling and green living.[8]

Affinity groups and employee networks offer incentives for employees to collaborate in order to address sustainability workplace issues. In addition, they empower employees to address social and environmental issues that are of personal concern to them. This is an example of the sustainability value mapping depicted in Figure 6.5. Providing this opportunity for employee impact is a meaningful way to celebrate employee diversity and to promote engagement.

15.4 Culture of Sustainability

As an organization matures in its sustainability journey, the concept of celebrating sustainability success becomes part of the performance culture of the organization. Sustainability goals are part of the compensation structure through performance scorecards for management, employees, and even suppliers. Policies and governance guidelines provide direction on ethics, inclusion, environmental goals, and community impacts. Sustainability success becomes synonymous with organizational success. This approach is reflected in core compensation benefits to employees such as paying a living wage, providing safe working conditions, offering health care plans, savings and spending accounts such as Health Savings Accounts, flexible spending accounts, dependent care, insurance plans, and retirement savings plans. Sustainability is reflected in board of director initiatives, senior management vision, human resource policies and strategy, organizational policy, leadership development, strategic initiatives, partnership alliances, and community outreach. Rather than focusing on specific campaigns or contests, sustainability becomes ingrained into traditional measures of success for leadership, the workforce, and key vendors. Celebration comes from being recognized by internal and external stakeholders as a sustainable organization and the benefits that accrue to an organization that has adopted sustainability as a core value. If you refer back to the Eccles et al. research discussed in Chapter 2, highly sustainability organizations outperformed their peer group.

As an organization moves into the transformation stage, celebrating success takes on more public recognition of achievement by third parties. Some of these measures of sustainability success include being invited to participate in the RobecoSAM DJSI, inclusion on Corporate Responsibilities' list of "100 Best Corporate Citizens," being named to Forbes' "Great Companies for Women," or a variety of other "Best of . . ." that celebrate sustainable accomplishments. Other forms of celebration come from industry group recognition as a leader or significant contributor in the sustainability field. Government agencies invite organizations with sustainability success stories to join think tanks, participate in consortia, and join public–private partnerships to address local and national challenges. Accepting leadership roles in UN sustainable development projects that promote collaboration among government, private, and societal partnerships to address global issues is another way to celebrate sustainability success.

15.5 Promoting Project Team Success

One of the most effective means of promoting success is allowing program and project managers and team members to share their sustainability project successes with others. Provide opportunities for employees to share their sustainability knowledge and expertise both inside and outside the organization. Internally, project managers and team members can present their sustainability stories at staff meetings, corporate town hall gatherings, steering committees, and more formal quarterly leadership meetings. This opportunity provides recognition for the project team and facilitates collaboration and sharing of best practices with other areas of the organization.

Another approach is to create a dedicated page on the organization's intranet for articles on sustainability projects, including profiles of the project team, their project story, and lessons learned. Depending on the organization's corporate policies, Facebook or other private collaboration sites can be an effective means of publicizing sustainability project success stories and building a community of internal and external stakeholders interested in monitoring the organization's progress. Providing a forum for recognition and an opportunity for project teams to share their areas of sustainability expertise is a significant motivational tool for project team members.

Formal communication channels are effective as well. Sharing project success stories in organizational newsletters or on blogs helps to promote the depth and breadth of sustainability programs within the organization. Some organizations provide blog sites for employee resource groups to highlight their own success stories.

Other project-related incentives include making pilot project team members into subject-matter experts (SMEs) on project rollouts across the organization, inviting business unit leaders who have embraced and successfully implemented programs to be part of a sustainability steering committee, and providing pilot team members with the opportunity to serve on cross-functional teams to promote new sustainability initiatives both internally and externally. Sharing success stories at industry events, customer forums, and supplier meetings facilitates celebrating sustainability program and project success with a broader stakeholder community, furthering the process of embedding sustainability into the organization's culture and perceived image.

15.6 Sonoma County Winegrowers Celebrate Sustainability

Sonoma County Winegrowers, in partnership with Sonoma County Vintners, are committed to becoming the nation's first 100% sustainable wine region by 2019. The organization has a marketing and education focus and is committed to the promotion and preservation of Sonoma County (California) as a premier grape-growing region. Their project goals are for land to be preserved for agriculture, communities and workers to be treated with respect, and business continuity to preserve and create a positive economic impact in the region. The project group reflects a collaboration among 1800 growers in 16 American Viticulture Areas. In early 2015 they published their first Sustainability Report, providing benchmarking statistics for the group as well as a series of articles on members highlighting success stories and best practices. The president, Karissa Kruse, announced their goal of becoming the first 100% sustainable wine region in the United States and laid out their project plan for sustainability assessments and third-party verification. While reaching this goal by 2019 is a challenge, the report indicates that 43% of vineyards have gone through the self-assessment sustainability process, with 33% obtaining third-party verification.[9] The organization has made significant progress toward its goals, with 58% of total acres sustainable just 15 months after the organization announced its commitment.[10]

The group has promoted their undertaking of this significant sustainability project and celebrated the success of their vineyard owners meeting this challenge. Promotional events include speaking about their sustainability initiatives to groups within Sonoma County and across California. Their program has been covered and celebrated by media sources such as the *San Francisco Chronicle,* CNN, NPR, and *Environmental Leader,* reaching readers worldwide with over 50 stories reaching over 50 million people.[11]

The project plan calls for:

1. **Supporting growers as they transition to sustainable practices, including land use, canopy management, energy efficiency, water quality, carbon emission, health care, and employee training.** In order to provide this support, a full-time sustainability manager has been hired to conduct workshops, seminars, and to help work with growers to complete the sustainability assessment.
2. **Facilitating third-party verification.** Support is coming from a partnership with the Sonoma Green Business Program, which currently offers a California Sustainable Winegrowing Alliance Code of Sustainability (CSWA) certification. Wineries will receive an annual report card, progress updates, and be able to track progress on the Winegrowers website.[12] The goal is to utilize existing funding in the Green Business Program to help defray the costs of the third-party verification process.
3. **Educating the local community.** A local advertising campaign highlighting the organization's commitment to sustainability has been launched. They have engaged regulatory agencies and the Agricultural Commissioner to discuss voluntary sustainability programs as a pathway to comply with concerns that may more traditionally be addressed through regulation and permits.
4. **Creating brand value for Sonoma Grapes as a 100% sustainable source, providing opportunities for growers to sell their produce at a premium price.** This initiative is supported by an advertising campaign linking Sonoma wines to sustainably produced wines, which ran in several highly recognized food and wine magazines such as *The Wine Spectator*, *Wine Enthusiast*, and *Food & Wine*. In addition, the organization will be part of Super Bowl 50, which is being hosted in the San Francisco Bay Area in 2016. The event is being showcased as the most sustainable Super Bowl yet. The event provides an opportunity to showcase Sonoma County wines as the sustainable choice.[13]

The organization's first Sustainability Report is a celebration of Sonoma County growers' and vintners' sustainability success stories. One "grower profile" highlights Balletto Vineyards' work in preserving land and continuing their farming legacy. The vineyard is certified by the Lodi Woodbridge Winegrape Commission (Lodi) for use of environmentally friendly, sustainable farming practices, as well as accredited by Protect Harvest, a not-for-profit that certifies farmers' use of 101 sustainable farming management practices in six categories: business, human resources, ecosystems, soil, water, and pest management.[14] Sustainable practices play a key role in Balletto Vineyards & Winery producing award-winning wines. A "sustainable case study" features the Jackson Family Wine sustainability programs, including energy and water conservation projects.[15] In addition, the grape growers and vintners who have conducted a sustainability self-assessment and received CSWA, Lodi, or Sustainable in Practice certifications are recognized on the "Sustainability Honor Roll."

While the Sonoma County Winegrowers' first Sustainability Report provides benchmarking and metrics about its 100% sustainable Sonoma goal, it also includes best practices and stories celebrating individual growers' and vintners' success. Participating in an industry consortium provides winegrowers and vineyards an opportunity to celebrate their own sustainability success while creating market value for their product as part of a uniquely sustainable wine region in the United States.

15.7 The Sustainability Continuum Impact

As an organization moves forward on its sustainability journey, the types of incentives and celebrations will change. In the exposure stage, incentives and campaigns are focused on engaging internal stakeholders and helping them relate to the concept of sustainability. The Walmart MSP program discussed earlier in the chapter falls into this category. "One associate asserts, 'with our new garden we're spending more time together. Cooking more. Eating fresher food. And saving more on groceries.'"[16] The personal sustainability impact of the program is reflected in this associate's comment.

Another example comes from University Hospitals (UH), a signatory to the Healthy Cleveland pledge, which focuses on improving the health of local citizens through improved nutrition, exercise, and behavior change. UH hosts an "Eat Real Food" photo contest for employees. Employees submit photographs showing how they are using real, unprocessed, and locally produced foods in their lives. Photos include home gardens, home-cooked dishes, and families cooking together. Winners are recognized on National Food Day.[17]

Another program designed to recognize individual employees' sustainability efforts is Green Health Heroes. These are peer-nominated employees who take leadership roles in helping UH advance environmental sustainability, promote health and wellness, and save money. Green Health Heroes are interviewed, and their efforts are shared on the external Greening UH web page. In addition, they are given a $50 credit at the online store, and they are formally recognized at the annual sustainability celebration in front of peers and senior leadership.[18] In these examples, the key to promoting the desired behavior is recognition and acknowledgement rather than a significant financial incentive.

As an organization moves along the sustainability continuum, celebrating sustainability success becomes more intertwined with external stakeholders and celebrating business success (see Figure 15.2). At Lush, a cosmetics company, management believes in creating effective products from organic fruits and vegetables, a supply chain free from animal testing, minimal packaging, ethical practices, and good products at good value.

Co-founder Mark Constantine says the "real art" is creating a product where customers don't need to think about sustainability when they make their choice. It is baked into the culture of the organization and its products.[19] Lush is committed to ethical and environmental standards. Rather than promoting customer overconsumption driven by marketing campaigns offering industry-standard promotions of three products for the price of two, Lush offers a "buy one, set one free" offer, to raise funds and awareness for a human rights organization, Reprieve. Lush's success is based on giving the best product to clients for a good value with good service. Lush celebrates their sustainability success by investing in their communities, paying living wages, paying full corporate taxes, and promoting environmental and social issues important to internal and external stakeholders. Part of the Lush story is their contributions of 10% of proceeds from Charity Pot product sales to small, grass-roots organizations that focus on environmental conservation, animal welfare, or human rights. Since its launch in 2007, the Charity Pot program has contributed $10 million to 850 grass-roots organizations in 42 countries.[20] One of the Charity Pot partners is 5Gyres Institute, which is working on eliminating microbeads that have become standard in most facial scrubs. The environmental impact is that these microbeads, which are made from nonbiodegradable plastic, are being washed into the waterways and then ingested by fish and other marine life. With the support of Lush, this organization has been able to raise awareness, resulting in several U.S. states banning the use of microbeads in products. At Lush, stakeholders' incentives comes through offering a sustainable product and the organization's support of customers' common interest in social and environmental programs.

As indicated in the Walmart example, vendor sustainability performance has become more of a driver in purchasing decisions for organizations. According to a recent poll, 54% of health-care professionals say that sustainability is incorporated into their purchasing decisions.[21] In order to address this growing trend and to provide products with sustainable features, Johnson & Johnson (J&J) developed the Earthwards program to design more sustainable solutions and to improve the sustainability impact across the product life cycle. The program considers impacts on product lines in seven categories of sustainability: material and packaging, energy, water and waste reduction, innovation, and social impact.[22] In order to be recognized under Earthwards classification, the product must meet regulatory compliance, core company standards, and at least three of the seven Earthwards categories. In order to meet

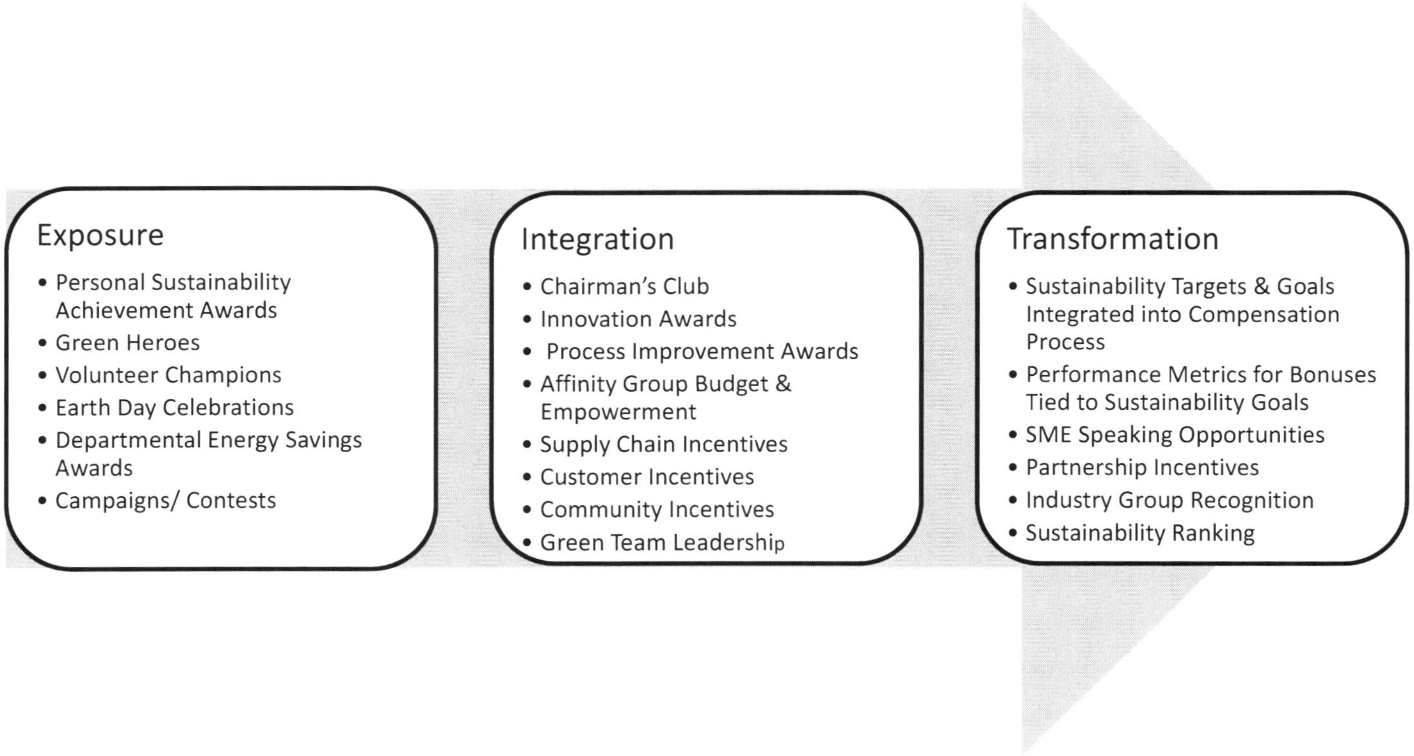

Exposure
- Personal Sustainability Achievement Awards
- Green Heroes
- Volunteer Champions
- Earth Day Celebrations
- Departmental Energy Savings Awards
- Campaigns/Contests

Integration
- Chairman's Club
- Innovation Awards
- Process Improvement Awards
- Affinity Group Budget & Empowerment
- Supply Chain Incentives
- Customer Incentives
- Community Incentives
- Green Team Leadership

Transformation
- Sustainability Targets & Goals Integrated into Compensation Process
- Performance Metrics for Bonuses Tied to Sustainability Goals
- SME Speaking Opportunities
- Partnership Incentives
- Industry Group Recognition
- Sustainability Ranking

Figure 15.2 Celebrating Success on the Sustainability Continuum

these cross-functional hurdles, project teams need to collaborate across business lines. Receiving an Earthwards recognition is a significant honor within J&J. Teams are congratulated and recognized for their innovative thinking and collaborative approach by J&J leadership.[23]

AVEENO PURE RENEWAL™ shampoo and conditioner received an Earthwards recognition as a result of several significant team contributions, such as changing ingredients to a sulfate-free formula, team members' contributions to community beautification, and energy savings from changing a transportation route from truck to train. By 2014, the Earthwards portfolio of products reflected 73 products, which exceeded the 2015 goals of 60 products. These products represent more than $8 billion in revenue generation for J&J.[24] The Earthwards program is another way in which J&J management has incorporated celebrating product sustainability into their culture and making it part of the strategy for organizational success.

As an organization moves further along on the continuum, sustainable strategy becomes part of the organizational culture. Celebrating sustainability moves from contests, campaigns, and special recognition awards to becoming an integral piece of total rewards. As a result, sustainability becomes entrenched in management's and employees' policies, processes, and daily functions. Strategic sustainability targets are cascaded down within the organization from senior management through business units to managers and employees. Sustainability goals are part of managers' balanced scorecards for measuring performance. (Balanced scorecards are discussed in Chapter 14.) Performance metrics and operational key performance indicators include measurements that track progress toward sustainability targets for the organization. Success and recognition within the organization require achievement of business goals, including sustainability goals. Sustainability achievement becomes an integral part of operations and overall strategic vision. It becomes part of the organizational decision making and is evidenced in the portfolio component process for selecting programs and projects to drive strategic goals.

At the transformative phase in the journey, sustainability is celebrated by allocation of senior leadership's time and attention to sustainable strategies, highlighting their commitment to its importance in the overall strategic vision of the business. Operationally, this means that sustainability goals and targets are part of the strategic planning process and that they are reviewed, evaluated, and monitored with the same frequency as other strategic initiatives. Achieving sustainable targets is celebrated in the same manner as meeting other key strategic targets such as revenue growth or product-specific targets. At this stage, sustainability is celebrated by having a designated Chief Sustainability Officer who has the same resources, power, and influence as the other C-suite members. The most powerful message that leadership can send about the importance of sustainability is creating a meaningful role for sustainability within the organization at the most senior level and empowering that role with access to key decision makers, organizational resources, and the same level of attention as other C-suite functions.

Establishing portfolio standards to embed sustainability into programs and projects demonstrates the importance of sustainability to an organization. Through this process, sustainability becomes part of all programs and projects, not just sustainability-driven portfolio components. Celebrating sustainability success by providing opportunities for program and project managers to share success stories and to develop best practice standards has a meaningful impact on driving change toward a sustainable culture. While internal sharing of success stories is an important change management process, sharing of these success stories with external stakeholders at industry events, think tank meetings, and stakeholder engagement forums provides recognition for program and project managers beyond the walls of the organization. Celebrate your program and project managers' successes and encourage their contributions to government, partnership, industry, and community forums. Academic institutions and government agencies frequently bring diverse groups together to share success stories and best practices. Other forums include global conferences such as those hosted by Sustainable Brands and GreenBiz. Providing an opportunity for a sustainability leader, project manager, or project team to attend these types of conferences and/or present their own success stories provides meaningful recognition and acknowledgment of their efforts and successes.

Unilever's Sustainable Living Plan reflects leadership's focus on sustainability as markets and consumers are increasingly concerned about sustainability issues. Pier Luigi Sigismondi, Chief Supply Chain Officer, predicts that "responsible consumption products will account for 70% of total grocery growth in the U.S. and Europe over the next 5 years."[25] In order to manage its 76,000 suppliers, Unilever launched its Partners to Win Platform, designed to create a network of partners to provide solutions that align with its strategic priorities. These strategic priorities—capability and capacity, quality and service, innovation, value, and responsible and sustainable living—drive innovation through R&D and joint business development plans with suppliers. While Unilever is making progress toward its 2020 goals in the areas of increasing sustainable sourcing and human rights in the supply chain, some challenges require broader collaborations to address issues such as climate change. External partnerships and collaboration are needed to address significant environmental and social challenges facing our global community. Unilever's leadership believes in exchanging best practices and aligning standards with other companies, including competitors, to promote a healthy planet. Challenges such as climate change caused by deforestation require a broad base of support to provide solutions. Working in concert with other companies, NGOs, and governments, traders, producers, manufacturers, and retailers that have signed on to zero or net-zero deforestation commitments, Unilever is driving toward limiting deforestation and the corresponding climate change impacts.[26] As an outgrowth of the UN Climate Summit, Unilever is working with governments, companies, civil society, and indigenous people on the development of the New York Declaration of Forests, a commitment to reduce deforestation 50% by 2020, end it by 2030, and to restore 350 million hectares of degraded land.[27] When asked about his legacy, Mr. Sigismondi hopes that he will be remembered for driving sustainable and profitable growth for the company as well as achieving results that will allow him to tell his family that he made a positive impact in our global society.[28]

In a fully transformed organization, sustainability success is celebrated by the impact that the organization's leadership, people, policies, and actions are having in terms of making our planet a healthier and better place. As organizations reach the transformative stage of sustainability, they realize that in order to address the challenges created by megatrends—climate change, population growth, and water quality and availability—they must build on networked solutions based on collaborations among competitors, governments, NGOs, and academia. At this stage in the journey, people within the organization are recognized and rewarded by being able to have the time and resources to be involved in addressing these global challenges in order to create impactful collaborative solutions.

15.8 Conclusion

Driving sustainable strategy into an organization to create a culture of sustainability requires senior management to communicate their vision clearly throughout the organization. The process begins with a vision but requires an effective implementation plan in order to bridge the performance gap between management vision and organizational action. Effecting change within an organization relies on selecting portfolio components that incorporate and align with sustainability goals in order to drive the organization toward its long-term sustainability vision. Creating sustainable organizational change means changing policies, people, and processes. Incorporating incentives into the process promotes desired behaviors and actions. Creating programs and time to celebrate progress, milestones, and achievements enhances the effectiveness of the change management process. Ultimately, rewarding desired behaviors increases program and project success rates and moves an organization forward on the sustainability continuum.

The types of incentives selected are based on the specific organization, culture, and stage of the sustainability journey. In the early stages, incentives built around campaigns and contests are useful for educating employees and encouraging desired behaviors. As an organization matures in its sustainability journey, the celebration process becomes more closely aligned with core business success. Opportunities for project team members to act as subject-matter experts speaking at events and serving on committees helps to reinforce their own commitment and helps to promote interest and engagement with other internal and external stakeholders. Ultimately, sustainability becomes part of the organization's culture, and incentives, and celebrations are incorporated into core compensation and bonuses rather than being separate campaigns.

Celebrating success is important for external stakeholders as well. Providing incentives for supplier engagement and compliance significantly impacts supply chain sustainability and overall vendor performance on core business initiatives. Providing clear guidelines to suppliers helps them to better serve the organization and to build a foundation for a long-term, mutually beneficially relationship that also helps to preserve the planet and protect human rights. Offering suppliers opportunities to engage with management to develop shared solutions to challenges and to open new opportunities is a meaningful way to reward vendors that share your organization's sustainability values.

Contest, games, and other incentive programs are effective ways to engage consumers on environmental and social issues. By engaging consumers in the sustainability process and making it fun and rewarding, organizations create a dual benefit of favorably impacting the environment and society while building brand loyalty. Moving beyond first-level incentives to create shared platforms that allow consumers to have meaningful impact on sustainable solution offerings engages customers in long-term value creation.

Celebration of internal and external stakeholders reinforces sustainable actions and behaviors that are desired by the leadership team. Moving toward a sustainable culture is a journey. Taking time to recognize and celebrate success by employees, partners, suppliers, and customers makes the journey more effective and more enjoyable for all. Acknowledging success provides leadership with opportunities both to confer appreciation and to solicit additional feedback for future sustainability growth.

Senior leadership's strategic vision drives organizational sustainability, creating business value while protecting natural and social capital. The journey to becoming a sustainable organization is complex and fraught with barriers that create a performance gap between sustainability vision and organizational behavior and action. In order to adopt sustainability as an organizational pillar and drive sustainability into the organization's culture, management must change the organization's policies, processes, and people. Project management professionals are skilled in managing programs and projects to effect these types of organizational change. With the additional knowledge, methodologies and techniques discussed in this book, project management professionals are well positioned to help close the sustainability performance gap and drive sustainable transformation. In order for Earth's resources to continue to meet the needs of society today and into the future, we all need to think and act more sustainably.

Notes

[1] Paul Baier, "5 Lessons from Walmart on Making Supplier Scorecards Work for You," *GreenBiz*, May 24, 2012, http://www.greenbiz.com/blog/2012/05/24/5-lessons-walmart-making-supplier-scorecards-work-your-business.

[2] Ibid.

[3] Ibid.

[4] Kristina Kohl, "Applying Lifelong Learning to Sustainability," *Wharton Magazine*, October 4, 2012, accessed July 14, 2015, http://whartonmagazine.com/blogs/applying-lifelong-learning-to-sustainability.

[5] "About Us," *Recyclebank*, accessed July 14, 2015, https://www.recyclebank.com/about-us.

[6] "Gamification of Environment," accessed July 14, 2015, https://badgeville.com/wiki/Gamification_of_Environment.

[7] "Bucks for Stars / Drink Sustainably / Betacup / Jovoto," accessed July 14, 2015, https://betacup.jovoto.com/ideas/4962.

[8] Hasbro, "Rewarding Employees," accessed July 9, 2015, http://csr.hasbro.com/employees/nurturing-networks.

9 Sonoma County Winegrowers, "Sonoma County Winegrowers 1st Annual Sustainability Report," January 2015, www.sonomawinegrape.org.
10 Ibid.
11 Ibid.
12 Sustainable Business, "Sonoma Strives to Become First 100 Percent Sustainable Wine Region," *GreenBiz*, January 23, 2014, http://www.greenbiz.com/blog/2014/01/23/sonoma-county-first-100-percent-sustainable-wine-region.
13 Sonoma County Winegrowers, "Sonoma County Winegrowers 1st Annual Sustainability Report."
14 Ibid.
15 Ibid.
16 Ellen Weinreb, "How Walmart Associates Put the 'U' and 'I' into Sustainability," *GreenBiz.com*, accessed September 12, 2014, http://www.greenbiz.com/blog/2013/01/09/walmart-associates-u-i-sustainability.
17 Business Ethics Health Economics Law, and Public Policy Operations Management North America, "Employees Can Be a Powerful Force in Sustainability," *Knowledge@Wharton*, March 3, 2015, accessed July 9, 2015, http://knowledge.wharton.upenn.edu/article/employees-powerful-force-sustainability.
18 "Employee Education and Engagement," accessed July 10, 2015, http://www.uhhospitals.org/about/greening-uh/progress-report-on-sustainability/progress-report-on-sustainability-issues/2013-progress-report-on-sustainability/education-and-outreach/employee-education-and-engagement.
19 Jessica Shankleman, "Lush Founder on Sustainable Business: Don't Worry, Be Profitable," *GreenBiz*, July 14, 2014, http://www.greenbiz.com/blog/2014/07/14/lush-founder-dont-worry-be-profitable.
20 Lush, "Lush: Charity Pot," accessed September 10, 2015, http://www.lushusa.com/on/demandware.store/Sites-Lush-Site/en_US/CharityPot-Featured.
21 "Earthwards®: Path to Innovation," *2014 Year in Review*, accessed July 13, 2015, http://2014yearinreview.jnj.com/stories/Earthwards-Path-to-Innovation.
22 Business Ethics Health Economics Law, and Public Policy Operations Management North America, "Employees Can Be a Powerful Force in Sustainability."
23 Johnson & Johnson, "Our Strategic Framework: Our Most Sustainable Products," accessed July 13, 2015, https://www.jnj.com/caring/citizenship-sustainability/strategic-framework/our-most-sustainable-products.
24 "Earthwards®: Path to Innovation."
25 Tom Idle, "How Unilever Is Creating a Web of Partnerships," *GreenBiz*, July 13, 2015, http://www.greenbiz.com/article/how-unilever-creating-web-partnerships.
26 Ibid.
27 "UN Climate Summit: Forests Results," *UNFCCC*, accessed July 14, 2015, http://newsroom.unfccc.int/nature-s-role/un-climate-summit-forests.
28 Idle, "How Unilever Is Creating a Web of Partnerships."

Index

A

Accenture, 22, 24, 41, 47, 52, 216
affinity groups, 236, 253, 260, 276, 277, 298, 324, 325
alignment of internal and external messaging, 243
alignment of strategy, goals, and metrics, 301
ambassadors, 66, 137, 212, 218, 220, 252, 262, 294
Apple, 36, 195, 196
Aqueduct Project, 182
assessment of an organization's sustainable cultural foundation, 123, 140

B

Ball Corporation, 54–58
barriers, 24, 43, 58, 59, 72–74, 82, 91, 114, 158, 188, 204, 233, 234, 236–238, 332
B Corp. *See* B Corporation
B Corporation (B Corp), 38, 127, 128
benchmarking organizational sustainability, 275
benefit corporations, 127
best practices for leveraging sustainable strategy, 262
BHCI. *See* Business Health Culture Index
B Lab, 128
Brundtland Commission, 2, 6–8
business case for sustainability, 21, 43, 47, 54, 94, 273
business function leaders, 68, 78, 83, 88, 113, 180, 187, 188, 191, 193, 198, 199, 201, 207, 264
Business Health Culture Index (BHCI), 218, 219
business value, 21, 22, 24, 25, 29, 35, 40, 44–46, 48, 49, 58, 59, 65, 66, 68, 76, 79, 96, 105, 107, 116, 126, 131, 140, 145, 149, 150, 157, 188, 200–204, 218, 231, 232, 254, 270, 273, 294, 296–298, 303, 308, 312, 325, 332

C

Carbon Disclosure Project (CDP), 14, 26, 27, 29, 35–37, 39, 139, 149, 156, 192, 200, 274, 303, 305, 313
cascading sustainable strategy, 78
Caterpillar Inc., 99
CDP. *See* Carbon Disclosure Project
celebrating success, 321, 323, 325, 329, 332
celebrating success on the sustainability continuum, 329
CEO's Top Drivers for Sustainability, 25
Ceres, 29, 31, 34, 139, 175, 200
Chameleon Studios, 38, 39
checklist for engaging internal stakeholders in sustainability, 246
Chief Sustainability Officer (CSO), 45, 68, 73, 74, 77, 78, 86, 88, 96, 123, 151, 153, 154, 158, 180, 181, 187, 188, 191–193, 206, 211, 213, 237, 264, 278, 330
circular economy, 98, 99, 111, 133, 149, 178, 189, 197, 198, 204, 249
Clean Cities program, 285, 287, 299
climate change risk, 22, 29–31, 81, 84, 132, 134, 191
Component Portfolio Selection Matrix, 157
creating business value, 21, 35, 40, 44, 59, 65, 218, 332

CSO. *See* Chief Sustainability Officer
culture of sustainability matrix, 131, 132

D

D&I. *See* Diversity & Inclusion
data quality and accuracy, 311
determine organizational readiness, 153, 291
Diversity & Inclusion (D&I), 74, 96, 100–102, 111, 132, 225, 296–298
DJSI. *See* Dow Jones Sustainability Index
DOE. *See* US Department of Energy
Dow Chemical, 44, 45, 52, 53, 60–62, 86, 87, 158, 172, 180, 305
Dow Jones Sustainability Index (DJSI), 4, 8, 35, 54, 136, 210, 304
Dow Sustainability Chemistry Index (SCI), 61, 86, 87
driver, 3, 4, 18, 24, 25, 27, 40, 45, 48, 51–55, 57, 62, 63, 86, 87, 97–99, 114, 126, 131, 132, 139, 148–150, 153–155, 161, 162, 188, 189, 196, 199, 201, 203, 206, 207, 216, 229–231, 245, 249, 255, 256, 264, 266, 269, 288, 292, 302, 309, 312, 324, 328

E

Earthwards, 328, 330
Electronic Product Environmental Assessment Tool (EPEAT), 195, 196
ELEEP. *See* Emerging Leaders in Energy and Environmental Policy
embedding sustainability into the culture, 248, 254
Emerging Leaders in Energy and Environmental Policy (ELEEP), 129
employee and team engagement, 257
employee engagement, 9, 22, 24, 25, 35, 40, 46, 53, 55, 57, 62, 111, 138, 149, 175, 180, 187, 190, 201, 204, 205, 211, 213, 215–218, 220, 222, 224, 234, 244, 246, 248, 251–254, 258, 260, 262, 263, 265, 266, 273, 277, 290, 292–294, 297, 299, 303, 309, 311, 312, 321, 324
employee resource groups, 296, 326
employee's life cycle, 261
energy management plan, 281–283, 309
Energy Star, 136, 196, 274, 279–281, 284, 299
engagement map, 171, 193–195

engagement plan, 132, 165, 168–170, 173, 182, 226
engaging function business leaders, 193
engaging stakeholders in the change process, 245
engaging the C-suite, 49, 50, 53, 58, 60, 62, 92, 94
EPEAT. *See* Electronic Product Environmental Assessment Tool
EV Everywhere Grand Challenge, 287
Executive Education for the Environment, 203
exposure, 22, 45, 46, 92, 103, 160, 189, 190, 274, 312, 327
extended leadership, 47
external stakeholder, 1, 5, 31, 44, 45, 47, 54, 61, 66, 69, 73, 76, 79, 85, 87, 92, 95, 96, 103, 106, 108, 111, 114, 116, 120, 122–124, 126, 134, 136, 137, 139, 144–146, 148, 157, 161, 163, 165, 169–171, 173, 177, 181, 182, 184, 187–189, 191–193, 196, 203, 211, 212, 214, 223, 225, 226, 233, 235, 239, 242, 243, 245, 248, 254, 261, 268, 273, 274, 277, 278, 280, 281, 284, 296, 301, 305, 307, 313, 315–318, 321–323, 325, 326, 328, 330, 332
external stakeholder management, 181, 192, 225

F

flexible work drives business value, 298
flexible work program, 220, 242, 289, 297–299
framework for sustainable culture change management, 236
framework for sustainable work, 257

G

gamification, 323, 324
GHG. *See* greenhouse gas emission
Global Reporting Initiative (GRI), 29, 97, 139, 149, 192, 210, 211, 274, 275, 303, 304, 313
greenhouse gas emission (GHG), 3, 4, 12, 13, 15, 16, 27, 29, 69, 71, 73, 77, 81, 84, 92, 99–102, 105, 107, 112, 132, 136, 138, 139, 147, 148, 156, 160, 167, 172, 195, 200, 215, 218, 222, 224, 225, 235–237, 243, 256, 266, 269, 278–281, 285, 287–290, 293, 294, 299, 302, 307, 309, 312, 317
greenhouse gas emission drivers by scope classification, 288
green team project plan, 278, 279, 299
greenwashing, 16, 39, 214, 243
GRI. *See* Global Reporting Initiative

Guidelines for Stakeholder Engagement, 179

H

HCM. *See* Human Capital Management
HCM survey to build a sustainable workforce, 247
Hewlett Packard (HP), 224, 225
HP. *See* Hewlett Packard
HR. *See* Human Resources
HR scorecard for sustainability, 294
Hubway, 267–270
Human Capital Management (HCM), 210–215, 218, 220–226, 232, 235, 247, 248, 282, 293, 294, 299
human capital relevance, 210
Human Resources (HR), 46, 53, 68, 80, 100–102, 110, 111, 137, 139, 195, 201, 209–214, 216–218, 220–225, 229, 230, 232, 239, 247–249, 253, 254, 255, 258, 261, 264, 265, 277, 282, 292–295, 298, 325, 327

I

Iconic Energy Consulting, 129
identify stakeholders, 74, 236
IKEA, 139, 245, 246, 248
impact of engagement, 166
impact of incentives, 322
incentives drive sustainable outcomes, 321, 322
Integrated Reporting (IR), 316
integrating sustainability into core HR functionality, 223
integration, 4, 46, 52, 60, 65, 92, 94, 96, 103, 119, 175, 187–190, 196, 273, 274, 312, 313
internal stakeholder communication tools, 241
internal stakeholder management, 191, 207
Internal Stakeholder Drivers for Sustainability, 196
Internal Stakeholder Engagement Techniques and Impacts, 189, 190
internal stakeholders, 43, 85, 88, 103, 132, 134, 137, 144, 165, 169–172, 187, 189–191, 193, 196, 201, 202, 206, 207, 209, 213, 214, 220, 222–224, 226, 229, 236, 241, 242, 246, 248, 254, 274, 280, 323, 327
International Standards Organization (ISO), 29, 303
IR. *See* Integrated Reporting
Irv & Shelly's Fresh Picks, 166, 167

ISO. *See* International Standards Organization

K

keys to creating an organization to support sustainability, 231

L

Leadership in Energy & Environmental Design (LEED), 73, 99, 126, 179, 275, 289
LEED. *See* Leadership in Energy & Environmental Design
lessons learned, 17, 53, 61, 76, 78, 86, 94, 105, 107, 108, 114, 127, 134, 140, 143, 153, 157–163, 170, 182, 189, 222, 235, 236, 245, 248, 265, 266, 268, 269, 277, 279–281, 284, 289, 293, 304, 309, 317, 326
Levi, 242, 248
Lush, 328

M

Malthusian catastrophe, 15
managing risk, 27, 145, 239
materiality, 54–57, 60, 63, 79–84, 99, 101–103, 105, 126, 130, 169, 175, 180, 211, 275, 301, 304, 306
materiality matrix, 54–56, 169
materiality project priority matrix, 102, 103, 105
matrix, 54–56, 100–103, 105, 107, 131, 132, 143, 146, 147, 157, 158, 169, 172, 173, 175, 176, 180, 181, 213, 231, 237, 241, 248, 253, 298, 309, 311
Matrix for Assessment of Material Issues by Function, 100
MDG. *See* Millennium Development Goals
meaningful metrics, 82, 222, 238, 302, 306, 308
measuring engagement, 206
megatrends, 1, 10, 11, 17, 18, 146, 147, 214, 331
metrics, 1, 3, 4, 13, 17, 21, 22, 24, 26, 27, 32, 35, 38, 44, 49, 51, 53, 54, 59, 74, 76–80, 82, 83, 86–88, 92, 95, 96, 103, 106–108, 110, 116, 124, 131, 132, 134, 136, 138, 156, 160, 162, 167, 170, 173, 175, 182, 188, 203, 206, 216–218, 220, 222, 232, 236, 238, 241, 242, 244–247, 254, 256, 257, 260, 263, 265, 267, 269, 270, 280, 281, 284, 291, 293, 294, 297–299, 301–309, 311, 313, 315–318, 321, 327, 330

metrics drive engagement, 308
Millennials, 11, 17, 60, 62, 147, 213, 221, 223, 234, 244, 255, 256, 258, 264, 270
Millennium Development Goals (MDG), 215
MSP. *See* Walmart My Sustainability Plan
Mud Jeans, 198

N

natural capital, 3, 7, 9, 13, 21, 26, 35, 44, 48, 71, 87, 99, 107, 124, 126, 128, 196
networked, 128, 151, 231, 232, 248, 253, 331
New Jersey Sustainable Business Registry, 38
NEWPCP. *See* Northeast Water Pollution Control Plant
Nike, 199, 214, 243
Northeast Water Pollution Control Plant (NEWPCP), 159, 160

O

organizational structure, 37, 73, 74, 77, 83, 123, 143, 150–152, 154, 203, 206, 211, 220, 229–234, 237, 238, 248, 253
organizational structure for change, 230

P

performance gap, 41, 251, 284, 331, 332
PEV. *See* plug-in vehicle
Philadelphia Water Department (PWD), 159, 160, 162
plug-in vehicle (PEV), 285, 287, 289–293
PMI. *See* Project Management Institute
portfolio component performance tracking matrix, 158
Portfolio Component Proposal Questionnaire, 155
portfolio management, 1, 40, 41, 43, 59, 76, 94, 106, 143, 145, 150, 151, 153, 154, 157, 158, 160, 163, 165, 207, 234, 252
portfolio management alignment, 150
PPP. *See* public-private partnership
product life-cycle assessment, 48, 70, 102, 103, 116, 189, 308
project checklist to promote a sustainable culture, 134, 135, 140
project management impact, 17, 178, 315
Project Management Institute (PMI), 68, 87, 106

project/program alignment with sustainable strategy, 89, 116
promoting project team success, 325
public-private partnership (PPP), 159
PWD. *See* Philadelphia Water Department

R

reporting standards and frameworks, 303
resources, 3, 4, 7, 9–13, 15–18, 24, 26, 31, 47–49, 52, 53, 58, 66, 68, 69, 71–74, 76, 77, 82, 86–89, 91, 94–96, 99–103, 105–108, 110, 111, 113, 116, 120, 123, 124, 138, 143–145, 148, 150, 151, 153–157, 160, 161, 163, 165, 168, 170, 171, 173, 177, 179–184, 188, 190, 193, 197, 199, 205–207, 209, 211, 224, 229–233, 235, 238, 239, 242, 248, 252, 253, 263, 265, 266, 273, 274, 276, 277, 279–281, 283–285, 287, 289, 292, 294, 297, 299, 303, 304, 307, 308, 313, 316, 318, 323, 327, 330–332
role for technology, 311
Rubicon Global, 128, 199

S

SC Johnson Greenlist™, 113, 114
SAP SE, 219
SASB. *See* Sustainability Accounting Standards Board
SCI. *See* Dow Sustainability Chemistry Index
shared value, 5, 6, 9, 47, 48, 51, 60, 66, 71, 126, 232, 253
Skanska AB, 124
SMART goals, 136, 173, 280, 281, 302
social capital, 1, 3, 8, 9, 15, 17, 27, 48, 59, 65, 80–85, 100, 107, 131, 132, 200, 332
Socially Responsible Investor (SRI), 32, 33, 127, 169, 190, 192
Sonoma County Winegrowers, 326, 327
SRI. *See* Socially Responsible Investor
stakeholder, 1, 3–5, 8, 9, 11, 17, 18, 21, 22, 25, 26, 29, 31–34, 36, 37, 39, 41, 43–49, 51, 54, 55, 57, 60, 61, 63, 66, 68, 69, 71–74, 76–79, 83, 85–88, 91, 92, 95, 96, 100, 101, 103, 105–108, 111, 113, 114, 116, 117, 120, 122–124, 126, 127, 130–132, 134–137, 139, 144–146, 148–151, 155, 157, 159, 161–163, 165–184, 187–193, 195, 196, 199, 201–203,

206, 207, 209, 211–214, 220, 222–226, 229, 233–248, 252–254, 257, 260, 261, 264, 266–270, 273–275, 277–287, 289, 294, 296, 297, 299, 301–303, 305–309, 311, 313, 315–318, 321–328, 330, 332
stakeholder communication strategies, 177
stakeholder engagement matrix, 176, 180, 181, 241
stakeholder engagement process, 165, 169, 170, 173, 174, 179, 184
stakeholder requirements matrix, 172, 173
STARS. *See* Sustainability Tracking Assessment & Rating Systems
sustainability, 1–6, 8, 9, 11, 16–18, 21, 22, 24–29, 31–41, 43–49, 51–55, 57–63, 65, 66, 68–74, 76–83, 85–89, 91–99, 101–114, 116, 117, 119, 121–124, 126–132, 134–140, 143–151, 153–161, 163, 165, 166, 169–173, 175–181, 183, 184, 187–193, 195, 196, 198–207, 209–218, 220–226, 229–239, 241–249, 251–266, 270, 273–279, 281, 283–285, 287, 289, 290, 292–297, 299, 301–318, 321–332
Sustainability Accounting Standards Board (SASB), 79–83, 169, 274, 275, 304
sustainability assessment, 5, 35, 99, 101, 326, 327
sustainability balanced scorecard, 295, 310
sustainability balanced scorecard for HR, 295
sustainability business model, 27, 28
sustainability champion, 6, 21, 40, 43, 44, 47, 48, 51–54, 58, 59, 62, 63, 68, 69, 74, 76, 77, 83, 86, 89, 92, 96, 101, 105, 159, 160, 180, 187, 189, 195, 199, 207, 211, 213, 222, 231, 246
sustainability change management plan, 237
sustainability continuum, 103, 104, 110, 113, 116, 124, 146, 158, 188, 193, 235, 246, 273, 285, 322, 327–329, 331
sustainability cycle for success, 108, 109
sustainability driver, 97–99, 155, 201
sustainability drives HCM benefit, 211, 212
sustainability engagement hierarchy, 259
sustainability framework, 83, 85, 86, 110, 263
sustainability journey, 3, 17, 24, 38, 39, 43, 45–48, 52, 53, 59, 61, 63, 65, 69, 73, 87, 88, 91–93, 99, 103, 105, 124, 128, 137, 140, 145, 146, 149, 153, 158, 165, 187, 188, 196, 207, 209, 220, 224, 230, 234, 235, 243, 245, 253, 255, 273, 275, 285, 299, 304, 305, 308, 315, 322, 325, 327, 331, 332
sustainability portfolio alignment matrix, 146, 147

sustainability portfolio assessmen, 147, 148
sustainability project charter, 94, 95, 108
sustainability project impact, 111
sustainability ranking indices, 4, 35
sustainability reporting journey, 315
sustainability software selection, 312–314
sustainability systems, 71
Sustainability Tipping Point, 87
Sustainability Tracking Assessment & Rating Systems (STARS), 79, 304, 316, 317, 324
sustainability value mapping, 134, 137, 325
sustainability vision, 9, 17, 59, 74, 76–79, 92, 113, 114, 119, 122, 123, 127, 131, 140, 143, 144, 148, 150, 151, 153, 159, 163, 165, 172, 180, 187, 191, 202, 203, 206, 211, 223, 225, 232–234, 236, 239, 241, 245, 248, 249, 251, 254, 255, 257, 258, 260–262, 264, 266, 270, 309, 318, 321, 331, 332
sustainable culture, 18, 60, 66, 76, 89, 119, 122, 124, 127, 128, 130, 134–137, 140, 212, 220, 229–232, 236, 238, 244, 248, 256, 258, 277, 293, 299, 302, 303, 308, 309, 330, 332
sustainable portfolio management cycle, 153, 154
sustainable strategy, 1–4, 6, 8, 11, 17, 18, 21–27, 31, 33, 37, 40, 41, 43–47, 49, 51–54, 58–61, 63, 65–70, 74, 76–79, 83, 86–89, 91, 94, 95, 99, 103, 107, 110, 111, 113, 114, 116, 117, 120, 122, 123, 127, 130, 138–140, 143–147, 149–154, 156–158, 170, 171, 173, 176, 177, 180, 184, 187–189, 191, 193, 195–197, 199–207, 209, 211–214, 216, 218, 220, 222, 223, 226, 229, 232–235, 237, 239, 246–248, 251–254, 256, 258, 260–262, 264–266, 274, 293, 294, 297, 299, 302, 308, 309, 311, 316, 318, 321, 324, 330, 331
sustainable strategy change management, 234
sustainable strategy portfolio alignment, 144
sustainable values, 124, 125, 231
sustainable workforce, 247–249, 251, 255, 257, 264, 265, 270, 276, 277, 293
sustainable workplace, 256, 257, 270
system thinking, 72

T

TBL. *See* triple bottom line
The City of Cambridge, Massachusetts, 266
The Valley of Despair, 237, 238
tiered stakeholder communication plan, 240

tragedy of the commons, 10
transformation, 6, 17, 21, 35, 47, 59, 63, 82, 83, 94, 103, 136, 155, 156, 187–191, 209, 223, 225, 226, 231, 234–238, 245–248, 266, 274, 275, 301, 322, 325, 332
transformative sustainable strategy, 67
triple bottom line (TBL), 2, 3, 7, 18, 25, 37, 40, 43, 55, 66, 70, 91, 126, 127, 129, 166, 167, 184, 209, 226, 229, 232, 233, 234, 262, 270, 289

U

UNGC. *See* United Nations Global Compact
United Nations Global Compact (UNGC), 3–5, 22, 24, 47, 52, 149, 216
United Nations Principles for Responsible Investing (UNPRI), 32, 275
University of Vermont (UVM), 79, 304, 305, 316, 317
UNPRI. *See* United Nations Principles for Responsible Investing
UN World Council for Economic Development, 2
U.S. Army, 193
US Department of Energy (DOE), 161, 162, 274, 285, 287, 289, 290, 292, 299
UVM. *See* University of Vermont

V

value chain stakeholders, 286, 299
value of engagement, 252
voluntarism drives employee engagement, 263

W

Walmart My Sustainability Plan (MSP), 220–222, 322, 323, 327
workplace charging, 287, 289–293
World Resource Institute (WRI), 182, 183
World Wildlife Fund (WWF), 39
WRI. *See* World Resource Institute
WWF. *See* World Wildlife Fund